BIRKHÄUSER

Operator Theory: Advances and Applications

Vol. 210

Founded in 1979 by Israel Gohberg

Recent Trends in Toeplitz and Pseudodifferential Operators

The Nikolai Vasilevskii Anniversary Volume

Roland Duduchava
Israel Gohberg
Sergei M. Grudsky
Vladimir Rabinovich
Editors

Birkhäuser

Editors:

Roland Duduchava
Department of Mathematical Physics
Andrea Razmadze Mathematical Institute
M. Alexidze str. 1
Tbilisi 0193
Georgia
e-mail: dudu@rmi.acnet.ge

Sergei M. Grudsky
Departamento de Matemáticas
CINVESTAV
Av. I.P.N. 2508
Col. San Pedro Zacatenco
Apartado Postal14-740
Mexico,D.F.
Mexico
e-mail: grudsky@math.cinvestav.mx

Israel Gohberg (Z"L)

Vladimir Rabinovich
National Polytechnic Institute
ESIME Zacatenco
Av. IPN
07738 Mexico, D.F.
Mexico
e-mail:vladimir.rabinovich@gmail.com

2010 Mathematics Subject Classification: 47-06

Library of Congress Control Number: 2010928340

Bibliographic information published by Die Deutsche Bibliothek.
Die Deutsche Bibliothek lists this publication in the Deutsche Nationalbibliografie; detailed bibliographic data is available in the Internet at http://dnb.ddb.de

ISBN 978-3-0346-0547-2

P.O. Box 133, CH-4010 Basel, Switzerland
Part of Springer Science+Business Media
Printed on acid-free paper produced from chlorine-free pulp. TCF∞
Printed in Germany

ISBN 978-3-0346-0547-2
e-ISBN 978-3-0346-0548-9

9 8 7 6 5 4 3 2 1

www.birkhauser.ch

Contents

Nikolai Vasilevski

Operator Theory:
Advances and Applications, Vol. 210, 1–14

The Life and Work of Nikolai Vasilevski

Sergei Grudsky, Yuri Latushkin and Michael Shapiro

Nikolai Leonidovich Vasilevski was born on January 21, 1948 in Odessa, Ukraine. His father, Leonid Semenovich Vasilevski, was a lecturer at Odessa Institute of Civil Engineering, his mother, Maria Nikolaevna Krivtsova, was a docent at the Department of Mathematics and Mechanics of Odessa State University.

In 1966 Nikolai graduated from Odessa High School Number 116, a school with special emphasis in mathematics and physics, that made a big impact at his creative and active attitude not only to mathematics, but to life in general. It was a very selective high school accepting talented children from all over the city, and famous for a high quality selection of teachers. A creative, nonstandard, and at the same time highly personal approach to teaching was combined at the school with a demanding attitude towards students. His mathematics instructor at the high school was Tatjana Aleksandrovna Shevchenko, a talented and dedicated teacher. The school was also famous because of its quite unusual by Soviet standards system of self-government by the students. Quite a few graduates of the school later became well-known scientists, and really creative researchers.

In 1966 Nikolai became a student at the Department of Mathematics and Mechanics of Odessa State University. Already at the third year of studies, he began his serious mathematical work under the supervision of the well-known Soviet mathematician Georgiy Semenovich Litvinchuk. Litvinchuk was a gifted teacher and scientific adviser. He, as anyone else, was capable of fascinating his students by new problems which have been always interesting and up-to-date. The weekly Odessa seminar on boundary value problems, chaired by Prof. Litvinchuk for more than 25 years, very much influenced Nikolai Vasilevski as well as others students of G.S. Litvinchuk.

N. Vasilevski started to work on the problem of developing the Fredholm theory for a class of integral operators with nonintegrable integral kernels. In essence, the integral kernel was the Cauchy kernel multiplied by a logarithmic factor. The integral operators of this type lie between the singular integral operators and the integral operators whose kernels have weak (integrable) singularities. A famous Soviet mathematician F.D. Gakhov posted this problem in early 1950ies, and it remained open for more than 20 years. Nikolai managed to provide a complete solution in the setting which was much more general than the original. Working on this problem, Nikolai has demonstrated one of the main traits of his mathematical talent: his ability to achieve a deep penetration in the core of the problem, and to see rather unexpected connections between different theories. For instance, in order to solve Gakhov's Problem, Nikolai utilized the theory of singular integral operators with coefficients having discontinuities of first kind, and the theory of operators whose integral kernels have fixed singularities – both theories just appeared at that time. The success of the young mathematician was well recognized by a broad circle of experts working in the area of boundary value problems and operator theory. In 1971 Nikolai was awarded the prestigious M. Ostrovskii Prize, given to the young Ukrainian scientists for the best research work. Due to his solution of the famous problem, Nikolai quickly entered the mathematical community, and became known to many prominent mathematicians of that time. In particular, he was very much influenced by the his regular interactions with such outstanding mathematicians as M.G. Krein and S.G. Mikhlin.

In 1973 N. Vasilevski defended his PhD thesis entitled "To the Noether theory of a class of integral operators with polar-logarithmic kernels". In the same year he became an Assistant Professor at the Department of Mathematica and Mechanics of Odessa State University, where he was later promoted to the rank of Associate Professor, and, in 1989, to the rank of Full Professor.

Having received the degree, Nikolai continued his active mathematical work. Soon, he displayed yet another side of his talent in approaching mathematical problems: his vision and ability to use general algebraic structures in operator theory, which, on one side, simplify the problem, and, on another, can be used in many other problems. We will briefly describe two examples of this.

The first example is the method of orthogonal projections. In 1979, studying the algebra of operators generated by the Bergman projection, and by the operators of multiplication by piece-wise continuous functions, N. Vasilevski gave a description of the C^*-algebra generated by two self-adjoint elements s and n satisfying the properties $s^2 + n^2 = \mathrm{e}$ and $sn + ns = 0$. A simple substitution $p = (\mathrm{e} + s - n)/2$ and $q = (\mathrm{e} - s - n)/2$ shows that this algebra is also generated by two self-adjoint idempotents (orthogonal projections) p and q (and the identity element e). During the last quarter of the past century, the latter algebra has been rediscovered by many authors all over the world. Among all algebras generated by orthogonal projections, the algebra generated by two projections is the only tame algebra (excluding the trivial case of the algebra with identity generated by one orthogonal projection). All algebras generated by three or more orthogonal projections are known to be wild, even when the projections satisfy some additional constrains. Many model algebras arising in operator theory are generated by orthogonal projections, and thus any information of their structure essentially broadens the set of operator algebras admitting a reasonable description. In particular, two and more orthogonal projections naturally appear in the study of various algebras generated by the Bergman projection and by piece-wise continuous functions having two or more different limiting values at a point. Although these projections, say, P, $Q_1, \dots, Q_n$, satisfy an extra condition $Q_1 + \cdots + Q_n = I$, they still generate, in general, a wild C^*-algebra. At the same time, it was shown that the structure of the algebra just mentioned is determined by the joint properties of certain positive injective contractions C_k, $k = 1, \dots, n$, satisfying the identity $\sum_{k=1}^n C_k = I$, and, therefore, the structure is determined by the structure of the C^*-algebra generated by the contractions. The principal difference between the case of two projections and the general case of a finite set of projections is now completely clear: for $n = 2$ (and the projections P and $Q + (I - Q) = I$) we have only one contraction, and the spectral theorem directly leads to the desired description of the algebra. For $n \geq 2$ we have to deal with the C^*-algebra generated by a finite set of noncommuting positive injective contractions, which is a wild problem. Fortunately, for many important cases related to concrete operator algebras, these projections have yet another special property: the operators $PQ_1P, \dots, PQ_nP$ mutually commute. This property makes the respective algebra tame, and thus it has a nice and simple description as the algebra of all $n \times n$ matrix-valued functions that are continuous on the joint spectrum Δ of the operators $PQ_1P, \dots, PQ_nP$, and have certain degeneration on the boundary of Δ.

Another notable example of the algebraic structures used and developed by N. Vasilevski is his version of the Local Principle. The notion of locally equivalent operators, and localization theory were introduced and developed by I. Simonenko in mid-sixtieth. According to the tradition of that time, the theory was focused on the study of individual operators, and on the reduction of the Fredholm properties of an operator to local invertibility. Later, different versions of the local principle have been elaborated by many authors, including, among others, G.R. Allan, R. Douglas, I.Ts. Gohberg and N.Ia. Krupnik, A. Kozak, B. Silbermann. In spite of the fact that many of these versions are formulated in terms of Banach- or C^*-algebras, the main result, as before, reduces invertibility (or the Fredholm property) to local invertibility. On the other hand, at about the same time, several papers on the description of algebras and rings in terms of continuous sections were published by J. Dauns and K.H. Hofmann, M.J. Dupré, J.M.G. Fell, M. Takesaki and J. Tomiyama. These two directions have been developed independently, with no known links between the two series of papers. N. Vasilevski was the one who proposed a local principle which gives the global description of the algebra under study in terms of continuous sections of a certain canonically defined C^*-bundle. This approach is based on general constructions of J. Dauns and K.H. Hofmann, and results of J. Varela. The main contribution consists of a deep re-comprehension of the traditional approach to the local principles unifying the ideas coming from both directions mentioned above, which results in a canonical procedure that provides the global description of the algebra under consideration in terms of continuous sections of a C^*-bundle constructed by means of local algebras.

In the eighties and even later, the main direction of the work of Nikolai Vasilevski has been the study of multi-dimensional singular integral operators with discontinuous coefficients. The main philosophy here was to study first algebras containing these operators, thus providing a solid foundation for the study of various properties (in particular, the Fredholm property) of concrete operators. The main tool has been the described above version of the local principle. This principle was not merely used to reduce the Fredholm property to local invertibility but also for a global description of the algebra as a whole based on the description of the local algebras. Using this methodology, Nikolai Vasilevski obtained deep results in the theory of operators with Bergman's kernel and piece-wise continuous coefficients, in the theory of multi-dimensional Toeplitz operators with pseudo-differential presymbols, in the theory of multi-dimensional Bitsadze operators, in the theory of multi-dimensional operators with shift, etc. In 1988 N. Vasilevski defended the Doctor of Sciences dissertation, based on these results, and entitled "Multi-dimensional singular integral operators with discontinuous classical symbols".

Besides being a very active mathematician, N. Vasilevski has been an excellent lecturer. His lectures are always clear, and sparkling, and full of humor, which so natural for someone who grew up in Odessa, a city with a longstanding tradition of humor and fun. He was the first at Odessa State University who designed and started to teach a class in general topology. Students happily attended his lectures in Calculus, Real Analysis, Complex Analysis, Functional Analysis. He has been one of the most popular professor at the Department of Mathematics and Mechanics of Odessa State University. Nikolai is a master of presentations, and his colleagues always enjoy his talks at conferences and seminars.

In 1992 Nikolai Vasilevski moved to Mexico. He started his career there as an Investigator (Full Professor) at the Mathematics Department of CINVESTAV (Centro de Investagacion y de Estudios Avansados). His appointment significantly strengthen the department which is one of the leading mathematical centers in Mexico. His relocation also visibly revitalized mathematical activity in the country in the field of operator theory. Actively pursuing his own research agenda, Nikolai also served as the organizer of several important conferences. For instance, let us mention the (regular since 1998) annual workshop "Análisis Norte-Sur", and the well-known international conference IWOTA-2009. He initiated the relocation to Mexico a number of active experts in operator theory such as Yu. Karlovich and S. Grudsky, among others.

During his tenure in Mexico, Nikolai Vasilevski produced a sizable group of students and younger colleagues; five of young mathematicians received PhD under his supervision.

The contribution of N. Vasilevski in the theory of multi-dimensional singular integral operators found its rather unexpected development in his work on quaternionic and Clifford analysis, published mainly with M. Shapiro in 1985–1995, starting still in the Soviet Union, with the subsequent continuation during the Mexican period of his life. Among others, the following topics have been considered: The settings for the Riemann boundary value problem for quaternionic functions that are taking into account both the noncommutative nature of quaternionic multiplication and the presence of a family of classes of hyperholomorphic functions, which adequately generalize the notion of holomorphic functions of one complex variable; algebras, generated by the singular integral operators with quaternionic Cauchy kernel and piece-wise continuous coefficients; operators with quaternion and Clifford Bergman kernels. The Toeplitz operators in quaternion and Clifford setting have been introduced and studied in the first time. This work found the most favorable response and initiated dozens of citations.

During his life in Mexico, the scientific interests of Nikolai Vasilevski mainly concentrated around the theory of Toeplitz operators on Bergman and Fock spaces. In the end of 1990ies, N. Vasilevski discovered a quite surprising phenomenon in the theory of Toeplitz operators on the Bergman space. Unexpectedly, there exists a rich family of commutative C^*-algebras generated by Toeplitz operators with non-trivial defining symbols. In 1995 B. Korenblum and K. Zhu proved that the Toeplitz operators with radial defining symbols acting on the Bergman space over the unit disk can be diagonalized with respect to the standard monomial basis in the Bergman space. The C^*-algebra generated by such Toeplitz operators is therefore obviously commutative. Four years later N. Vasilevski also showed the commutativity of the C^*-algebra generated by the Toeplitz operators acting on the Bergman space over the upper half-plane and with defining symbols depending only on $\operatorname{Im} z$. Furthermore, he discovered the existence of a rich family of commutative C^*-algebras of Toeplitz operators. Moreover, it turned out that the smoothness properties of the symbols do not play any role in commutativity: the symbols can be merely measurable. Surprisingly, everything is governed by the geometry of the underlying manifold, the unit disk equipped with the hyperbolic metric. The precise description of this phenomenon is as follows. Each pencil of hyperbolic geodesics determines the set of symbols which are constant on the corresponding cycles, the orthogonal trajectories to geodesics forming the pencil. The C^*-algebra generated by the Toeplitz operators with such defining symbols is commutative. An important feature of such algebras is that they remain commutative for the Toeplitz operators acting on each of the commonly considered weighted Bergman spaces. Moreover, assuming some natural conditions on "richness" of the classes of symbols, the following complete characterization has been obtained: A C^*-algebra generated by the Toeplitz operators is commutative on each weighted Bergman space if and only if the corresponding defining symbols are constant on cycles of some pencil of hyperbolic geodesics. Apart from its own beauty, this result reveals an extremely deep influence of the geometry of the underlying manifold on the properties of the Toeplitz operators over the manifold. In each of the mentioned above cases, when the algebra is commutative, a certain unitary operator has been constructed. It reduces the corresponding Toeplitz operators to certain multiplication operators, which also allows one to describe their representations of spectral type. This gives a powerful research tool for the subject, in particular, yielding direct access to the majority of the important properties such as boundedness, compactness, spectral properties, invariant subspaces, of the Toeplitz operators under study.

The results of the research in this directions became a part of the monograph "Commutative Algebras of Toeplitz Operators on the Bergman Space" published by N. Vasilevski in Birkhäuser in 2008.

Nikolai Leonidovich Vasilevski passed his sixties birthday on full speed, and being in excellent shape. We, his friends, students, and colleagues, wish him further success and, above all, many new interesting and successfully solved problems.

Principal publications of Nikolai Vasilevski

Book

1. N.L. Vasilevski. *Commutative Algebras of Toeplitz Operators on the Bergman Space*, Operator Theory: Advances and Applications, Vol. 183, Birkhäuser Verlag, Basel-Boston-Berlin, 2008, XXIX, 417 p.

Articles

1. N.L. Vasilevski. *On the Noether conditions and a formula for the index of a class of integral operators.* Doklady Akad. Nauk SSSR, 1972, v. 202, N° 4, p. 747–750 (Russian).
   English translation: Soviet Math. Dokl., v. 13, no. 1, 1972, p. 175–179.
2. N.L. Vasilevski. *On properties of a class of integral operators in the space L_p.* Matemat. Zametki, 1974, v. 16, N° 4, p. 529–535 (Russian).
   English translation: Math. Notes, v. 16, no. 4, 1974, p. 905–909.
3. N.L. Vasilevski. *The Noether theory of a class of potential type integral operators.* Izvestija VUZov. Matematika, 1974, N° 7, p. 12–20 (Russian).
   English translation: Soviet Math. (Izv. VUZ), v. 18, no. 7, 1974, p. 8–15.
4. N.L. Vasilevski. *On the Noetherian theory of integral operators with a polar logarithmic kernel.* Doklady Akad. Nauk SSSR, 1974, v. 215, N° 3, p. 514–517 (Russian).
   English translation: Soviet Math. Dokl., v. 15, no. 2, 1974, p. 522–527.
5. N.L. Vasilevski, E.V. Gutnikov. *On the symbol of operators forming finite-dimensional algebras.* Doklady Akad. Nauk SSSR, 1975, v. 221, N° 1, p. 18–21 (Russian).
   English translation: Soviet Math. Dokl., v. 16, no. 2, 1975, p. 271–275.
6. N.L. Vasilevski, G.S. Litvinchuk. *Theory of solvability of a class of singular integral equations with involution.* Doklady Akad. Nauk SSSR, 1975, v. 221, N° 2, p. 269–271 (Russian).
   English translation: Soviet Math. Dokl., v. 16, no. 2, 1975, p. 318–321.
7. N.L. Vasilevski, M.V. Shapiro. *On an algebra generated by singular integral operators with the Carleman shift and in the case of piece-wise continuous coefficients.* Ukrainski Matematicheski Zurnal, 1975, v. 27, N° 2, p. 216–223 (Russian).
   English translation: Ukrainian Math. J., v. 27, no. 2, 1975, p. 171–176.

8. N.L. Vasilevski. *On a class of singular integral operators with kernels of polar-logarithmic type.* Izvestija Akad. Nauk SSSR, ser. matem., 1976, v. 40, N° 1, p. 131–151 (Russian).
English translation: Math. USSR Izvestija, v. 10, no. 1, 1976, p. 127–143.

9. N.L. Vasilevski, E.V. Gutnikov. *On the structure of the symbol of operators forming finite-dimensional algebras.* Doklady Akad. Nauk SSSR, 1976, v. 230, N° 1, p. 11–14 (Russian).
English translation: Soviet Math. Dokl., v. 17, no. 5, 1976, p. 1225–1229.

10. N.L. Vasilevski, A.A. Karelin, P.V. Kerekesha, G.S. Litvinchuk. *On a class of singular integral equations with shift and its applications in the theory of boundary value problems for partial differential equations.* I. Differentsialnye Uravnenija, 1977, v. 13, N° 9, p. 1692–1700 (Russian).
English translation: Diff. Equations, v. 13, no. 9, 1977, p. 1180–1185.

11. N.L. Vasilevski, A.A. Karelin, P.V. Kerekesha, G.S. Litvinchuk. *On a class of singular integral equations with shift and its applications in the theory of boundary value problems for partial differential equations.* II. Differentsialnye Uravnenija, 1977, v. 13, N° 11, p. 2051–2062 (Russian).
English translation: Diff. Equations, v. 13, no. 11, 1977, p. 1430–1438.

12. N.L. Vasilevski. *Symbols of operator algebras.* Doklady Akad. Nauk SSSR, 1977, v. 235, N° 1, p. 15–18 (Russian).
English translation: Soviet Math. Dokl., v. 18, no. 4, 1977, p. 872–876.

13. N.L. Vasilevski, A.A. Karelin. *An investigation of a boundary value problem for the partial differential equation of the mixed type with the help of reduction to the singular integral equation with Carleman shift.* Izvestija VUZov. Matematika, 1978, N° 3, p. 15–19, (Russian).
English translation: Soviet Math. (Izv. VUZ), v. 22, no. 3, 1978, p. 11–15.

14. N.L. Vasilevski, R. Trujillo. *On Φ_R-operators in matrix algebras of operators.* Doklady Akad. Nauk SSSR, 1979, v. 245, N° 6, p. 1289–1292 (Russian).
English translation: Soviet Math. Dokl., v. 20, no. 2, 1979, p. 406–409.

15. N.L. Vasilevski, R. Trujillo. *On the theory of Φ_R-operators in matrix algebras of operators.* Linejnye Operatory, Kishinev, 1980, p. 3–15 (Russian).

16. N.L. Vasilevski. *On an algebra generated by some two-dimensional integral operators with continuous coefficients in a subdomain of the unit disc.* Journal of Integral Equations, 1980, v. 2, p. 111–116.

17. N.L. Vasilevski. *Banach algebras generated by some two-dimensional integral operators.* I. Math. Nachr., 1980, b. 96, p. 245–255 (Russian).

18. N.L. Vasilevski. *Banach algebras generated by some two-dimensional integral operators.* II. Math. Nachr., 1980, b. 99, p. 136–144 (Russian).

19. N.L. Vasilevski. *On the symbol theory for Banach operator algebras which generalizes algebras of singular integral operators.* Differentsialnye Uravnenija, 1981, v. 17, N° 4, p. 678–688 (Russian).
English translation: Diff. Equations, v. 17, no. 4, 1981, p. 462–469.

20. N.L. Vasilevski, I.M. Spitkovsky. *On an algebra generated by two projectors.* Doklady Akad. Nauk UkSSR, Ser. “A”, 1981, N° 8, p. 10–13 (Russian).

21. N.L. Vasilevski. *On the algebra generated by two-dimensional integral operators with Bergman kernel and piece-wise continuous coefficients.* Doklady Akad. Nauk SSSR, 1983, v. 271, N° 5, p. 1041–1044 (Russian).
English translation: Soviet Math. Dokl., v. 28, no. 1, 1983, p. 191–194.

22. N.L. Vasilevski. *On certain algebras generated by a space analog of the singular operator with Cauchy kernel.* Doklady Akad. Nauk SSSR, 1983, v. 273, N° 3, p. 521–524 (Russian).
English translation: Soviet Math. Dokl., v. 28, no. 3, 1983, p. 654–657.

23. N.L. Vasilevski. *On an algebra generated by abstract singular operators and Carleman shift.* Soobshchenija Akad. Nauk GSSR, 1984, v. 115, N° 3, p. 473–476 (Russian).

24. N.L. Vasilevski. *On an algebra generated by multivariable Wiener-Hopf operators.* Reports of Enlarged Session of Seminars of the I.N. Vekua Institute of Applied Mathematics. Tbilisi, 1985, v. 1, p. 59–62 (Russian).

25. N.L. Vasilevski, M.V. Shapiro. *On an analogy of monogenity in the sense of Moisil-Teodoresko and some applications in the theory of boundary value problems.* Reports of Enlarged Sessions of Seminars of the I.N. Vekua Institute of Applied Mathematics. Tbilisi, 1985, v. 1, p. 63–66 (Russian).

26. N.L. Vasilevski. *Algebras generated by multivariable Toeplitz operators with piece-wise continuous presymbols.* Scientific Proceedings of the Boundary Value Problems Seminar Dedicated to 75th birthday of Academician BSSR Academy of Sciences F.D. Gahov. Minsk, 1985, p. 149–150 (Russian).

27. N.L. Vasilevski. *Two-dimensional Mikhlin-Calderon-Zygmund operators and bisingular operators.* Sibirski Matematicheski Zurnal, 1986, v. 27, N° 2, p. 23–31 (Russian).
English translation: Siberian Math. J., v. 27, no. 2, 1986, p. 161–168.

28. N.L. Vasilevski. *Banach algebras generated by two-dimensional integral operators with Bergman Kernel and piece-wise continuous coefficients.* I. Izvestija VUZov, Matematika, 1986, N° 2, p. 12–21 (Russian).
English translation: Soviet Math. (Izv. VUZ), v. 30, no. 2, 1986, p. 14–24.

29. N.L. Vasilevski. *Banach algebras generated by two-dimensional integral operators with Bergman Kernel and piece-wise continuous coefficients.* II. Izvestija VUZov, Matematika, 1986, N° 3, p. 33–38 (Russian).
English translation: Soviet Math. (Izv. VUZ), v. 30, no. 3, 1986, p. 44–50.

30. N.L. Vasilevski. *Algebras generated by multidimensional singular integral operators and by coefficients admitting discontinuities of homogeneous type.* Matematicheski Sbornik, 1986, v. 129, N° 1, p. 3–19 (Russian).
English translation: Math. USSR Sbornik, v. 57, no. 1, 1987, p. 1–19.

31. N.L. Vasilevski. *On an algebra generated by Toeplitz operators with zero-order pseudodifferential presymbols.* Doklady Akad. Nauk SSSR, 1986, v. 289, N° 1, p. 14–18 (Russian).
English translation: Soviet Math. Dokl., v. 34, no. 1, 1987, p. 4–7.

32. N.L. Vasilevski, M.V. Shapiro. *On quaternion Ψ-monogenic function.* "Methods of solving of the direct and inverse geoelectrical problems". 1987, p. 54–65 (Russian).

33. N.L. Vasilevski. *On an algebra connected with Toeplitz operators on the tube domains.* Izvestija Akad. Nauk SSSR, ser. matem., 1987, v. 51, N° 1, p. 79–95 (Russian).
English translation: Math. USSR Izvestija, v. 30, no.1, 1988, p. 71–87.

34. N.L. Vasilevski, R. Trujillo. *On C^*-algebra generated by almost-periodic two-dimensional singular integral operators with discontinuous presymbols.* Funkcionalny Analiz i ego Prilogenija, 1987, v. 21, N° 3, p. 75–76 (Russian).
English translation: Func. Analysis and its Appl., v. 21, no. 3, 1987, p. 235–236.

35. N.L. Vasilevski. *Toeplitz operators associated with the Siegel domains.* Matematicki Vesnik, 1988, v. 40, p. 349–354.

36. N.L. Vasilevski. *Hardy spaces associated with the Siegel domains.* Reports of Enlarged Sessions of Seminars of the I.N. Vekua Institute of Applied Mathematics. Tbilisi, 1988, v. 3, N° 1, p. 48–51 (Russian).

37. N.L. Vasilevski, M.V. Shapiro. *Holomorphy, hyperholomorphy Toeplitz operators.* Uspehi Matematicheskih Nauk, 1989, v. 44, N° 4 (268), p. 226–227 (Russian).
English translation: Russian Math. Surveys, v. 44, no. 4, 1989, p. 196–197.

38. N.L. Vasilevski, M.V. Shapiro. *Some questions of hypercomplex analysis* "Complex Analysis and Applications '87", Sofia, 1989, p. 523–531.

39. N.L. Vasilevski. *Non-classical singular integral operators and algebras generated by them.* Integral Equations and Boundary Value Problems. World Scientific. 1991, p. 210–215.

40. M.V. Shapiro, N.L. Vasilevski. *Singular integral operator in Clifford analysis,* Clifford Algebras and Their Applications in Mathematical Physics, Kluwer Academic Publishers, Netherlands, 1992, p. 271–277.

41. N.L. Vasilevski. *On an algebra generated by abstract singular operators and a shift operator.* Math. Nachr., v. 162, 1993, p. 89–108.

42. R.M. Porter, M.V. Shapiro, N.L. Vasilevski. *On the analogue of the $\overline{\partial}$-problem in quaternionic analysis.* Clifford Algebras and Their Applications in Mathematical Physics, F. Brackx et al., eds., Kluwer Academic Publishers, Netherlands, 1993, p. 167–173.

43. E. Ramírez de Arellano, M.V. Shapiro, N.L. Vasilevski. *Hurwitz pairs and Clifford algebra representations.* Clifford Algebras and Their Applications in Mathematical Physics, F. Brackx et al., eds., Kluwer Academic Publishers, Netherlands, 1993, p. 175–181.

44. M.V. Shapiro, N.L. Vasilevski. *On the Bergman kernel function in the Clifford analysis.* Clifford Algebras and Their Applications in Mathematical Physics, F. Brackx et al., eds., Kluwer Academic Publishers, Netherlands, 1993, p. 183–192.

45. N.L. Vasilevski. *On "discontinuous" boundary value problems for pseudodifferential operators.* International Conference on Differential Equations, Vol. 1, 2, (Barcelona, 1991), World Sci. Publishing, River Edge, NJ, 1993, p. 953–958.

46. N.L. Vasilevski, R. Trujillo. *Convolution operators on standard CR-manifolds. I. Structural Properties.* Integral Equations and Operator Theory, v. 19, no. 1, 1994, p. 65–107.

47. N.L. Vasilevski. *Convolution operators on standard CR-manifolds. II. Algebras of convolution operators on the Heisenberg group.* Integral Equations and Operator Theory, v. 19, no. 3, 1994, p. 327–348.

48. N.L. Vasilevski. *On an algebra generated by two-dimensional singular integral operators in plane domains.* Complex Variables, v. 26, 1994, p. 79–91.

49. R.M. Porter, M. Shapiro, N. Vasilevski. *Quaternionic differential and integral operators and the $\overline{\partial}$-problem.* Journal of Natural Geometry, v. 6, no. 2, 1994, p. 101–124.

50. E. Ramírez de Arellano, M.V. Shapiro, N.L. Vasilevski. *Two types of analysis associated to the notion of Hurwitz pairs.* Differential Geometric Methods in Theoretical Physics, Ed. J. Keller, Z. Oziewich, Advances in Applied Clifford Algebras, v. 4 (S1), 1994, p. 413–422.

51. M.V. Shapiro, N.L. Vasilevski. *Quaternionic Ψ-hyperholomorphic functions, singular integral operators and boundary value problems. I. Ψ-hyperholomorphic function theory.* Complex Variables, v. 27, 1995, p. 17–46.

52. M.V. Shapiro, N.L. Vasilevski. *Quaternionic Ψ-hyperholomorphic functions, singular integral operators and boundary value problems. II. Algebras of singular integral operators and Riemann type boundary value problems.* Complex Variables, v. 27, 1995, p. 67–96.

53. N. Vasilevski, V. Kisil, E. Ramirez de Arellano, R. Trujillo. *Toeplitz operators with discontinuous presymbols on the Fock space.* Russian Math. Doklady, v. 345, no. 2, 1995, p. 153–155 (Russian).
English translation: Russian Math. Doklady.

54. E. Ramírez de Arellano, N.L. Vasilevski. *Toeplitz operators on the Fock space with presymbols discontinuous on a thick set,* Mathematische Nachrichten, v. 180, 1996, p. 299–315.

55. E. Ramírez de Arellano, N.L. Vasilevski. *Algebras of singular integral operators generated by three orthogonal projections,* Integral Equations and Operator Theory, v. 25, no. 3, 1996, p. 277–288.

56. N. Vasilevski, E. Ramirez de Arellano, M. Shapiro. *Hurwitz classical problem and associated function theory.* Russian Math. Doklady, v. 349, no. 5, 1996, p. 588–591 (Russian).
English translation: Russian Math. Doklady

57. E. Ramírez de Arellano, M.V. Shapiro, N.L. Vasilevski. *The hyperholomorphic Bergman projector and its properties.* In: *Clifford Algebras and Related Topics,* J. Ryan, Ed. CRC Press, Chapter 19, 1996, p. 333–344.

58. E. Ramírez de Arellano, M.V. Shapiro, N.L. Vasilevski. *Hurwitz analysis: basic concepts and connection with Clifford analysis.* In: *Generalizations of Complex Analysis and their Applications in Physics,* J. Lawrynowicz, Ed. Banach Center Publications, V. 37, Warszawa, 1996, p. 209–221.

59. M.V. Shapiro, N.L. Vasilevski. *On the Bergman kernel function in hyperholomorphic analysis,* Acta Applicandae Mathematicae, v. 46, 1997, p. 1–27.

60. E. Ramírez de Arellano, N.L. Vasilevski. *Bargmann projection, three-valued functions and corresponding Toeplitz operators,* Contemporary Mathematics, v. 212, 1998, p. 185–196.

61. N.L. Vasilevski. *C*-algebras generated by orthogonal projections and their applications.* Integral Equations and Operator Theory, v. 31, 1998, p. 113–132.

62. N.L. Vasilevski, M.V. Shapiro. *On the Bergman kern-function on quaternionic analysis.* Izvestiia VUZov, Matematika, no. 2, 1998, p. 84–88 (Russian).
English translation: Russian Math. (Izvestiia VUZ), v. 42, no. 2, 1998, p. 81–85.

63. N.L. Vasilevski, *On the structure of Bergman and poly-Bergman spaces,* Integral Equations and Operator Theory, v. 33, 1999, p. 471–488.

64. N.L. Vasilevski, *On Bergman-Toeplitz operators with commutative symbol algebras,* Integral Equations and Operator Theory, v. 34, no. 1, 1999, p. 107–126.

65. E. Ramírez de Arellano, M.V. Shapiro, N.L. Vasilevski. *The hyperholomorphic Bergman projector and its properties.* In: *Clifford Algebras and Related Topics,* J. Ryan, Ed. CRC Press, Chapter 19, 1996, p. 333–344.

66. E. Ramírez de Arellano, M.V. Shapiro, N.L. Vasilevski. *Hurwitz analysis: basic concepts and connection with Clifford analysis.* In: *Generalizations of Complex Analysis and their Applications in Physics,* J. Lawrynowicz, Ed. Banach Center Publications, v. 37, Warszawa, 1996, p. 209–221.

67. N.L. Vasilevski, *On quaternionic Bergman and poly-Bergman spaces,* Complex Variables, v. 41, 2000, p. 111–132.

68. V.S. Rabinovish, N.L. Vasilevski, *Bergman-Toeplitz and pseudodifferential operators,* Operator Theory. Advances and Applications v. 114, 2000, p. 207–234.

69. N.L. Vasilevski, *Poly-Fock Spaces,* Operator Theory. Advances and Applications v. 117, 2000, p. 371–386.

70. V.V. Kucherenko, N.L. Vasilevski, *A shift operator generated by a trigonometric system,* Mat. Zametki, v. 67, no. 4, 2000, p. 539–548 (Russian). English translation: Mat. Notes.

71. N.L. Vasilevski, *The Bergman space in tube domains, and commuting Toeplitz operators,* Doklady RAN, v. 372, no. 1, 2000, p. 9–12 (Russian). English translation: Doklady Mathematics, v. 61, no. 3, 2000.

72. N.L. Vasilevski. *Bergman space on tube domains and commuting Toeplitz operators.* In: *Proceedings of the Second ISAAC Congress, Volume* 2, H.G.W. Begehr et al. (eds.), Kluwer Academic Publishers, The Netherlands, Chapter 163, 2000, p. 1523–1537.

73. S. Grudsky, N. Vasilevski, *Bergman-Toeplitz operators: Radial component influence,* Integral Equations and Operator Theory, v. 40, no. 1, 2001, p. 16–33.

74. A.N. Karapetyants, V.S. Rabinovich, N.L. Vasilevski, *On algebras of two-dimensional singular integral operators with homogeneous discontinuities in symbols,* Integral Equations and Operator Theory, v. 40, no. 3, 2001, p. 278–308.

75. N.L. Vasilevski. *Toeplitz Operators on the Bergman Spaces: Inside-the-Domain Effects,* Contemporary Mathematics, v. 289, 2001, p. 79–146.

76. N.L. Vasilevski. *Bergman spaces on the unit disk.* In: *Clifford Analysis and Its Applications* F. Brackx et al. (eds.), Kluwer Academic Publishers, The Netherlands, 2001, p. 399–409.

77. S. Grudsky, N. Vasilevski. *Toeplitz operators on the Fock space: Radial component effects,* Integral Equations and Operator Theory, v. 44, no. 1, 2002, p. 10–37.

78. N.L. Vasilevski. *Commutative algebras of Toeplitz operators and hyperbolic geometry.* In: *Proceedings of the Ukranian Mathematical Congress* – 2001, *Functional Analysis, Section* 11, Institute of Mathematics of the National Academy of Sciences, Ukraine, 2002, p. 22–35.

79. N.L. Vasilevski. *Bergman Space Structure, Commutative Algebras of Toeplitz Operators and Hyperbolic Geometry,* Integral Equations and Operator Theory, v. 46, 2003, p. 235–251.

80. S. Grudsky, A. Karapetyants, N. Vasilevski. *Toeplitz Operators on the Unit Ball in* $\mathbb{C}^n$ *with Radial Symbols,* J. Operator Theory, v. 49, 2003, p. 325–346.

81. J. Ramírez Ortega, N. Vasilevski, E. Ramírez de Arellano *On the algebra generated by the Bergman projection and a shift operator. I.* Integral Equations and Operator Theory, v. 46, no. 4, 2003, p. 455–471.

82. N.L. Vasilevski. *Toeplitz operators on the Bergman space.* In: *Factorization, Singular Operators and Related Problems*, Edited by S. Samko, A. Lebre, A.F. dos Santos, Kluwer Academic Publishers, 2003, p. 315–333.

83. J. Ramírez Ortega, E. Ramírez de Arellano, N. Vasilevski *On the algebra generated by the Bergman projection and a shift operator. II.* Bol. Soc. Mat. Mexicana (3), v. 10, 2004, p. 105–117.

84. S. Grudsky, A. Karapetyants, N. Vasilevski. *Dynamics of properties of Toeplitz operators on the upper half-plane: Hyperbolic case,* Bol. Soc. Mat. Mexicana (3), v. 10, 2004, p. 119–138.

85. S. Grudsky, A. Karapetyants, N. Vasilevski. *Dynamics of properties of Toeplitz operators on the upper half-plane: Parabolic case,* J. Operator Theory, v. 52, no. 1, 2004, p. 185–204.

86. S. Grudsky, A. Karapetyants, N. Vasilevski. *Dynamics of properties of Toeplitz operators with radial Symbols,* Integral Equations and Operator Theory, v. 50, no. 2, 2004, p. 217–253.

87. N.L. Vasilevski. *On a general local principle for C^*-algebras,* Izv.VUZ North-Caucasian Region, Natural Sciences, Special Issue, "Pseudodifferential operators and some problems of mathematical physics", 2005, p. 34–42 (Russian).

88. S. Grudsky, N. Vasilevski. *Dynamics of Spectra of Toeplitz Operators,* Advances in Analysis. Proceedings of the 4th International ISAAC Congress. (York University, Toronto, Canada 11–16 August 2003), World Scientific, New Jersey London Singapore, 2005, p. 495–504.

89. S. Grudsky, R. Quiroga-Barranco, N. Vasilevski. *Commutative C^*-algebras of Toeplitz operators and quantization on the unit disk,* J. Functional Analysis, v. 234, 2006, p. 1–44.

90. N.L. Vasilevski, S.M. Grudsky, A.N. Karapetyants. *Dynamics of properties of Toeplitz operators on weighted Bergman spaces,* Siberian Electronic Math. Reports, v. 3, 2006, p. 362–383 (Russian).

91. N. Vasilevski. *On the Toeplitz operators with piecewise continuous symbols on the Bergman space,* In: "Modern Operator Theory and Applications", Operator Theory: Advances and Applications, v. 170, 2007, p. 229–248.

92. N. Vasilevski. *Poly-Bergman spaces and two-dimensional singular integral operators,* Operator Theory: Advances and Applications, v. 171, 2007, p. 349–359.

93. N. Tarkhanov, N. Vasilevski. *Microlocal analysis of the Bochner-Martinelli integral,* Integral Equations and Operator Theory, v. 57, 2007, p. 583–592.

94. R. Quiroga-Barranco, N. Vasilevski. *Commutative algebras of Toeplitz operators on the Reinhardt domains*, Integral Equations and Operator Theory, v. 59, no. 1, 2007, p. 67–98.

95. R. Quiroga-Barranco, N. Vasilevski. *Commutative C^*-algebras of Toeplitz operators on the unit ball, I. Bargmann-type transforms and spectral representations of Toeplitz operators*, Integral Equations and Operator Theory, v. 59, no. 3, 2007, p. 379–419.
96. R. Quiroga-Barranco, N. Vasilevski. *Commutative C^*-algebras of Toeplitz operators on the unit ball, II. Geometry of the level sets of symbols*, Integral Equations and Operator Theory, v. 60, no. 1, 2008, p. 89–132.
97. N. Vasilevski. *Commutative algebras of Toeplitz operators and Berezin quantization*, Contemporary Mathematics, v. 462, 2008, p. 125–143.
98. S. Grudsky, N. Vasilevski. *On the structure of the C^*-algebra generated by Toeplitz operators with piece-wise continuous symbols*, Complex Analysis and Operator Theory, v. 2, no. 4, 2008, p. 525–548.

Ph. D. dissertations directed by Nikolai Vasilevski

1. Rafael Trujillo, *Fredholm Theory of Tensor Product of Operator Algebras*, Odessa State University, 1986.
2. Zhelko Radulovich, *Algebras of Multidimensional Singular Integral Operators with Discontinuous Symbols with Respect to Dual Variable*, Odessa State University, 1991.
3. Vladimir Kisil, *Algebras of Pseudodifferential Operators Associated with the Heisenberg Group*, Odessa State University, 1992.
4. Josué Ramírez Ortega, *Algebra generada por la proyección de Bergman y un operador de translación*, CINVESTAV del I.P.N., Mexico City, 1999.
5. Maribel Loaiza Leyva, *Algebra generada por la proyección de Bergman y por los operadores de multiplicación por funciones continuas a trozos*, CINVESTAV del I.P.N., Mexico City, 2000.
6. Ernesto Prieto Sanabría, *Operadores de Toeplitz en la 2-esfera en los espacios de Bergman con peso*, CINVESTAV del I.P.N., Mexico City, 2007.
7. Armando Sánchez Nungaray, *Super operadores de Toeplitz en la dos-esfera*, CINVESTAV del I.P.N., Mexico City, 2008.
8. Carlos Moreno Muñoz, *Operadores de Toeplitz en el espacio de Bergman con peso: Caso parabólico*, CINVESTAV del I.P.N., Mexico City, 2009.

Operator Theory:
Advances and Applications, Vol. 210, 15–36

On the Structure of the Eigenvectors of Large Hermitian Toeplitz Band Matrices

Albrecht Böttcher, Sergei M. Grudsky and Egor A. Maksimenko

For Nikolai Vasilevski on His 60th Birthday

Abstract. The paper is devoted to the asymptotic behavior of the eigenvectors of banded Hermitian Toeplitz matrices as the dimension of the matrices increases to infinity. The main result, which is based on certain assumptions, describes the structure of the eigenvectors in terms of the Laurent polynomial that generates the matrices up to an error term that decays exponentially fast. This result is applicable to both extreme and inner eigenvectors.

Mathematics Subject Classification (2000). Primary 47B35; Secondary 15A18, 41A25, 65F15.

Keywords. Toeplitz matrix, eigenvector, asymptotic expansions.

1. Introduction and main results

Given a function a in L^1 on the complex unit circle $\mathbf{T}$, we denote by a_ℓ the ℓth Fourier coefficient,

$$a_\ell = \frac{1}{2\pi}\int_0^{2\pi} a(e^{ix})e^{-i\ell x}dx \quad (\ell \in \mathbf{Z}),$$

and by $T_n(a)$ the $n \times n$ Toeplitz matrix $(a_{j-k})_{j,k=1}^n$. We assume that a is real-valued, in which case the matrices $T_n(a)$ are all Hermitian. Let

$$\lambda_1^{(n)} \le \lambda_2^{(n)} \le \cdots \le \lambda_n^{(n)}$$

be the eigenvalues of $T_n(a)$ and let

$$\{v_1^{(n)}, v_2^{(n)}, \ldots, v_n^{(n)}\}$$

This work was partially supported by CONACYT project 80503, Mexico.

be an orthonormal basis of eigenvectors such that $T_n(a)v_j^{(n)} = \lambda_j^{(n)} v_j^{(n)}$. The present paper is dedicated to the asymptotic behavior of the eigenvectors $v_j^{(n)}$ as $n \to \infty$.

To get an idea of the kind of results we will establish, consider the function $a(e^{ix}) = 2 - 2\cos x$. The range $a(\mathbf{T})$ is the segment $[0, 4]$. It is well known that the eigenvalues and eigenvectors of $T_n(a)$ are given by

$$\lambda_j^{(n)} = 2 - 2\cos\frac{\pi j}{n+1}, \quad x_j^{(n)} = \sqrt{\frac{2}{n+1}} \left(\sin \frac{m\pi j}{n+1} \right)_{m=1}^{n}. \tag{1.1}$$

(We denote the eigenvectors in this reference case by $x_j^{(n)}$ and reserve the notation $v_j^{(n)}$ for the general case.) Let φ be the function

$$\varphi : [0, 4] \to [0, \pi], \quad \varphi(\lambda) = \arccos\frac{2-\lambda}{2}.$$

We have $\varphi(\lambda_j^{(n)}) = \pi j/(n+1)$ and hence, apart from the normalization factor $\sqrt{2/(n+1)}$, $x_{j,m}^{(n)}$ is the value of $\sin(m\varphi(\lambda))$ at $\lambda = \lambda_j^{(n)}$. In other words, an eigenvector for λ is given by $(\sin(m\varphi(\lambda)))_{m=1}^n$. A speculative question is whether in the general case we can also find functions Ω_m such that, at least asymptotically, $(\Omega_m(\lambda))_{m=1}^n$ is an eigenvector for λ. It turns out that this is in general impossible but that after a slight modification the answer to the question is in the affirmative. Namely, we will prove that, under certain assumptions, there are functions Ω_m, Φ_m and real-valued functions σ, η such that an eigenvector for $\lambda = \lambda_j^{(n)}$ is always of the form

$$\left(\Omega_m(\lambda)+\Phi_m(\lambda)+(-1)^{j+1}e^{-i(n+1)\sigma(\lambda)}e^{-i\eta(\lambda)}\overline{\Phi_{n+1-m}(\lambda)}+\text{error term}\right)_{m=1}^{n}. \tag{1.2}$$

The error term will be shown to decrease to zero exponentially fast and uniformly in j and m as $n \to \infty$. Moreover, we will show that $\Omega_m(\lambda)$ is an oscillating function of m for each fixed λ and that $\Phi_m(\lambda)$ decays exponentially fast to zero as $m \to \infty$ for each λ (which means that $\Phi_{n+1-m}(\lambda)$ is an exponentially increasing function of m for each λ). Finally, it will turn out that

$$\sum_{m=1}^{n} |\Phi_m(\lambda)|^2 \Big/ \sum_{m=1}^{n} |\Omega_m(\lambda)|^2 = O\left(\frac{1}{n}\right)$$

as $n \to \infty$, uniformly in λ. Thus, the dominant term in (1.2) is $\Omega_m(\lambda)$, while the terms containing $\Phi_m(\lambda)$ and $\Phi_{n+1-m}(\lambda)$ may be viewed as twin babies.

If a is also an even function, $a(e^{ix}) = a(e^{-ix})$ for all x, then all the matrices $T_n(a)$ are real and symmetric. In [4], we conjectured that then, again under additional but reasonable assumptions, the appropriately rotated extreme eigenvectors $v_j^{(n)}$ are all close to the vectors $x_j^{(n)}$. To be more precise, we conjectured that if $n \to \infty$ and j (or $n - j$) remains fixed, then there are complex numbers $\tau_j^{(n)}$ of modulus 1 such that

$$\left\| \tau_j^{(n)} v_j^{(n)} - x_j^{(n)} \right\|_2 = o(1), \tag{1.3}$$

where $\|\cdot\|_2$ is the ℓ^2 norm. Several results related to this conjecture were established in [3] and [4]. We here prove this conjecture under assumptions that will be specified in the following paragraph. We will even be able to show that the $o(1)$ in (1.3) is $O(j/n)$ if $j/n \to 0$ and $O(1-j/n)$ if $j/n \to 1$.

Throughout what follows we assume that a is a Laurent polynomial

$$a(t) = \sum_{k=-r}^{r} a_k t^k \quad (t = e^{ix} \in \mathbf{T})$$

with $r \geq 2$, $a_r \neq 0$, and $\overline{a_k} = a_{-k}$ for all k. The last condition means that a is real-valued on $\mathbf{T}$. We assume without loss of generality that $a(\mathbf{T}) = [0, M]$ with $M > 0$ and that $a(1) = 0$ and $a(e^{i\varphi_0}) = M$ for some $\varphi_0 \in (0, 2\pi)$. We require that the function $g(x) := a(e^{ix})$ is strictly increasing on $(0, \varphi_0)$ and strictly decreasing on $(\varphi_0, 2\pi)$ and that the second derivatives of g at $x = 0$ and $x = \varphi_0$ are nonzero. Finally, we denote by $[\alpha, \beta] \subset [0, M]$ a segment such that if $\lambda \in [\alpha, \beta]$, then the $2r - 2$ zeros of the Laurent polynomial $a(z) - \lambda$ that lie in $\mathbf{C} \setminus \mathbf{T}$ are pairwise distinct.

Note that we exclude the case $r = 1$, because in this case the eigenvalues and eigenvectors of $T_n(a)$ are explicitly available. Also notice that if $r = 2$, which is the case of pentadiagonal matrices, then for every $\lambda \in [0, M]$ the polynomial $a(z) - \lambda$ has two zeros on $\mathbf{T}$, one zero outside $\mathbf{T}$, and one zero inside $\mathbf{T}$. Thus, in this situation the last requirement of the previous paragraph is automatically satisfied for $[\alpha, \beta] = [0, M]$.

The asymptotic behavior of the extreme eigenvalues and eigenvectors of $T_n(a)$, that is, of $\lambda_j^{(n)}$ and $v_j^{(n)}$ when j or $n - j$ remain fixed, has been studied by several authors. As for extreme eigenvalues, the pioneering works are [7], [9], [11], [12], [18], while recent papers on the subject include [3], [6], [8], [10], [13], [14], [15], [19], [20]. See also the books [1] and [5]. Much less is known about the asymptotics of the eigenvectors. Part of the results of [4] and [19] may be interpreted as results on the behavior of the eigenvectors "in the mean" on the one hand and as insights into what happens if eigenvectors are replaced by pseudomodes on the other. In [3], we investigated the asymptotics of the extreme eigenvectors of certain Hermitian (and not necessarily banded) Toeplitz matrices. Our paper [2] may be considered as a first step to the understanding of the asymptotic behavior of individual inner eigenvalues of Toeplitz matrices. In the same vein, this paper intends to understand the nature of individual eigenvectors as part of the whole, independently of whether they are extreme or inner ones.

To state our main results, we need some notation. Let $\lambda \in [0, M]$. Then there are uniquely defined $\varphi_1(\lambda) \in [0, \varphi_0]$ and $\varphi_2(\lambda) \in [\varphi_0 - 2\pi, 0]$ such that

$$g(\varphi_1(\lambda)) = g(\varphi_2(\lambda)) = \lambda;$$

recall that $g(x) := a(e^{ix})$. We put

$$\varphi(\lambda) = \frac{\varphi_1(\lambda) - \varphi_2(\lambda)}{2}, \quad \sigma(\lambda) = \frac{\varphi_1(\lambda) + \varphi_2(\lambda)}{2}.$$

We have

$$\begin{aligned} a(z) - \lambda &= z^{-r}\Big(a_r z^{2r} + \cdots + (a_0 - \lambda)z^r + \cdots + a_{-r}\Big) \\ &= a_r z^{-r} \prod_{k=1}^{2r} (z - z_k(\lambda)), \end{aligned}$$

and our assumptions imply that we can label the zeros $z_k(\lambda)$ so that the collection $\mathcal{Z}(\lambda)$ of the zeros may be written as

$$\begin{aligned} &\{z_1(\lambda), \dots, z_{r-1}(\lambda), z_r(\lambda), z_{r+1}(\lambda), z_{r+2}(\lambda), \dots, z_{2r}(\lambda)\} \\ &= \{u_1(\lambda), \dots, u_{r-1}(\lambda), e^{i\varphi_1(\lambda)}, e^{i\varphi_2(\lambda)}, 1/\overline{u}_1(\lambda), \dots, 1/\overline{u}_{r-1}(\lambda)\} \end{aligned} \tag{1.4}$$

where $|u_\nu(\lambda)| > 1$ for $1 \le \nu \le r-1$ and each $u_\nu(\lambda)$ depends continuously on $\lambda \in [0, M]$. Here and in similar places below we write $\overline{u}_k(\lambda) := \overline{u_k(\lambda)}$. We define $\delta_0 > 0$ by

$$e^{\delta_0} = \min_{\lambda \in [0,M]} \min_{1 \le \nu \le r-1} |u_\nu(\lambda)|.$$

Throughout the following, δ stands for any number in $(0, \delta_0)$. Further, we denote by h_λ the function

$$h_\lambda(z) = \prod_{\nu=1}^{r-1} \left(1 - \frac{z}{u_\nu(\lambda)}\right).$$

The function $\Theta(\lambda) = h_\lambda(e^{i\varphi_1(\lambda)})/h_\lambda(e^{i\varphi_2(\lambda)})$ is continuous and nonzero on $[0, M]$ and we have $\Theta(0) = \Theta(M) = 1$. In [2], it was shown that the closed curve

$$[0, M] \to \mathbf{C} \setminus \{0\}, \quad \lambda \mapsto \Theta(\lambda)$$

has winding number zero. Let $\theta(\lambda)$ be the continuous argument of $\Theta(\lambda)$ for which $\theta(0) = \theta(M) = 0$.

In [2], we proved that if n is large enough, then the function

$$f_n : [0, M] \to [0, (n+1)\pi], \quad f_n(\lambda) = (n+1)\varphi(\lambda) + \theta(\lambda)$$

is bijective and increasing and that if $\lambda_{j,*}^{(n)}$ is the unique solution of the equation $f_n(\lambda_{j,*}^{(n)}) = \pi j$, then the eigenvalues $\lambda_j^{(n)}$ satisfy

$$|\lambda_j - \lambda_{j,*}^{(n)}| \le K e^{-\delta n}$$

for all $j \in \{1, \dots, n\}$, where K is a finite constant depending only on a. Thus, we have

$$(n+1)\varphi(\lambda_j^{(n)}) + \theta(\lambda_j^{(n)}) = \pi j + O(e^{-\delta n}), \tag{1.5}$$

uniformly in $j \in \{1, \dots, n\}$.

Now take λ from (α, β). For $j \in \{1, \dots, n\}$ and $\nu \in \{1, \dots, r-1\}$, we put

$$A(\lambda) = \frac{e^{i\sigma(\lambda)}}{2i\, h_\lambda(e^{i\varphi_1(\lambda)})}, \quad B(\lambda) = \frac{e^{i\sigma(\lambda)}}{2i\, h_\lambda(e^{i\varphi_2(\lambda)})},$$

$$D_\nu(\lambda) = \frac{e^{2i\sigma(\lambda)} \sin\varphi(\lambda)}{(u_\nu(\lambda) - e^{i\varphi_1(\lambda)})(u_\nu(\lambda) - e^{i\varphi_2(\lambda)}) h'_\lambda(u_\nu(\lambda))},$$

$$F_\nu(\lambda) = \frac{\sin\varphi(\lambda)}{(\overline{u}_\nu(\lambda) - e^{-i\varphi_1(\lambda)})(\overline{u}_\nu(\lambda) - e^{-i\varphi_2(\lambda)})\overline{h'_\lambda(u_\nu(\lambda))}} \times \frac{|h_\lambda(e^{i\varphi_1(\lambda)}) h_\lambda(e^{i\varphi_2(\lambda)})|}{h_\lambda(e^{i\varphi_1(\lambda)}) h_\lambda(e^{i\varphi_2(\lambda)})}$$

and define the vector $w_j^{(n)}(\lambda) = (w_{j,m}^{(n)}(\lambda))_{m=1}^n$ by

$$\begin{aligned} w_{j,m}^{(n)}(\lambda) &= A(\lambda)e^{-im\varphi_1(\lambda)} - B(\lambda)e^{-im\varphi_2(\lambda)} \\ &+ \sum_{\nu=1}^{r-1} \left(D_\nu(\lambda) \frac{1}{u_\nu(\lambda)^m} + F_\nu(\lambda) \frac{(-1)^{j+1} e^{-i(n+1)\sigma(\lambda)}}{\overline{u}_\nu(\lambda)^{n+1-m}} \right). \end{aligned}$$

The assumption that zeros $u_\nu(\lambda)$ are all simple guarantees that $h'(u_\nu) \neq 0$. We denote by $\|\cdot\|_2$ and $\|\cdot\|_\infty$ the ℓ^2 and ℓ^∞ norms on $\mathbf{C}^n$, respectively.

Here are our main results.

Theorem 1.1. *As $n \to \infty$ and if $\lambda_j^{(n)} \in (\alpha, \beta)$,*

$$\|w_j^{(n)}(\lambda_j^{(n)})\|_2^2 = \frac{n}{4} \left(\frac{1}{|h_\lambda(e^{i\varphi_1(\lambda)})|^2} + \frac{1}{|h_\lambda(e^{i\varphi_2(\lambda)})|^2} \right)\Bigg|_{\lambda=\lambda_j^{(n)}} + O(1),$$

uniformly in j.

Theorem 1.2. *Let $n \to \infty$ and suppose $\lambda_j^{(n)} \in (\alpha, \beta)$. Then the eigenvectors $v_j^{(n)}$ are of the form*

$$v_j^{(n)} = \tau_j^{(n)} \left(\frac{w_j^{(n)}(\lambda_j^{(n)})}{\|w_j^{(n)}(\lambda_j^{(n)})\|_2} + O_\infty(e^{-\delta n}) \right)$$

where $\tau_j^{(n)} \in \mathbf{T}$ and $O_\infty(e^{-\delta n})$ denotes vectors $\xi_j^{(n)} \in \mathbf{C}^n$ such that $\|\xi_j^{(n)}\|_\infty \leq Ke^{-\delta n}$ for all j and n with some finite constant K independent of j and n.

Note that the previous theorem gives (1.2) with

$$\Omega_m(\lambda) = A(\lambda)e^{-im\varphi_1(\lambda)} - B(\lambda)e^{-im\varphi_2(\lambda)}, \quad \Phi_m(\lambda) = \sum_{\nu=1}^{r-1} \frac{D_\nu(\lambda)}{u_\nu(\lambda)^m},$$

$$e^{-i\eta(\lambda)} = \frac{|h_\lambda(e^{i\varphi_1(\lambda)}) h_\lambda(e^{i\varphi_2(\lambda)})|}{h_\lambda(e^{i\varphi_1(\lambda)}) h_\lambda(e^{i\varphi_2(\lambda)})}.$$

Things can be a little simplified for symmetric matrices. Thus, suppose all a_k are real and $a_k = a_{-k}$ for all k. We will show that then $\{u_1(\lambda), \dots, u_{r-1}(\lambda)\} = \{\overline{u}_1(\lambda), \dots, \overline{u}_{r-1}(\lambda)\}$. Put

$$Q_\nu(\lambda) = \frac{|h_\lambda(e^{i\varphi(\lambda)})| \sin\varphi(\lambda)}{(u_\nu(\lambda) - e^{i\varphi(\lambda)})(u_\nu(\lambda) - e^{-i\varphi(\lambda)})h'_\lambda(u_\nu(\lambda))}$$

and let $y_j^{(n)}(\lambda) = (y_{j,m}^{(n)}(\lambda))_{m=1}^n$ be given by

$$y_{j,m}^{(n)}(\lambda) = \sin\left(m\varphi(\lambda) + \frac{\theta(\lambda)}{2}\right) - \sum_{\nu=1}^{r-1} Q_\nu(\lambda)\left(\frac{1}{u_\nu(\lambda)^m} + \frac{(-1)^{j+1}}{u_\nu(\lambda)^{n+1-m}}\right). \tag{1.6}$$

Theorem 1.3. *Let $n \to \infty$ and suppose $\lambda_j^{(n)} \in (\alpha, \beta)$. If $a_k = a_{-k}$ for all k, then*

$$\|y_j^{(n)}(\lambda_j^{(n)})\|_2^2 = \frac{n}{2} + O(1)$$

uniformly in j, and the eigenvectors $v_j^{(n)}$ are of the form

$$v_j^{(n)} = \tau_j^{(n)}\left(\frac{y_j^{(n)}(\lambda_j^{(n)})}{\|y_j^{(n)}(\lambda_j^{(n)})\|_2} + O_\infty(e^{-\delta n})\right)$$

where $\tau_j^{(n)} \in \mathbf{T}$ and $O_\infty(e^{-\delta n})$ is as in the previous theorem.

Let J be the $n \times n$ matrix with ones on the counterdiagonal and zeros elsewhere. Thus, $(Jv)_m = v_{n+1-m}$. A vector v is called symmetric if $Jv = v$ and skew-symmetric if $Jv = -v$. Trench [17] showed that the eigenvectors $v_1^{(n)}, v_3^{(n)}, \dots$ are all symmetric and that the eigenvectors $v_2^{(n)}, v_4^{(n)}, \dots$ are all skew-symmetric. From (1.5) we infer that

$$\begin{aligned} &\sin\left((n+1-m)\varphi(\lambda_j^{(n)}) + \frac{\theta(\lambda_j^{(n)})}{2}\right) \\ &= (-1)^{j+1}\sin\left(m\varphi(\lambda_j^{(n)}) + \frac{\theta(\lambda_j^{(n)})}{2}\right) + O(e^{-\delta n}) \end{aligned}$$

and hence (1.6) implies that

$$(Jy_j^{(n)}(\lambda_j^{(n)}))_m = (-1)^{j+1} y_{j,m}^{(n)}(\lambda_j^{(n)}) + O(e^{-\delta n}).$$

Consequently, apart from the term $O(e^{-\delta n})$, the vectors $y_j^{(n)}(\lambda_j^{(n)})$ are symmetric for $j = 1, 3, \dots$ and skew-symmetric for $j = 2, 4, \dots$. This is in complete accordance with Trench's result.

Due to (1.5), we also have

$$\sin\left(m\varphi(\lambda_j^{(n)}) + \frac{\theta(\lambda_j^{(n)})}{2}\right) = \sin\left(\left(m - \frac{n+1}{2}\right)\varphi(\lambda_j^{(n)})\right) + O(e^{-\delta n}).$$

Thus, Theorem 1.3 remains valid with (1.6) replaced by

$$\begin{aligned} y_{j,m}^{(n)}(\lambda) &= \sin\left(\left(m - \frac{n+1}{2}\right)\varphi(\lambda) + \frac{\pi j}{2}\right) \\ &\quad - \sum_{\nu=1}^{r-1} Q_\nu(\lambda)\left(\frac{1}{u_\nu(\lambda)^m} + \frac{(-1)^{j+1}}{u_\nu(\lambda)^{n+1-m}}\right). \end{aligned} \tag{1.7}$$

In this expression, the function θ has disappeared.

Define $y_j^{(n)}$ again by (1.6). The following theorem in conjunction with Theorem 1.3 proves (1.3).

Theorem 1.4. *Let $n \to \infty$ and suppose $\lambda_j^{(n)} \in (\alpha, \beta)$. If $a_k = a_{-k}$ for all k, then*

$$\left\| \frac{y_j^{(n)}(\lambda_j^{(n)})}{\|y_j^{(n)}(\lambda_j^{(n)})\|_2} - x_j^{(n)} \right\|_2 = O\left(\frac{j}{n}\right).$$

The rest of the paper is as follows. We approach eigenvectors by using the elementary observation that if λ is an eigenvalue of $T_n(a)$, then every nonzero column of the adjugate matrix of $T_n(a) - \lambda I = T_n(a - \lambda)$ is an eigenvector for λ. In Section 2 we employ "exact" formulas by Trench and Widom for the inverse and the determinant of a banded Toeplitz matrix to get a representation of the first column of the adjugate matrix of $T_n(a-\lambda)$ that will be convenient for asymptotic analysis. This analysis is carried out in Section 3. On the basis of these results, Theorems 1.1 and 1.2 are proved in Section 4, while the proofs of Theorems 1.3 and 1.4 are given in Section 4. Section 6 contains numerical results.

2. The first column of the adjugate matrix

The adjugate matrix $\operatorname{adj} B$ of an $n \times n$ matrix $B = (b_{jk})_{j,k=1}^n$ is defined by

$$(\operatorname{adj} B)_{jk} = (-1)^{j+k} \det M_{kj}$$

where M_{kj} is the $(n-1)\times(n-1)$ matrix that results from B by deleting the kth row and the jth column. We have

$$(A - \lambda I)\operatorname{adj}(A - \lambda I) = (\det(A - \lambda I))I.$$

Thus, if λ is an eigenvalue of A, then each nonzero column of $\operatorname{adj}(A - \lambda I)$ is an eigenvector. For an invertible matrix B,

$$\operatorname{adj} B = (\det B)B^{-1}. \tag{2.1}$$

Formulas for $\det T_n(b)$ and $T_n^{-1}(b)$ were established by Widom [18] and Trench [16], respectively. The purpose of this section is to transform Trench's formula for the first column of $T_n^{-1}(b)$ into a form that will be convenient for further analysis.

Theorem 2.1. *Let*

$$b(t) = \sum_{k=-p}^{q} b_k t^k = b_p t^{-q} \prod_{j=1}^{p+q} (t - z_j) \quad (t \in \mathbf{T})$$

where $p \geq 1$, $q \geq 1$, $b_p \neq 0$, *and* $z_1, \dots, z_{p+q}$ *are pairwise distinct nonzero complex numbers. If* $n > p+q$ *and* $1 \leq m \leq n$, *then the* m*th entry of the first column of* $\operatorname{adj} T_n(b)$ *is*

$$[\operatorname{adj} T_n(b)]_{m,1} = \sum_{J \subset \mathcal{Z}, |J|=p} C_J W_J^n \sum_{z \in J} S_{m,J,z} \tag{2.2}$$

where $\mathcal{Z} = \{z_1, \dots, z_{p+q}\}$, *the sum is over all sets* $J \subset \mathcal{Z}$ *of cardinality* p, *and, with* $\overline{J} := \mathcal{Z} \setminus J$,

$$C_J = \prod_{z \in J} z^q \prod_{z \in J, w \in \overline{J}} \frac{1}{z - w}, \qquad W_J = (-1)^p b_p \prod_{z \in J} z,$$

$$S_{m,J,z} = -\frac{1}{b_p} \frac{1}{z^m} \prod_{w \in J \setminus \{z\}} \frac{1}{z - w}.$$

Proof. It suffices to prove (2.2) under the assumption that $\det T_n(b) \neq 0$ because both sides of (2.2) are continuous functions of $z_1, \dots, z_{p+q}$. Thus, let $\det T_n(b) \neq 0$. We will employ (2.1) with $B = T_n(b)$.

Trench [16] proved that $[T_n^{-1}(b)]_{m,1}$ equals

$$-\frac{1}{b_p} \frac{D_{\{1,\dots,p+q\}}(0, \dots, q-1, q+n, \dots, q+n+p-2, q+n-m)}{D_{\{1,\dots,p+q\}}(0, \dots, q-1, q+n, \dots, q+n+p-1)} \tag{2.3}$$

where $D_{\{j_1,\dots,j_k\}}(s_1, \dots, s_k)$ denotes the determinant

$$\det \begin{pmatrix} z_{j_1}^{s_1} & z_{j_1}^{s_2} & \dots & z_{j_1}^{s_k} \\ z_{j_2}^{s_1} & z_{j_2}^{s_2} & \dots & z_{j_2}^{s_k} \\ \vdots & \vdots & & \vdots \\ z_{j_k}^{s_1} & z_{j_k}^{s_2} & \dots & z_{j_k}^{s_k} \end{pmatrix}.$$

Note that

$$D_J(s_1 + \xi, \dots, s_k + \xi) = \left(\prod_{j \in J} z_j^{\xi} \right) D_J(s_1, \dots, s_k),$$

$$D_{\{1,2,\dots,k\}}(0, 1, \dots, k-1) = \prod_{\substack{j,\ell \in J \\ \ell > j}} (z_\ell - z_j).$$

We first consider the denominator of (2.3). Put $Z = \{1, \dots, p+q\}$. Laplace expansion along the last p columns gives

$$\begin{aligned}
&D_Z(0, \dots, q-1, q+n, \dots, q+n+p-1) \\
&= \sum_{J \subset Z, |J|=p} (-1)^{\mathrm{inv}(\overline{J},J)} D_J(q+n, \dots, q+n+p-1) D_{\overline{J}}(0, \dots, q-1) \\
&= \sum_{J \subset Z, |J|=p} (-1)^{\mathrm{inv}(\overline{J},J)} \prod_{k \in J} z_k^{q+n} \prod_{\substack{k,\ell \in J \\ \ell > k}} (z_\ell - z_k) \prod_{\substack{k,\ell \in \overline{J} \\ \ell > k}} (z_\ell - z_k),
\end{aligned}$$

where $\mathrm{inv}(\overline{J}, J)$ is the number of inversions in the permutation of length $p+q$ whose first q elements are the elements of the set $\overline{J}$ in increasing order and whose last p elements are the elements of the set J in increasing order. A little thought reveals that $\mathrm{inv}(\overline{J}, J)$ is just the number of pairs (k, ℓ) with $k \in J$, $\ell \in \overline{J}$, $k < \ell$. We have

$$\begin{aligned}
\prod_{j \in J, s \in \overline{J}} (z_j - z_s) &= \prod_{\substack{\ell \in J, k \in \overline{J} \\ \ell > k}} (z_\ell - z_k) \prod_{\substack{k \in J, \ell \in \overline{J} \\ \ell > k}} (z_k - z_\ell) \\
&= (-1)^{\mathrm{inv}(\overline{J},J)} \prod_{\substack{\ell \in J, k \in \overline{J} \\ \ell > k}} (z_\ell - z_k) \prod_{\substack{\ell \in \overline{J}, k \in J \\ \ell > k}} (z_\ell - z_k)
\end{aligned} \tag{2.4}$$

and hence the denominator is equal to

$$R_n \sum_{J \subset \mathcal{Z}, |J|=p} C_J W_J^n \quad \text{with} \quad R_n := \frac{(-1)^{pn}}{b_p^n} \prod_{\ell > k} (z_\ell - z_k).$$

A formula by Widom [18], which can also be found in [1], says that

$$\det T_n(b) = \sum_{J \subset \mathcal{Z}, |J|=p} C_J W_J^n.$$

Consequently, the denominator of (2.3) is nothing but $R_n \det T_n(b)$.

Let us now turn to the numerator of (2.3). This time Laplace expansion along the last p columns yields

$$\begin{aligned}
&D_Z(0, \dots, q-1, q+n, \dots, q+n+p-2, q+n-m) \\
&= \sum_{J \subset Z, |J|=p} (-1)^{\mathrm{inv}(\overline{J},J)} D_J(q+n, \dots, q+n+p-1, q+n-m) D_{\overline{J}}(0, \dots, q-1) \\
&= \sum_{J \subset Z, |J|=p} (-1)^{\mathrm{inv}(\overline{J},J)} D_{\overline{J}}(0, \dots, q-1) \left(\prod_{j \in J} z_j^{q+n} \right) D_J(0, \dots, p-2, -m).
\end{aligned}$$

Expanding $D_J(0, \dots, p-2, -m)$ by its last column we get

$$D_J(0, \dots, p-2, -m) = \sum_{j \in J} (-1)^{\mathrm{inv}(J \setminus \{j\}, j)} z_j^{-m} D_{J \setminus \{j\}}(0, \dots, p-2) \tag{2.5}$$

with $\operatorname{inv}(J\setminus\{j\},j)$ being the number of $s\in J\setminus\{j\}$ such that $s>j$. Thus, (2.5) is

$$\sum_{j\in J}(-1)^{\operatorname{inv}(J\setminus\{j\},j)}z_j^{-m}\prod_{\substack{k,\ell\in J\setminus\{j\}\\ \ell>k}}(z_\ell-z_k)$$

$$=\sum_{j\in J}z_j^{-m}\prod_{\substack{k,\ell\in J\\ \ell>k}}(z_\ell-z_k)\prod_{s\in J\setminus\{j\}}\frac{1}{z_j-z_s}.$$

This in conjunction with (2.4) shows that the numerator of (2.2) equals

$$-b_pR_n\sum_{J\subset\mathcal{Z},|J|=p}C_JW_J^n\sum_{z\in J}S_{m,J,z}.$$

In summary, from (2.3) we obtain that

$$[T_n^{-1}(b)]_{m,1}=\frac{1}{\det T_n(b)}\sum_{J\subset\mathcal{Z},|J|=p}C_JW_J^n\sum_{z\in J}S_{m,J,z},$$

which after multiplication by $\det T_n(b)$ becomes (2.2). □

3. The main terms of the first column

We now apply Theorem 2.1 to

$$b(t)=a(t)-\lambda=a_rt^{-r}\prod_{k=1}^{2r}(t-z_k(\lambda)) \tag{3.1}$$

where $\lambda\in(\alpha,\beta)$. The set $\mathcal{Z}=\mathcal{Z}(\lambda)$ is given by (1.4). Let

$$d_0(\lambda)=(-1)^ra_re^{i\sigma(\lambda)}\prod_{k=1}^{r-1}u_k(\lambda). \tag{3.2}$$

In [2], we showed that $d_0(\lambda)>0$ for all $\lambda\in(0,M)$. The dependence on λ will henceforth frequently be suppressed in notation. Let

$$J_1=\{u_1,\dots,u_{r-1},e^{i\varphi_1}\},\quad J_2=\{u_1,\dots,u_{r-1},e^{i\varphi_2}\}$$

and for $\nu\in\{1,\dots,r-1\}$, put

$$J_\nu^0=\{u_1,\dots,u_{r-1},1/\overline{u}_\nu\}.$$

Lemma 3.1. *If $J\subset\mathcal{Z}$, $|J|=r$, $J\notin\{J_1,J_2,J_1^0,\dots,J_{r-1}^0\}$, then*

$$|C_JW_j^nS_{m,J,z}|\le K\frac{d_0^n}{\sin\varphi}e^{-\delta n}$$

for all $z\in J$, $n\ge1$, $1\le m\le n$, $\lambda\in(\alpha,\beta)$ with some finite constant K that does not depend on z,n,m,λ.

Proof. If both $e^{i\varphi_1}$ and $e^{i\varphi_2}$ belong to J, then

$$J = \{u_{\nu_1}, \dots, u_{\nu_k}, e^{i\varphi_1}, e^{i\varphi_2}, 1/\overline{u}_{s_1}, \dots, 1/\overline{u}_{s_\ell}\}$$

with $k + \ell = r - 2$. Since

$$\min_{\lambda \in [\alpha,\beta]} \min_{j_1 \neq j_2} |u_{j_1}(\lambda) - u_{j_2}(\lambda)| > 0,$$

we conclude that $|C_J| \le K_1$. Here and in the following K_i denotes a finite constant that is independent of $\lambda \in [\alpha, \beta]$. We have $k \le r - 2$ and thus

$$|W_J| = |a_r| \frac{|u_{\nu_1} \dots u_{\nu_k}|}{|u_{s_1} \dots u_{s_\ell}|} \le \frac{d_0 e^{-\delta}}{|u_{s_1} \dots u_{s_\ell}|}. \tag{3.3}$$

If $z \in \{u_{\nu_1}, \dots, u_{\nu_k}, e^{i\varphi_1}, e^{i\varphi_2}\}$, then obviously $|S_{m,J,z}| \le K_2/\sin\varphi$ and hence

$$|C_J W_J^n S_{m,J,z}| \le K_1 K_2 \frac{d_0^n e^{-\delta n}}{\sin\varphi}.$$

In case $z \in \{1/\overline{u}_{s_1}, \dots, 1/\overline{u}_{s_\ell}\}$, say $z = 1/\overline{u}_{s_1}$, we have $|S_{m,J,z}| \le K_3 |u_{\nu_1}|^m$, which gives

$$|C_J W_J^n S_{m,J,z}| \le K_1 K_3 d_0^n e^{-\delta n} \frac{|u_{\nu_1}|^m}{|u_{\nu_1}|^n} \le K_1 K_3 d_0^n e^{-\delta n} \le K_1 K_3 \frac{d_0^n e^{-\delta n}}{\sin\varphi}.$$

The only other possibility for J is to be of the type

$$J = \{u_{\nu_1}, \dots, u_{\nu_k}, e^{i\varphi_1}, 1/\overline{u}_{s_1}, \dots, 1/\overline{u}_{s_\ell}\}$$

with $k + \ell \le r - 1$, $k \le r - 2$, $\ell \ge 1$. (The case where $e^{i\varphi_1}$ is replaced by $e^{i\varphi_2}$ is completely analogous.) This time, $|C_J| \le K_4/\sin\varphi$ and (3.3) holds again. For $z \in \{u_{\nu_1}, \dots, u_{\nu_k}, e^{i\varphi_1}\}$ we have $|S_{m,J,z}| \le K_5$ and thus get the assertion. If $z = 1/\overline{u}_s$ for some $s \in \{s_1, \dots, s_\ell\}$, say $s = s_1$, then $|S_{m,J,z}| \le K_6 |u_{s_1}|^m$, and the assertion follows as above, too. □

Let

$$d_1(\lambda) = \frac{1}{|h_\lambda(e^{i\varphi_1(\lambda)}) h_\lambda(e^{i\varphi_2(\lambda)})|} \prod_{k,s=1}^{r-1} \left(1 - \frac{1}{u_k(\lambda)\overline{u}_s(\lambda)}\right)^{-1}.$$

It is easily seen that $d_1(\lambda) > 0$ for all $\lambda \in [0, M]$.

Lemma 3.2. *If $\lambda = \lambda_j^{(n)} \in (0, M)$, then*

$$C_{J_1} W_{J_1}^n S_{m,J_1,e^{i\varphi_1}} = \frac{d_1 d_0^{n-1}}{\sin\varphi} \left[(-1)^j A e^{-im\varphi_1} + O(e^{-\delta n})\right],$$

$$C_{J_2} W_{J_2}^n S_{m,J_2,e^{i\varphi_2}} = \frac{d_1 d_0^{n-1}}{\sin\varphi} \left[(-1)^{j+1} B e^{-im\varphi_2} + O(e^{-\delta n})\right]$$

uniformly in m and λ.

Proof. We abbreviate $\prod_{k=1}^{r-1}$ to $\prod_k$. Clearly,

$$W_{j_1} = (-1)^r a_r \left(\prod_k u_k\right) e^{i\varphi_1} = (-1)^r a_r \left(\prod_k u_k\right) e^{i\sigma} e^{i\varphi} = d_0 e^{i\varphi}.$$

We have

$$\begin{aligned} C_{J_1} &= \frac{\left(\prod_k u_k^r\right) e^{ir\varphi_1}}{(e^{i\varphi_1} - e^{i\varphi_2}) \prod_{k,s}\left(u_k - \frac{1}{\overline{u}_s}\right) \prod_k (u_k - e^{i\varphi_2}) \prod_k \left(e^{i\varphi_1} - \frac{1}{\overline{u}_k}\right)} \\ &= \frac{e^{ir\varphi_1}}{e^{i\sigma}(e^{i\varphi} - e^{-i\varphi}) \prod_{k,s}\left(1 - \frac{1}{u_k \overline{u}_s}\right) \prod_k \left(1 - \frac{e^{i\varphi_2}}{u_k}\right) e^{i(r-1)\varphi_1} \prod_k \left(1 - \frac{e^{-i\varphi_1}}{\overline{u}_k}\right)} \\ &= \frac{e^{i\varphi}}{2i \sin\varphi \prod_{k,s}\left(1 - \frac{1}{u_k \overline{u}_s}\right) h(e^{i\varphi_2}) \overline{h(e^{i\varphi_1})}} \end{aligned}$$

and because

$$h(e^{i\varphi_2}) \overline{h(e^{i\varphi_1})} = |h(e^{i\varphi_1}) h(e^{i\varphi_2})| e^{-i\theta}, \tag{3.4}$$

it follows that

$$C_{J_1} = \frac{d_1 e^{i(\varphi+\theta)}}{2i \sin\varphi}.$$

Furthermore,

$$\begin{aligned} S_{m,J_1,e^{i\varphi_1}} &= -\frac{1}{a_r} \frac{1}{e^{im\varphi_1} \prod_k (e^{i\varphi_1} - u_k)} \\ &= -\frac{1}{a_r} \frac{e^{-im\varphi_1}}{(-1)^{r-1} \left(\prod_k u_k\right) \prod_k (1 - e^{i\varphi_1}/u_k)} \\ &= \frac{e^{-im(\sigma+\varphi)}}{(-1)^r a_r \left(\prod_k u_k\right) h(e^{i\varphi_1})} = \frac{e^{-im(\sigma+\varphi)} e^{i\sigma}}{d_0 h(e^{i\varphi_1})}. \end{aligned}$$

Putting things together we arrive at the formula

$$C_{J_1} W_{J_1}^n S_{m,J_1,e^{i\varphi_1}} = \frac{d_1 d_0^{n-1}}{\sin\varphi} A\, e^{-im(\sigma+\varphi)} e^{i((n+1)\varphi+\theta)}.$$

Obviously, $\sigma + \varphi = \varphi_1$. By virtue of (1.5),

$$e^{i((n+1)\varphi+\theta)} = e^{i\pi j}(1 + O(e^{-\delta n})) = (-1)^j (1 + O(e^{-\delta n})).$$

This proves the first of the asserted formulas. Analogously,

$$W_{J_2} = d_0 e^{-i\varphi}, \quad C_{J_2} = -\frac{d_1 e^{-i(\varphi+\theta)}}{2i \sin\varphi}, \quad S_{m,J_2,e^{i\varphi_2}} = \frac{e^{-im(\sigma-\varphi)} e^{i\sigma}}{d_0 h(e^{i\varphi_2})},$$

which gives the second formula. □

Lemma 3.3. *If* $1 \le \nu \le r-1$ *and* $\lambda = \lambda_j^{(n)} \in (\alpha, \beta)$, *then, uniformly in* m *and* λ,

$$C_{J_1} W_{J_1}^n S_{m,J_1,u_\nu} + C_{J_2} W_{J_2}^n S_{m,J_2,u_\nu} = \frac{d_1 d_0^{n-1}}{\sin\varphi} \left[(-1)^j D_\nu \frac{1}{u_\nu^m} + O(e^{-\delta n})\right].$$

Proof. By definition,

$$\begin{aligned} S_{m,J_1,u_\nu} &= -\frac{1}{a_r} \frac{1}{u_\nu^m (u_\nu - e^{i\varphi_1}) \prod_{s\neq\nu}(u_\nu - u_s)} \\ &= \frac{u_\nu^{-m}}{(-1)^{r-1} \left(\prod_k u_k\right) a_r (u_\nu - e^{i\varphi_1}) h'(u_\nu))} \end{aligned}$$

Since $-h'(z)$ equals

$$\frac{1}{u_1}\left(1-\frac{z}{u_2}\right)\dots\left(1-\frac{z}{u_{r-1}}\right)+\dots+\frac{1}{u_{r-1}}\left(1-\frac{z}{u_1}\right)\dots\left(1-\frac{z}{u_{r-2}}\right),$$

we obtain that

$$h'(u_\nu) = -\frac{1}{u_\nu}\prod_{s\neq\nu}\left(1-\frac{u_\nu}{u_s}\right).$$

Thus,

$$S_{m,J_1,u_\nu} = \frac{u_\nu^{-m}}{(-1)^r a_r \left(\prod_k u_k\right)(u_\nu - e^{i\varphi_1})h'(u_\nu)} = \frac{u_\nu^{-m} e^{i\sigma}}{d_0(u_\nu - e^{i\varphi_1})h'(u_\nu)}.$$

Changing φ_1 to φ_2 we get

$$S_{m,J_2,u_\nu} = \frac{u_\nu^{-m} e^{i\sigma}}{d_0(u_\nu - e^{i\varphi_2})h'(u_\nu)}.$$

These two expressions along with the expressions for $C_{J_1}, W_{J_1}, C_{J_2}, W_{J_2}$ derived in the proof of Lemma 3.2 show that the sum under consideration is

$$\frac{d_1 d_0^{n-1}}{2i\sin\varphi}\,\frac{u_\nu^{-m}e^{i\sigma}}{h'(u_\nu)}\left[\frac{e^{i((n+1)\varphi+\theta)}}{u_\nu - e^{i\varphi_1}} - \frac{e^{-i((n+1)\varphi+\theta)}}{u_\nu - e^{i\varphi_2}}\right].$$

Because of (1.5), the term in brackets equals

$$\begin{aligned} &(-1)^j\left[\frac{1}{u_\nu - e^{i\varphi_1}} - \frac{1}{u_\nu - e^{i\varphi_2}} + O(e^{-\delta n})\right] \\ &= (-1)^j \frac{e^{i\sigma} 2i\sin\varphi}{(u_\nu - e^{i\varphi_1})(u_\nu - e^{i\varphi_2})} + O(e^{-\delta n}). \end{aligned}$$

□

Lemma 3.4. *For $1 \le \nu \le r-1$ and $\lambda \in (\alpha, \beta)$,*

$$C_{J_\nu^0} W_{J_\nu^0}^n S_{m,J_\nu^0,1/\overline{u}_\nu} = -\frac{d_1 d_0^{n-1}}{\sin\varphi} F_\nu \frac{e^{-i(n+1)\sigma}}{\overline{u}_\nu^{n+1-m}}.$$

Proof. We have $C_{J^0_\nu} = \left(\prod_k u_k^r\right)/(\overline{u}_\nu^r P_1P_2P_3)$ with

$$\begin{aligned}
P_1 &= \left(\frac{1}{\overline{u}_\nu} - e^{i\varphi_1}\right)\left(\frac{1}{\overline{u}_\nu} - e^{i\varphi_2}\right) = \frac{(\overline{u}_\nu - e^{-i\varphi_1})(\overline{u}_\nu - e^{-i\varphi_2})}{\overline{u}_\nu^2 e^{-2i\sigma}},\\
P_2 &= \prod_k (u_k - e^{i\varphi_1})\prod_k (u_k - e^{i\varphi_2}) = \left(\prod_k u_k^2\right) h(e^{i\varphi_1})h(e^{i\varphi_2}),\\
P_3 &= \prod_{s\neq\nu}\left(\frac{1}{\overline{u}_\nu} - \frac{1}{\overline{u}_s}\right)\prod_k\prod_{s\neq\nu}\left(u_k - \frac{1}{\overline{u}_s}\right)\\
&= \frac{1}{\overline{u}_\nu^{r-2}}\prod_{s\neq\nu}\left(1 - \frac{\overline{u}_\nu}{\overline{u}_s}\right)\left(\prod_k u_k^{r-2}\right)\prod_k\prod_{s\neq\nu}\left(1 - \frac{1}{u_k\overline{u}_s}\right)\\
&= -\frac{1}{\overline{u}_\nu^{r-3}}\overline{h'(u_\nu)}\left(\prod_k u_k^{r-2}\right)\frac{1}{d_1|h(e^{i\varphi_1})h(e^{i\varphi_2})|}\frac{1}{\prod_k(1 - 1/(u_k\overline{u}_\nu))}.
\end{aligned}$$

Thus, $C_{J^0_\nu}$ equals

$$-\frac{d_1 e^{-2i\sigma}|h(e^{i\varphi_1})h(e^{i\varphi_2})|}{\overline{u}_\nu(\overline{u}_\nu - e^{-i\varphi_1})(\overline{u}_\nu - e^{-i\varphi_2})\overline{h'(u_\nu)}h(e^{i\varphi_1})h(e^{i\varphi_2})}\prod_k\left(1 - \frac{1}{u_k\overline{u}_\nu}\right).$$

Since $W_{J^0_\nu} = d_0 e^{-i\sigma}/\overline{u}_\nu$ and

$$\begin{aligned}
S_{m,J^0_\nu,1/\overline{u}_\nu} &= -\frac{1}{a_r}\overline{u}_\nu^m\frac{1}{\prod_k(1/\overline{u}_\nu - u_k)}\\
&= \frac{\overline{u}_\nu^m}{(-1)^r a_r\left(\prod_k u_k\right)\prod_k\left(1 - \frac{1}{u_k\overline{u}_\nu}\right)}\\
&= \frac{\overline{u}_\nu^m e^{i\sigma}}{d_0\prod_k\left(1 - \frac{1}{u_k\overline{u}_\nu}\right)},
\end{aligned}$$

we obtain that $C_{J^0_\nu}W^n_{J^0_\nu}S_{m,J^0_\nu,1/\overline{u}_\nu}$ is equal to

$$-\frac{d_1 d_0^{n-1}}{\overline{u}^{n+1-m}}\frac{e^{-i\sigma}e^{-in\sigma}|h(e^{i\varphi_1})h(e^{i\varphi_2})|}{(\overline{u}_\nu - e^{-i\varphi_1})(\overline{u}_\nu - e^{-i\varphi_2})\overline{h'(u_\nu)}h(e^{i\varphi_1})h(e^{i\varphi_2})}. \qquad \square$$

Lemma 3.5. *If* $1 \le k \le r-1$ *and* $\lambda \in (\alpha,\beta)$,

$$C_{J^0_\nu}W^n_{J^0_\nu}S_{m,J^0_\nu,u_k} = \frac{d_1 d_0^{n-1}}{\sin\varphi}O(e^{-\delta n})$$

uniformly in m *and* λ.

Proof. This time

$$\begin{aligned} S_{m,J_\nu^0,u_k} &= -\frac{1}{a_r}\frac{1}{u_k^m}\frac{1}{(u_k-1/\overline{u}_\nu)\prod_{s\neq k}(u_k-u_s)} \\ &= \frac{u_k^{-m}}{(-1)^{r-1}a_r u_k\left(1-\frac{1}{u_k\overline{u}_\nu}\right)\left(\prod_{s\neq k}u_s\right)\prod_{s\neq k}\left(1-\frac{u_k}{u_s}\right)} \\ &= -\frac{1}{d_0u_k^m}\frac{1}{\left(1-\frac{1}{u_k\overline{u}_\nu}\right)\prod_{s\neq k}\left(1-\frac{u_k}{u_s}\right)}. \end{aligned}$$

Expressions for $C_{J_\nu^0}$ and $W_{J_\nu^0}$ were given in the proof of Lemma 3.4. It follows that

$$C_{J_\nu^0}W_{J_\nu^0}^n S_{m,J_\nu^0,u_k} = G_{\nu,k}\frac{d_1d_0^{n-1}}{\sin\varphi}\frac{1}{\overline{u}_\nu^{n+1}u_k^m}$$

where $G_{\nu,k}$ equals

$$\frac{e^{2i\sigma}e^{-in\sigma}|h(e^{i\varphi_1})h(e^{i\varphi_2})|\sin\varphi}{(\overline{u}_\nu-e^{-i\varphi_1})(\overline{u}_\nu-e^{-i\varphi_2})\overline{h'(u_\nu)}h(e^{i\varphi_1})h(e^{i\varphi_2})}\frac{\prod_{s\neq k}\left(1-\frac{1}{u_s\overline{u}_\nu}\right)}{\prod_{s\neq k}\left(1-\frac{u_k}{u_s}\right)}.$$

Since

$$\overline{h'(u_\nu)} = -\frac{1}{\overline{u}_\nu}\prod_{s\neq\nu}\left(1-\frac{\overline{u}_\nu}{\overline{u}_s}\right),$$

we see that $G_{\nu,k}$ remains bounded on $[\alpha,\beta]$. Finally,

$$\frac{1}{|\overline{u}_\nu^{n+1}u_k^m|}\leq\frac{1}{|u_\nu|^n}\leq e^{-\delta n}.$$ □

Corollary 3.6. *If $\lambda=\lambda_j^{(n)}\in(\alpha,\beta)$, then*

$$[\operatorname{adj}T_n(a-\lambda)]_{m,1} = (-1)^j\frac{d_1(\lambda)d_0^{n-1}(\lambda)}{\sin\varphi(\lambda)}[w_{j,m}(\lambda)+O(e^{-\delta n})]$$

uniformly in m and λ.

Proof. This follows from Theorem 2.1 and Lemmas 3.1 to 3.5 along with the fact that d_1 is bounded and bounded away from zero on $[\alpha,\beta]$. □

4. The asymptotics of the eigenvectors

We now prove Theorem 1.1. There is a finite constant K_1 such that $|D_\nu|\leq K_1$ and $|F_\nu|\leq K_1$ for all ν and all $\lambda\in(\alpha,\beta)$. Thus, summing up two finite geometric series, we get

$$\sum_{m=1}^n\left|D_\nu\frac{1}{u_\nu^m}+F_\nu\frac{(-1)^{j+1}e^{-i(n+1)\sigma}}{\overline{u}_\nu^{n+1-m}}\right|^2\leq 2K_1^2\frac{1}{|u_\nu|^2}\frac{1-1/|u_\nu|^{2(n+1)}}{1-1/|u_\nu|^2}\leq K_2$$

for all ν, n, λ. We further have

$$\sum_{m=1}^{n} \left|Ae^{-im\varphi_1} - Be^{-im\varphi_2}\right|^2 = \sum_{m=1}^{n} \left|\frac{e^{-im\varphi_1}}{2h(e^{i\varphi_1})} - \frac{e^{-im\varphi_2}}{2h(e^{i\varphi_2})}\right|^2$$

$$= \sum_{m=1}^{n} \left(\frac{1}{4|h(e^{i\varphi_1})|^2} + \frac{1}{4|h(e^{i\varphi_1})|^2}\right)$$

$$- \sum_{m=1}^{n} \left(\frac{e^{-2im\varphi}}{4h(e^{i\varphi_1})\overline{h(e^{i\varphi_2})}} + \frac{e^{2im\varphi}}{4\overline{h(e^{i\varphi_1})}h(e^{i\varphi_2})}\right).$$

The first sum is of the form $\sum_{m=1}^{n}(\gamma/4)$ and therefore equals $(n/4)\gamma$. Hence, because of (3.4) we are left to prove that

$$\left|\sum_{m=1}^{n} e^{i\theta(\lambda_j^{(n)})} e^{2im\varphi(\lambda_j^{(n)})}\right| \le K_3 \tag{4.1}$$

for all n and j such that $\lambda_j^{(n)} \in (\alpha, \beta)$. The sum in (4.1) is

$$e^{i[(n+1)\varphi(\lambda_j^{(n)})+\theta(\lambda_j^{(n)})]} \frac{\sin n\varphi(\lambda_j^{(n)})}{\sin \varphi(\lambda_j^{(n)})}.$$

Thus, (4.1) will follow as soon as we have shown that

$$\left|\frac{\sin n\varphi(\lambda_j^{(n)})}{\sin \varphi(\lambda_j^{(n)})}\right| \le K_3$$

for all n and j in question. From (1.5) we infer that

$$n\varphi(\lambda_j^{(n)}) = \pi j - \varphi(\lambda_j^{(n)}) - \theta(\lambda_j^{(n)}) + O(e^{-\delta n}),$$

which implies that

$$\sin n\varphi(\lambda_j^{(n)}) = (-1)^{j+1} \sin\left(\varphi(\lambda_j^{(n)}) + \theta(\lambda_j^{(n)})\right) + O(e^{-\delta n}).$$

Suppose first that $0 < \varphi(\lambda_j^{(n)}) \le \pi/2$. Then

$$\left|\frac{\sin\left(\varphi(\lambda_j^{(n)}) + \theta(\lambda_j^{(n)})\right)}{\sin \varphi(\lambda_j^{(n)})}\right| \le \frac{\pi}{2} \frac{|\varphi(\lambda_j^{(n)}) + \theta(\lambda_j^{(n)})|}{|\varphi(\lambda_j^{(n)})|} \le \frac{\pi}{2}\left(1 + \frac{|\theta(\lambda_j^{(n)})|}{|\varphi(\lambda_j^{(n)})|}\right). \tag{4.2}$$

In [2], we proved that $|\theta/\varphi|$ is bounded on $(0, M)$. Thus, the right-hand side of (4.2) is bounded by some K_3 for all n and j. If $\pi/2 < \varphi(\lambda_j^{(n)}) < \pi$, we may replace (4.2) by the upper bound

$$\frac{\pi}{2}\left(1 + \frac{|\theta(\lambda_j^{(n)})|}{|\pi - \varphi(\lambda_j^{(n)})|}\right).$$

We know again from [2] that $|\theta/(\pi - \varphi)|$ is bounded on $(0, M)$. This completes the proof of Theorem 1.1.

Here is the proof of Theorem 1.2. By virtue of Theorem 1.1, $\|w_j(\lambda_j^{(n)})\|_2 > 1$ whenever n is sufficiently large. Corollary 3.6 therefore implies that the first column of $\operatorname{adj} T_n(a - \lambda_j^{(n)})$ is nonzero and thus an eigenvector for $\lambda_j^{(n)}$ for all $n \ge n_0$ and all $1 \le j \le n$ such that $\lambda_j^{(n)} \in (\alpha, \beta)$. Again by Corollary 3.6, the mth entry of this column is

$$\frac{d_1(\lambda) d_0^{n-1}(\lambda)}{\sin \varphi(\lambda)} \left[w_{j,m}(\lambda) + \xi_{j,m}^{(n)} \right] \Big|_{\lambda = \lambda_j^{(n)}}$$

where $|\xi_{j,m}^{(n)}| \le K e^{-\delta n}$ for all n and j under consideration and K does not depend on m, n, j. It follows that

$$w_j(\lambda_j^{(n)}) + \left(\xi_{j,m}^{(n)} \right)_{m=1}^{n} = w_j(\lambda_j^{(n)}) + O_\infty(e^{-\delta n})$$

is also an eigenvector for $\lambda_j^{(n)}$. Consequently,

$$\frac{w_j(\lambda_j^{(n)}) + O_\infty(e^{-\delta n})}{\|w_j(\lambda_j^{(n)}) + O_\infty(e^{-\delta n})\|_2} = \frac{w_j(\lambda_j^{(n)})}{\|w_j(\lambda_j^{(n)})\|_2} + O_\infty(e^{-\delta n}) \tag{4.3}$$

is a normalized eigenvector for $\lambda_j^{(n)}$. From (1.5) we deduce that all eigenvalues of $T_n(a)$ are simple. Thus, $v_j^{(n)}$ is a scalar multiple of modulus 1 of (4.3). This completes the proof of Theorem 1.2.

5. Symmetric matrices

The matrices $T_n(a)$ are all symmetric if and only if all a_k are real and $a_k = a_{-k}$ for all k. Obviously, this is equivalent to the requirement that the real-valued function $g(x) := a(e^{ix})$ be even, that is, $g(x) = g(-x)$ for all x. Thus, suppose g is even. In that case

$$\varphi_0 = \pi, \quad \varphi_1(\lambda) = -\varphi_2(\lambda) = \varphi(\lambda), \quad \sigma(\lambda) = 0.$$

Moreover, for $t \in \mathbf{T}$ we have

$$\begin{aligned} a_r t^{-r} \prod_{k=1}^{2r} (t - z_k(\lambda)) &= a(t) - \lambda = a(1/t) - \lambda \\ &= a_r t^r \prod_{k=1}^{2r} (1/t - z_k(\lambda)) = a_r \left(\prod_{k=1}^{2r} z_k(\lambda) \right) t^{-r} \prod_{k=1}^{2r} (t - 1/z_k(\lambda)), \end{aligned}$$

which in conjunction with (1.4) implies that

$$\{u_1(\lambda), \ldots, u_{r-1}(\lambda)\} = \{\overline{u}_1(\lambda), \ldots, \overline{u}_{r-1}(\lambda)\}. \tag{5.1}$$

The coefficients of the polynomial $h_\lambda(t)$ are symmetric functions of

$$1/u_1(\lambda), \ldots, 1/u_{r-1}(\lambda).$$

From (5.1) we therefore see that these coefficients are real. It follows in particular that $h_\lambda(e^{-i\varphi(\lambda)}) = \overline{h_\lambda(e^{i\varphi(\lambda)})}$, which gives $\theta(\lambda) = 2\arg h_\lambda(e^{i\varphi(\lambda)})$ and thus

$$h_\lambda(e^{i\varphi(\lambda)}) = |h_\lambda(e^{i\varphi(\lambda)})|e^{i\theta(\lambda)/2}, \quad h_\lambda(e^{-i\varphi(\lambda)}) = |h_\lambda(e^{i\varphi(\lambda)})|e^{-i\theta(\lambda)/2}.$$

We are now in a position to prove Theorem 1.3. To do so, we use Theorem 1.2. Consider the vector $w_j(\lambda_j^{(n)})$. We now have

$$\begin{aligned} Ae^{-im\varphi_1} - Be^{-im\varphi_2} &= \frac{e^{-im\varphi}}{2ih(e^{i\varphi})} - \frac{e^{im\varphi}}{2ih(e^{-i\varphi})} \\ &= \frac{1}{2i|h(e^{i\varphi})|}\left(\frac{e^{-im\varphi}}{e^{i\theta/2}} - \frac{e^{im\varphi}}{e^{-i\theta/2}}\right) = -\frac{1}{|h(e^{i\varphi})|}\sin\left(m\varphi + \frac{\theta}{2}\right). \end{aligned}$$

Furthermore,

$$\begin{aligned} D_\nu &= \frac{\sin\varphi}{(u_\nu - e^{i\varphi})(u_\nu - e^{-i\varphi})h'(u_\nu)} = \frac{Q_\nu}{|h(e^{i\varphi})|}, \\ F_\nu &= \frac{\sin\varphi}{(\overline{u}_\nu - e^{i\varphi})(\overline{u}_\nu - e^{-i\varphi})\overline{h'(u_\nu)}}\,\frac{|h(e^{i\varphi})h(e^{-i\varphi})|}{h(e^{i\varphi})h(e^{-i\varphi})} \\ &= \frac{\sin\varphi}{(\overline{u}_\nu - e^{i\varphi})(\overline{u}_\nu - e^{-i\varphi})h'(\overline{u}_\nu)}. \end{aligned}$$

Consequently, from (5.1) we infer that

$$\sum_{\nu=1}^{r-1} \frac{F_\nu}{\overline{u}_\nu^{\,n+1-m}} = \sum_{\nu=1}^{r-1} \frac{D_\nu}{u_\nu^{n+1-m}} = \sum_{\nu=1}^{r-1} \frac{Q_\nu}{|h(e^{i\varphi})|\,\overline{u}_\nu^{\,n+1-m}}.$$

In summary, it follows that

$$w_j(\lambda_j^{(n)}) = -\frac{1}{|h(e^{i\varphi(\lambda_j^{(n)})})|}\, y_j(\lambda_j^{(n)}). \tag{5.2}$$

Thus, the representation

$$v_j^{(n)} = \tau_j^{(n)}\left[\frac{y_j(\lambda_j^{(n)})}{\|y_j(\lambda_j^{(n)})\|_2} + O_\infty(e^{-\delta n})\right]$$

is immediate from Theorem 1.2. Finally, put $h_{j,n} = |h(e^{i\varphi(\lambda_j^{(n)})})| = |h(e^{-i\varphi(\lambda_j^{(n)})})|$. Theorem 1.1 shows that

$$\|w_j(\lambda_j^{(n)})\|_2^2 = \frac{n}{4}\left(\frac{1}{|h(e^{i\varphi})|^2} + \frac{1}{|h(e^{-i\varphi})|^2}\right)\Bigg|_{\lambda=\lambda_j^{(n)}} + O(1) = \frac{n}{2h_{j,n}^2} + O(1),$$

whence, by (5.2), $\|y_j(\lambda_j^{(n)})\|_2^2 = h_{j,n}^2\|w_j(\lambda_j^{(n)})\|_2^2 = n/2 + O(1)$. The proof of Theorem 1.3 is complete.

Here is the proof of Theorem 1.4. We first estimate the "small terms" in $y_j^{(n)}$. Summing up finite geometric series and using the assumption that $|u_\nu(\lambda)|$

are separated from 1 we come to

$$\sum_{m=1}^{n}\left|\sum_{\nu=1}^{r-1} Q_\nu(\lambda)\left(\frac{1}{u_\nu(\lambda)^m}+\frac{(-1)^{j+1}}{u_\nu(\lambda)^{n+1-m}}\right)\right|^2$$
$$\leq \sum_{\nu=1}^{r-1}\frac{4(r-1)|Q_\nu(\lambda)|^2}{1-|u_\nu(\lambda)|^2}\leq K\sin^2\varphi(\lambda)$$

where K is some positive number depending only on a. since $\varphi(\lambda_j^{(n)})=O\left(j/n\right)$, it follows that

$$\left\|\left(\sum_{\nu=1}^{r-1} Q_\nu(\lambda)\left(\frac{1}{u_\nu(\lambda)^m}+\frac{(-1)^{j+1}}{u_\nu(\lambda)^{n+1-m}}\right)\right)_{m=1}^{n}\right\|_2=O\left(\frac{j}{n}\right). \tag{5.3}$$

We next consider the difference between the "main term" of $y_j^{(n)}$ and $\sin\frac{mj\pi}{n+1}$. Using the elementary estimate

$$|\sin A-\sin B|^2=4\sin^2\frac{A-B}{2}\cos^2\frac{A+B}{2}$$
$$\leq 4\sin^2\frac{A-B}{2}=2-2\cos(A-B),$$

we get

$$\sum_{m=1}^{n}\left|\sin\left(m\varphi(\lambda_j^{(n)})+\theta(\lambda_j^{(n)})\right)-\sin\frac{mj\pi}{n+1}\right|^2$$
$$\leq 2n-2\sum_{m=1}^{n}\cos\left(m\left(\varphi(\lambda_j^{(n)})-\frac{\pi j}{n+1}\right)+\theta(\lambda_j^{(n)})\right).$$

To simplify the last sum, we use that

$$\begin{aligned}\sum_{m=1}^{n}\cos(m\xi+\omega) &= \frac{\sin\frac{n\xi}{2}\cos\left(\frac{(n+1)\xi}{2}+\omega\right)}{\sin\frac{\xi}{2}}\\ &= n\left(1+O(n^2\xi^2)\right)\left(1+O\left(\frac{(n+1)\xi}{2}+\omega\right)^2\right).\end{aligned}$$

In our case

$$\omega=\theta(\lambda_j^{(n)})=O\left(\sqrt{\lambda_j^{(n)}}\right)=O\left(\frac{j}{n}\right),$$

$$\xi=\varphi(\lambda_j^{(n)})-\frac{\pi j}{n+1}=-\frac{\theta(\lambda_j^{(n)})}{n+1}+O(e^{-n\delta})=O\left(\frac{j}{n^2}\right).$$

Consequently,

$$\sum_{m=1}^{n}\left|\sin\left(m\varphi(\lambda_j^{(n)})+\theta(\lambda_j^{(n)})\right)-\sin\frac{mj\pi}{n+1}\right|^2=O\left(\frac{j^2}{n}\right),$$

that is,

$$\left\| \left(\sin\left(m\varphi(\lambda_j^{(n)}) + \theta(\lambda_j^{(n)}) \right) - \sin\frac{mj\pi}{n+1} \right)_{m=1}^{n} \right\|_2 = O\left(\frac{j}{\sqrt{n}}\right). \tag{5.4}$$

Combining (5.3) and (5.4)we obtain that

$$\left\| y_j^{(n)} - \sqrt{\frac{n+1}{2}}\, x_j^{(n)} \right\|_2 = O\left(\frac{j}{n}\right) + O\left(\frac{j}{\sqrt{n}}\right) = O\left(\frac{j}{\sqrt{n}}\right), \tag{5.5}$$

which implies in particular that

$$\|y_j^{(n)}\|_2 = \sqrt{\frac{n+1}{2}} \left(1 + O\left(\frac{j}{n}\right)\right). \tag{5.6}$$

Clearly, estimates (5.5) and (5.6) yield the asserted estimate. This completes the proof of Theorem 1.4.

6. Numerical results

Given $T_n(a)$, determine the approximate eigenvalue $\lambda_{j,*}^{(n)}$ from the equation

$$(n+1)\varphi(\lambda_{j,*}^{(n)}) + \theta(\lambda_{j,*}^{(n)}) = \pi j.$$

In [2], we proposed an exponentially fast iteration method for solving this equation. Let $w_j^{(n)}(\lambda) \in \mathbf{C}^n$ be as in Section 1 and put

$$w_{j,*}^{(n)} = \frac{w_j^{(n)}(\lambda_{j,*}^{(n)})}{\|w_j^{(n)}(\lambda_{j,*}^{(n)})\|_2}.$$

We define the distance between the normalized eigenvector $v_j^{(n)}$ and the normalized vector $w_{j,*}^{(n)}$ by

$$\varrho(v_j^{(n)}, w_{j,*}^{(n)}) := \min_{\tau\in\mathbf{T}} \|\tau v_j^{(n)} - w_{j,*}^{(n)}\|_2 = \sqrt{2 - 2\langle v_j^{(n)}, w_{j,*}^{(n)}\rangle}$$

and put

$$\begin{aligned}
\Delta_*^{(n)} &= \max_{1\le j\le n} |\lambda_j^{(n)} - \lambda_{j,*}^{(n)}|,\\
\Delta_{v,w}^{(n)} &= \max_{1\le j\le n} \varrho(v_j^{(n)}, w_{j,*}^{(n)}),\\
\Delta_r^{(n)} &= \max_{1\le j\le n} \|T_n(a)w_{j,*}^{(n)}) - \lambda_{j,*}^{(n)} w_{j,*}^{(n)}\|_2.
\end{aligned}$$

The tables following below show these errors for three concrete choices of the generating function a.

For $a(t) = 8 - 5t - 5t^{-1} + t^2 + t^{-2}$ we have

	$n = 10$	$n = 20$	$n = 50$	$n = 100$	$n = 150$
$\Delta_*^{(n)}$	$5.4 \cdot 10^{-7}$	$1.1 \cdot 10^{-11}$	$5.2 \cdot 10^{-25}$	$1.7 \cdot 10^{-46}$	$9.6 \cdot 10^{-68}$
$\Delta_{v,w}^{(n)}$	$2.0 \cdot 10^{-6}$	$1.1 \cdot 10^{-10}$	$2.0 \cdot 10^{-23}$	$1.9 \cdot 10^{-44}$	$2.0 \cdot 10^{-65}$
$\Delta_r^{(n)}$	$8.0 \cdot 10^{-6}$	$2.7 \cdot 10^{-10}$	$3.4 \cdot 10^{-23}$	$2.2 \cdot 10^{-44}$	$1.9 \cdot 10^{-65}$

If $a(t) = 8 + (-4 - 2i)t + (-4 - 2i)t^{-1} + it - it^{-1}$ then

	$n = 10$	$n = 20$	$n = 50$	$n = 100$	$n = 150$
$\Delta_*^{(n)}$	$3.8 \cdot 10^{-8}$	$2.8 \cdot 10^{-13}$	$2.9 \cdot 10^{-30}$	$5.9 \cdot 10^{-58}$	$1.6 \cdot 10^{-85}$
$\Delta_{v,w}^{(n)}$	$1.8 \cdot 10^{-7}$	$4.7 \cdot 10^{-13}$	$2.0 \cdot 10^{-29}$	$7.0 \cdot 10^{-57}$	$2.4 \cdot 10^{-84}$
$\Delta_r^{(n)}$	$5.4 \cdot 10^{-7}$	$1.3 \cdot 10^{-12}$	$2.7 \cdot 10^{-29}$	$6.7 \cdot 10^{-57}$	$1.9 \cdot 10^{-84}$

In the case where $a(t) = 24 + (-12 - 3i)t + (-12 + 3i)t^{-1} + it^3 - it^{-3}$ we get

	$n = 10$	$n = 20$	$n = 50$	$n = 100$	$n = 150$
$\Delta_*^{(n)}$	$6.6 \cdot 10^{-6}$	$1.2 \cdot 10^{-10}$	$7.6 \cdot 10^{-24}$	$1.4 \cdot 10^{-45}$	$3.3 \cdot 10^{-67}$
$\Delta_{v,w}^{(n)}$	$1.9 \cdot 10^{-6}$	$1.3 \cdot 10^{-10}$	$2.0 \cdot 10^{-23}$	$7.2 \cdot 10^{-45}$	$2.8 \cdot 10^{-66}$
$\Delta_r^{(n)}$	$2.5 \cdot 10^{-5}$	$8.6 \cdot 10^{-10}$	$7.3 \cdot 10^{-23}$	$1.9 \cdot 10^{-44}$	$5.9 \cdot 10^{-66}$

References

[1] A. Böttcher and S.M. Grudsky, *Spectral Properties of Banded Toeplitz Matrices,* SIAM, Philadelphia 2005.

[2] A. Böttcher, S.M. Grudsky, E.A. Maksimenko, *Inside the eigenvalues of certain Hermitian Toeplitz band matrices.* Journal of Computational and Applied Mathematics **233** (2010), 2245–2264.

[3] A. Böttcher, S.M. Grudsky, E.A. Maksimenko, and J. Unterberger, *The first order asymptotics of the extreme eigenvectors of certain Hermitian Toeplitz matrices.* Integral Equations and Operator Theory **63** (2009), 165–180.

[4] A. Böttcher, S.M. Grudsky, and J. Unterberger, *Asymptotic pseudomodes of Toeplitz matrices,* Operators and Matrices **2** (2008), 525–541.

[5] A. Böttcher and B. Silbermann, *Introduction to Large Truncated Toeplitz Matrices,* Universitext, Springer-Verlag, New York 1999.

[6] C. Estatico and S. Serra Capizzano, *Superoptimal approximation for unbounded symbols,* Linear Algebra Appl. **428** (2008), 564–585.

[7] U. Grenander and G. Szegö, *Toeplitz Forms and Their Applications,* University of California Press, Berkeley 1958.

[8] C.M. Hurvich and Yi Lu, *On the complexity of the preconditioned conjugate gradient algorithm for solving Toeplitz systems with a Fisher-Hartwig singularity,* SIAM J. Matrix Anal. Appl. **27** (2005), 638–653.

[9] M. Kac, W.L. Murdock, and G. Szegö, *On the eigenvalues of certain Hermitian forms,* J. Rational Mech. Anal. **2** (1953), 767–800.

[10] A.Yu. Novosel'tsev and I.B. Simonenko, *Dependence of the asymptotics of extreme eigenvalues of truncated Toeplitz matrices on the rate of attaining an extremum by a symbol,* St. Petersburg Math. J. **16** (2005), 713–718.

[11] S.V. Parter, *Extreme eigenvalues of Toeplitz forms and applications to elliptic difference equations,* Trans. Amer. Math. Soc. **99** (1961), 153–192.

[12] S.V. Parter, *On the extreme eigenvalues of Toeplitz matrices,* Trans. Amer. Math. Soc. **100** (1961), 263–276.

[13] S. Serra Capizzano, *On the extreme spectral properties of Toeplitz matrices generated by L^1 functions with several minima/maxima,* BIT **36** (1996), 135–142.

[14] S. Serra Capizzano, *On the extreme eigenvalues of Hermitian (block) Toeplitz matrices,* Linear Algebra Appl. **270** (1998), 109–129.

[15] S. Serra Capizzano and P. Tilli, *Extreme singular values and eigenvalues of non-Hermitian block Toeplitz matrices,* J. Comput. Appl. Math. **108** (1999), 113–130.

[16] W.F. Trench, *Explicit inversion formulas for Toeplitz band matrices,* SIAM J. Algebraic Discrete Methods **6** (1985), 546–554.

[17] W.F. Trench, *Interlacement of the even and odd spectra of real symmetric Toeplitz matrices,* Linear Algebra Appl. **195** (1993), 59–68.

[18] H. Widom, *On the eigenvalues of certain Hermitian operators,* Trans. Amer. Math. Soc. **88** (1958), 491–522.

[19] N.L. Zamarashkin and E.E. Tyrtyshnikov, *Distribution of the eigenvalues and singular numbers of Toeplitz matrices under weakened requirements on the generating function,* Sb. Math. **188** (1997), 1191–1201.

[20] P. Zizler, R.A. Zuidwijk, K.F. Taylor, and S. Arimoto, *A finer aspect of eigenvalue distribution of selfadjoint band Toeplitz matrices,* SIAM J. Matrix Anal. Appl. **24** (2002), 59–67.

Albrecht Böttcher
Fakultät für Mathematik
TU Chemnitz
D-09107 Chemnitz, Germany
e-mail: `aboettch@mathematik.tu-chemnitz.de`

Sergei M. Grudsky
Departamento de Matemáticas, CINVESTAV del I.P.N.
Apartado Postal 14-740
07000 México, D.F., México
e-mail: `grudsky@math.cinvestav.mx`

Egor A. Maksimenko
Departamento de Matemáticas, CINVESTAV del I.P.N.
Apartado Postal 14-740
07000 México, D.F., México
e-mail: `emaximen@math.cinvestav.mx`

Operator Theory:
Advances and Applications, Vol. 210, 37–42

Complete Quasi-wandering Sets and Kernels of Functional Operators

Victor D. Didenko

To Nikolai Vasilevski on the occasion of his 60th birthday

Abstract. Kernels of functional operators generated by mapping that possess complete quasi-wandering sets are studied. It is shown that the kernels of the operators under consideration either consist of a zero element or contain a subset isomorphic to a space $L_\infty(\mathbb{S})$, where $\mathbb{S} \subset \mathbb{R}^n$ has a positive Lebesgue measure. Consequently, such operators are Fredholm if and only if they are invertible.

Mathematics Subject Classification (2000). Primary 47B33, 39A33, 45E10; Secondary 42C40, 39B42.

Keywords. Quasi-wandering set, homogeneous equation, solution.

1. Introduction

Let X be a domain in $\mathbb{R}^n$ provided with the Lebesgue measure μ, and let $\alpha : X \mapsto X$ be a measurable mapping satisfying the compatibility condition. Thus if E is a Lebesgue measurable subset of X and if $\mu(E) = 0$, then $\mu(\alpha^{-1}(E)) = 0$. Assume that $A_0, A_1, \ldots, A_s : X \mapsto \mathbb{C}^{d\times d}$ are matrix functions, and consider the functional equation

$$A_0(x)\varphi(x) + A_1(x)\varphi(\alpha(x)) + \cdots + A_s(x)\varphi(\alpha^s(x)) = f(x) \tag{1.1}$$

where $\varphi : X \mapsto \mathbb{C}^d$ is an unknown vector function and $f : X \mapsto \mathbb{C}^d$ is a given vector function. Equations of the form (1.1) arise in various fields of mathematics and its applications, and there is vast literature where different properties of such equations and the corresponding operators are studied. For details, the reader can consult [1, 2, 4, 5, 9, 10, 12] and literature therein.

Let us mention a few functional operators relevant to our following discussion. Let $A = A(x), x \in \mathbb{R}^n$ be a matrix function and $M \in \mathbb{R}^{n\times n}$ a non-singular integer expansive matrix – i.e., all eigenvalues $\lambda_j, j = 1, 2, \ldots, n$ of the matrix M satisfy the inequality $|\lambda_j| > 1$. Equations of the form

$$\varphi(x) = A(x)\varphi(Mx) \tag{1.2}$$

appear in most publications concerning wavelets or their applications. In particular, equation (1.2) arises in discussions of the discrete refinement equation

$$\psi(x) = \sum_{k\in\mathbb{Z}^n} a_k\psi(Mx-k), \quad x \in \mathbb{R}^n\,, \tag{1.3}$$

or the continuous refinement equation

$$\psi(x) = \int_{\mathbb{R}^n} c(x-My)\psi(x)\,dy, \quad x \in \mathbb{R}^n. \tag{1.4}$$

Note that the refinement equation (1.3) is associated with the functional equation (1.2), where the matrix A has a uniformly convergent Fourier series; whereas equation (1.4) leads to the equation (1.2), with the matrix A that is the Fourier image of the matrix $c \in L_1^{d\times d}(\mathbb{R}^n)$.

Let us also recall another class of functional equations. Consider $X = \mathbb{R}^n$ and choose an $h \in \mathbb{R}^n, h \neq 0$. The difference equations

$$A_0(x)\varphi(x) + A_1(x)\varphi(x+h) + \cdots + A_s(x)\varphi(x+sh) = f(x),$$

that often arise in applications have been investigated by various authors.

In the present paper, the kernel spaces and the Fredholm properties of the equations mentioned are studied from a unified point of view.

Let

$$U_\alpha := \sum_{j=0}^{s} A_j T_\alpha^j \tag{1.5}$$

be the operator defined by the left-hand side of equation (1.1), where

$$T_\alpha\varphi(x) := \varphi(\alpha(x)),$$

and $A_j, j = 0,1,\ldots,s$ are the operators of multiplication by the matrices $A_j(x)$. In the present paper, the operator U_α is considered on the space $L_2^d(X)$ of all Lebesgue measurable square summable vector functions. We study the kernel space of the functional operators for mappings α that possess the so-called complete quasi-wandering set. For such mappings, the kernel space of the above-mentioned operator U_α has a distinctive property – viz. it either consists of the single element $\varphi_0 = 0$ or it contains a subset isomorphic to a space $L_\infty(\mathbb{S})$, where $\mathbb{S}$ is a subset of $\mathbb{R}^n$ with a positive Lebesgue measure. Note that it was conjectured in [3] that the kernel of any operator (1.5) on L_p space is either infinite-dimensional or only has the zero element. However, an example from [11] shows that this conjecture is not true in general. On the other hand, the problem mentioned is closely connected to the hypothesis that any operator (1.5) is Fredholm if and only if it is invertible. Although valid for a number of operator algebras [1, 8], this hypothesis is also not true in general [6]. Thus, this paper presents another class of operators when both hypotheses are true. Moreover, our approach shows that the kernel space of the operators under consideration can include subspaces as 'massive' as $L_\infty(\mathbb{S})$.

2. The kernel space of the operator U_α

Now let us assume that the mapping $\alpha : X \mapsto X$ is invertible, and the inverse mapping α^{-1} also satisfies the compatibility condition.

Definition 2.1. A Lebesgue measurable set $E \subset X$ is a quasi-wandering set for the mapping α if, for any $j, k \in \mathbb{Z}, j \neq k$, either $\alpha^j(E) \cap \alpha^k(E) = \emptyset$, or if $\alpha^{j_1}(E) \cap \alpha^{k_1}(E) \neq \emptyset$ for some $j_1, k_1 \in \mathbb{Z}$, then $\alpha^{j_1}(x) = \alpha^{k_1}(x)$ for all $x \in E$.

If $\alpha^j(E) \cap \alpha^k(E) = \emptyset$ for all $j, k \in \mathbb{Z}, j \neq k$, the set E is called the wandering set for α.

Definition 2.2. A quasi-wandering set E is called complete if

$$X = \bigcup_{j \in \mathbb{Z}} \alpha^j(E).$$

It is clear that if a mapping $\alpha : X \mapsto X$ possesses a complete quasi-wandering set E and for some indices j, k, $j \neq k$, $\alpha^j(x) = \alpha^k(x)$ for all $x \in E$, then

$$X = \bigcup_{j=0}^{N} \alpha^j(E)$$

for an $N \in \mathbb{N}$.

Let us now consider a few mappings with complete quasi-wandering sets.

Example 1. Suppose $X = \mathbb{R}$ and $\alpha : \mathbb{R} \mapsto \mathbb{R}$ is the shift operator

$$\alpha(x) := x - 1.$$

The mapping α satisfies all of the conditions mentioned. It is invertible and the interval $[0, 1)$ is a complete wandering set for α.

Example 2. Let $X = \mathbb{R}^2$, and let Λ be a diagonal matrix

$$\Lambda = \begin{pmatrix} \lambda_1 & 0 \\ 0 & \lambda_2 \end{pmatrix},$$

where $\lambda_1 \neq 0$ and $\lambda_2 > 1$. Consider the mapping

$$\alpha_\Lambda(x) := \Lambda x.$$

Then the set

$$E_\Lambda := \{(x, y) \in \mathbb{R}^2, x \in \mathbb{R} \text{ and } y \in [-1, -1/\lambda_2) \cup (1/\lambda_2, 1]\}$$

is a complete wandering set for the mapping α_Λ.

Example 3. Let $X = \mathbb{R}^3$, $m \in \mathbb{N}$, $m \geq$ and let

$$\alpha_R(x) := Rx,$$

where R is the rotation matrix

$$R = \begin{pmatrix} 1 & 0 & 0 \\ 0 & \cos(\pi/m) & -\sin(\pi/m) \\ 0 & \sin(\pi/m) & \cos(\pi/m) \end{pmatrix}.$$

Then the set

$$E_R := \{(x,y,z) \in \mathbb{R}^3 : x \in \mathbb{R}, y > 0, z \geq 0, \text{ and } 0 \leq \arctan(z/y) < \pi/m\}$$

is a complete quasi-wandering set for the mapping α_R and

$$\mathbb{R}^3 = \bigcup_{j=0}^{2m-1} \alpha_R^j(E_R).$$

Example 4. Let $X = E_1 \cup E_2$, $E_1 \cap E_2 = \emptyset$ and $\mu(E_1) \neq 0$ and $\mu(E_2) \neq 0$. Consider a homomorphism $\alpha : X \mapsto X$ satisfying the compatibility condition and such that

- $\alpha(E_1) = E_2$, $\alpha(E_2) = E_1$.
- $\alpha^2(x) = x$ for all $x \in X$.

Then E_1 and E_2 are complete quasi-wandering sets for the mapping α.

Now we can study the kernel spaces of the operators (1.1) when the corresponding mapping α possesses a complete quasi-wandering set.

Theorem 2.3. *Let $A_1, A_2, \ldots, A_s \in L_\infty^{d\times d}(X)$ and $\alpha : X \mapsto X$ be a mapping such that $T_\alpha : L_2(X) \mapsto L_2(X)$ is a continuous operator. If the mapping α has a complete quasi-wandering set, then either*

$$\ker U_\alpha = 0$$

or there is a subspace $\mathfrak{S} \subset \ker U_\alpha$ and a set $\mathbb{S}_\alpha \subset X$, the Lebesgue measure of which is positive such that $\mathfrak{S}$ is isomorphic to the space $L_\infty(\mathbb{S}_\alpha)$.

Proof. Let E_α be a complete quasi-wandering set for the mapping α. Assume that the kernel of the operator U_α contains a non-zero element φ_0. For simplicity, also suppose that $d = 1$. Thus

$$\int_X |\varphi_0(x)|^2 \, dx > 0 \,,$$

so there is at least one index $j_0 \in \mathbb{Z}$ and an $\varepsilon > 0$ such that the set

$$\mathbb{S}_\alpha = \mathbb{S}_\alpha^\varepsilon := \{x \in \alpha^{j_0}(E_\alpha) : |\varphi_0(x)| \geq \varepsilon\}$$

has a positive Lebesgue measure. Consider now the space $L_\infty(\alpha^{j_0}(E_\alpha))$. For any $m \in L_\infty(\alpha^{j_0}(E_\alpha))$, such that the restriction of m on the set $\mathbb{S}_\alpha^\varepsilon$ is a non-zero element of $L_\infty(\mathbb{S}_\alpha^\varepsilon)$, define an extension $\widetilde{m}$ of the element m on the whole X by

$$\widetilde{m}(x) := m(\alpha^{-j}(x)) \text{ if } x \in \alpha^{j+j_0}(E_\alpha).$$

Note that the element $\widetilde{m}$ is well defined and belongs to the space $L_\infty(X)$. Moreover, it satisfies the equation

$$\widetilde{m}(x) = \widetilde{m}(\alpha(x)), \quad x \in X. \tag{2.1}$$

Consider now the element $\widetilde{m}\varphi_0$. Taking into account equation (2.1), one can easily check that $\widetilde{m}\varphi_0 \in \ker U_\alpha$. It remains to show that $\widetilde{m}\varphi_0 \neq 0$. However, if the restriction of $\widetilde{m}$ on the set $\mathbb{S}_\alpha$ is a non-zero element of $L_\infty(\mathbb{S}_\alpha)$, then

$$\int_X |\widetilde{m}(x)\varphi_0(x)|^2\,dx = \int_{\cup_{j\in\mathbb{Z}}\alpha^j(E_\alpha)} |\widetilde{m}(x)\varphi_0(x)|^2\,dx \geq \int_{\alpha^{j_0}(E_\alpha)} |\widetilde{m}(x)\varphi_0(x)|^2\,dx$$
$$\geq \int_{\mathbb{S}_\alpha} |\widetilde{m}(x)\varphi_0(x)|^2\,dx \geq \varepsilon^2 \int_{\mathbb{S}_\alpha} |\widetilde{m}(x)|^2\,dx > 0.$$

Thus the kernel of the operator U_α contains the subset

$$\mathfrak{S} := \{\widetilde{m}\varphi_0 : m \in L_\infty(\mathbb{S}_\alpha)\}$$

and the proof is complete. □

The connection between some sets related to the operators of multiplication by expansive matrices M and the kernel spaces of the corresponding refinement operators was first noted in [7]. It turns out that these relations have a universal nature and can be extended to general mappings α having quasi-wandering sets. This allows us to characterize the Fredholm properties of the functional operators.

Corollary 2.4. *Let $U_\alpha : L_2(X) \mapsto L_2(X)$ be a Φ^+-operator. If the mapping $\alpha : X \mapsto X$ possesses a complete quasi-wandering set, then*

$$\ker U_\alpha = 0.$$

Now consider functional operators generated by diffeomorphisms.

Theorem 2.5. *Let $A_1, A_2, \ldots, A_s \in L_\infty^{d\times d}(X)$ and let $\alpha : X \mapsto X$ be a differentiable mapping such that the Jacobian J_α of α satisfies the inequality*

$$0 < r_1 \leq J_\alpha(x) \leq r_2, \quad r_1, r_2 \in \mathbb{R} \tag{2.2}$$

for all $x \in X$. If the mapping α has a complete quasi-wandering set, then the operator $U_\alpha : L_2^d(X) \mapsto L_2^d(X)$ is Fredholm if and only if it is invertible.

Proof. If α has a complete quasi-wandering set E, then E is also a complete quasi-wandering set for the inverse mapping α^{-1}. The proof of Theorem 2.5 therefore follows from Theorem 2.3 and from condition (2.2), which implies that the structure of the adjoint operator U_α^* is similar to the structure of the operator U_α. □

Remark 2.6. This result is known for some classes of the functional operators [1, 8]. One way to prove it is to show that the algebra generated by the corresponding operator T_α contains no non-trivial compact operators – see [1, Theorem 8.3]. On the other hand, the approach used here allows us to obtain certain additional information concerning the kernels of the operators under consideration.

Remark 2.7. In theory of dynamical systems, the systems with complete wandering sets are called complete dissipative systems [13]. Thus, complete dissipative systems are Fredholm if and only if they are invertible.

Acknowledgment

The author would like to thank Alexei Karlovich for pointing out reference [6].

References

[1] A. Antonevich, *Linear Functional Equations. Operator Approach.* Operator Theory. Advances and Applications, Vol. 83, Birkhäuser, Basel-Boston-Berlin, 1996.

[2] A. Antonevich, A. Lebedev, *Functional Differential Equations: C^*-Theory.* Pitman Monographs and Surveys in Pure and Applied Mathematics, Vol. 70. Longman, Harlow, 1994.

[3] N.V. Azbelev, G.G. Islamov, A certain class of functional-differential equations. *Differencial'nye Uravnenija,* **12** (1976), 417–427.

[4] G. Belitskii, V. Tkachenko, *One-dimensional functional equations.* Operator Theory: Advances and Applications, Vol. 144, Birkhäuser, Basel, 2003.

[5] E. Castillo, A. Iglesias, R. Ruíz-Cobo, *Functional equations in applied sciences.* Mathematics in Science and Engineering, Vol. 199, Elsevier, Amsterdam, 2005.

[6] A.V. Chistyakov, A pathological counterexample to the non-Fredholmness conjecture in algebras of weighted shift operators, *Izv. Vyssh. Uchebn. Zaved. Mat.,* No. 10 (1995), 76–86.

[7] V.D. Didenko, Fredholm properties of multivariate refinement equations. *Math. Meth. Appl. Sci.,* **30** (2007), 1639–1644

[8] Yu.I. Karlovich, V.G. Kravchenko, G.S. Litvinchuk, Invertibility of functional operators in Banach spaces. In: *Functional-differential equations,* Perm. Politekh. Inst., Perm', 1990, 18–58.

[9] V.G. Kravchenko, G.S. Litvinchuk, *Introduction to the theory of singular integral operators with shift.* Mathematics and its Applications, Vol. 289, Kluwer, Dordrecht, 1994.

[10] M. Kuczma, *Functional equations in single variable.* Monografie Matematyczne, Vol. 46, PWN, Warszawa, 1968.

[11] V.G. Kurbatov, A conjecture in the theory of functional-differential equations, *Differentsial'nye Uravnenija,* **14** (1978), 2074–2075.

[12] G.S. Litvinchuk, *Solvability theory of boundary value problems and singular integral equations with shift.* Mathematics and its Applications, Vol. 523, Kluwer, Dordrecht, 2000.

[13] P.J. Nicholls, *The Ergodic Theory of Discrete Groups.* London Mathematical Society Lecture Note Series, Vol. 143, Cambridge University press, Cambridge, 1989.

Victor D. Didenko
Department of Mathematics
University Brunei Darussalam
Bandar Seri Begawan
BE1410 Brunei
e-mail: `diviol@gmail.com`

Operator Theory:
Advances and Applications, Vol. 210, 43–77

Lions' Lemma, Korn's Inequalities and the Lamé Operator on Hypersurfaces

Roland Duduchava

Dedicated to my friend and colleague Nikolai Vasilevski on the occasion of his 60th birthday anniversary

Abstract. We investigate partial differential equations on hypersurfaces written in the Cartesian coordinates of the ambient space. In particular, we generalize essentially Lions' Lemma, prove Korn's inequality and establish the unique continuation property from the boundary for Killing's vector fields, which are analogues of rigid motions in the Euclidean space. The obtained results, the Lax-Milgram lemma and some other results are applied to the investigation of the basic Dirichlet and Neumann boundary value problems for the Lamé equation on a hypersurface.

Mathematics Subject Classification (2000). 35J57, 74J35, 58J32.

Keywords. Lions's Lemma, Korn's inequality, Killing's fields, Lax-Milgram lemma, Lamé equation, Boundary value problems.

Introduction

Partial differential equations (PDEs) on hypersurfaces and corresponding boundary value problems (BVPs) appear rather often in applications: see [Ha1, §72] for the heat conduction by surfaces, [Ar1, §10] for the equations of surface flow, [Ci1], [Ci3],[Ci4], [Ko2], [Go1] for thin flexural shell problems in elasticity, [AC1] for the vacuum Einstein equations describing gravitational fields, [TZ1, TW1] for the Navier-Stokes equations on spherical domains and spheres, [MM1] for minimal surfaces, [AMM1] for diffusion by surfaces, as well as the references therein. Furthermore, such equations arise naturally while studying the asymptotic behavior of solutions to elliptic boundary value problems in a neighborhood of conical points (see the classical reference [Ko1]).

The investigation was supported by the grant of the Georgian National Science Foundation GNSF/ST07/3-175.

By a classical approach differential equations on surfaces are written with the help of covariant and contravariant frames, metric tensors and Christoffel symbols. To demonstrate a difference between a classical and the present approaches, let us consider an example. A surface $\mathscr{S}$ can be given by a local immersion

$$\Theta : \omega \to \mathscr{S}, \quad \omega \subset \mathbb{R}^{n-1}, \tag{0.1}$$

which means that the derivatives $\{\boldsymbol{g}_k := \partial_k\Theta\}_{k=1}^{n-1}$, constituting the *covariant frame* in the space of tangent vector fields to the surface $\mathscr{V}(\mathscr{S})$, are linearly independent. In equivalent formulation that means the Gram matrix $G_{\mathscr{S}}(\mathscr{X}) = [g_{jk}(\mathscr{X})]_{n-1\times n-1}$, $g_{jk} := \langle \boldsymbol{g}_j, \boldsymbol{g}_k\rangle$ has the inverse $G_{\mathscr{S}}^{-1}(\mathscr{X}) = [g^{jk}(\mathscr{X})]_{n-1\times n-1}$, $g^{jk} := \langle \boldsymbol{g}^j, \boldsymbol{g}^k\rangle$, where $\{\boldsymbol{g}^k\}_{k=1}^{n-1}$ is the contravariant frame and is biorthogonal to the covariant frame $\langle \boldsymbol{g}_j, \boldsymbol{g}^k\rangle = \delta_{jk}$, $j,k = 1,\dots,n-1$. Hereafter

$$f\langle \boldsymbol{U}, \boldsymbol{V}\rangle := \sum_{j=1}^{n} U_j^0 V_j^0, \qquad \boldsymbol{U} = (U_1^0,\dots,U_n^0)^\top \in \mathbb{R}^n, \quad \boldsymbol{V} = (V_1^0,\dots,V_n^0)^\top \in \mathbb{R}^n$$

denotes the scalar product. The Gram matrix $G_{\mathscr{S}}(\mathscr{X})$ is also called *covariant metric tensor* and is responsible for the *Riemannian metric* on $\mathscr{S}$.

The surface divergence and gradients in classical differential geometry (in intrinsic parameters of the surface $\mathscr{S}$) read as follows:

$$\begin{aligned} \operatorname{div}_{\mathscr{S}} \boldsymbol{U} &:= \big[\det G_{\mathscr{S}}\big]^{-1/2} \sum_{j=1}^{n} \partial_j \Big\{ \big[\det G_{\mathscr{S}}\big]^{1/2} U^j \Big\}, \\ \nabla_{\mathscr{S}} f &= \sum_{j,k=1}^{n-1} (g^{jk}\partial_j f)\,\partial_k, \qquad \boldsymbol{U} = \sum_{j=1}^{n-1} U^j \boldsymbol{g}_j \end{aligned} \tag{0.2}$$

(see [Ta2, Ch. 2, § 3]). The intrinsic parameters enable generalization to arbitrary manifolds, not necessarily immersed in the Euclidean space $\mathbb{R}^n$.

A derivative $\partial_{\boldsymbol{U}}^{\mathscr{S}} : C^1(\mathscr{S}) \to C^1(\mathscr{S})$ along some tangential vector field $\boldsymbol{U} \in \mathscr{V}(\mathscr{S})$ is called *covariant* if it is a linear automorphism of the space of tangential vector fields

$$\partial_{\boldsymbol{U}}^{\mathscr{S}} : \mathscr{V}(\mathscr{S}) \longrightarrow \mathscr{V}(\mathscr{S}). \tag{0.3}$$

The covariant derivative of a tangential vector field $\boldsymbol{V} = \sum_{j=1}^{n-1} V^j \boldsymbol{g}_j \in \mathscr{V}(\mathscr{S})$ along a tangential vector field $\boldsymbol{U} = \sum_{j=1}^{n-1} U^j \boldsymbol{g}_j \in \mathscr{V}(\mathscr{S})$ is defined by the formula

$$\partial_{\boldsymbol{U}}^{\mathscr{S}} \boldsymbol{V} := \pi_{\mathscr{S}} \partial_{\boldsymbol{U}} \boldsymbol{V} := \sum_{j,k,m=1}^{n-1} \left[U^j V^k \Gamma_{jk}^m + \delta_{jk} U^j \partial_j V^m\right] \boldsymbol{g}_m, \tag{0.4}$$

where $\Gamma_{jk}^m(x)$ are the *Christoffel symbols*

$$\begin{aligned} \Gamma_{jk}^m(x) &:= \langle \partial_k \boldsymbol{g}_j(x), \boldsymbol{g}^m(x)\rangle = \sum_{q=1}^{n-1} \frac{g^{mq}}{2}\left[\partial_k g_{jq}(x) + \partial_j g_{kq}(x) - \partial_q g_{jk}(x)\right] \\ &:= \Gamma_{kj}^m(x). \end{aligned} \tag{0.5}$$

The calculus of differential operators on hypersurfaces presented here is based on Günter's derivatives. The definition applies the natural basis

$$\boldsymbol{e}^1 = (1,0,\dots,0)^\top, \dots, \boldsymbol{e}^n = (0,\dots,0,1)^\top \tag{0.6}$$

in the ambient Euclidean space $\mathbb{R}^n$ and the field of unit normal vectors to the surface $\mathscr{S}$

$$\boldsymbol{\nu}(\mathscr{X}) := \pm \frac{\boldsymbol{g}_1(\Theta^{-1}(\mathscr{X})) \wedge \cdots \wedge \boldsymbol{g}_{n-1}(\Theta^{-1}(\mathscr{X}))}{\left|\boldsymbol{g}_1(\Theta^{-1}(\mathscr{X})) \wedge \cdots \wedge \boldsymbol{g}_{n-1}(\Theta^{-1}(\mathscr{X}))\right|}, \qquad \mathscr{X} \in \mathscr{S}, \tag{0.7}$$

where $\boldsymbol{U}^{(1)} \wedge \cdots \wedge \boldsymbol{U}^{(n-1)}$ (or also $\boldsymbol{U}^{(1)} \times \cdots \times \boldsymbol{U}^{(n-1)}$) denotes the vector product of vectors $\boldsymbol{U}^{(1)}, \dots, \boldsymbol{U}^{(n-1)} \in \mathbb{R}^n$. If a hypersurface $\mathscr{S}$ in $\mathbb{R}^n$ is defined implicitly

$$\mathscr{S} = \left\{\mathscr{X} \in \omega \,:\, \Psi_{\mathscr{S}}(\mathscr{X}) = 0\right\}, \tag{0.8}$$

where $\Psi_{\mathscr{S}} : \omega \to \mathbb{R}$ is a C^k-mapping (or is a Lipschitz mapping) which is regular $\nabla\,\Psi(\mathscr{X}) \neq 0$, then the normalized gradient

$$\boldsymbol{\nu}(\mathscr{X}) := \pm \frac{\nabla\Psi_{\mathscr{S}}(\mathscr{X})}{\left|\nabla\Psi_{\mathscr{S}}(\mathscr{X})\right|}, \qquad \mathscr{X} \in \mathscr{S} \tag{0.9}$$

coincides with the outer unit normal vector provided the sign $\pm$ is chosen appropriately.

The collection of the tangential *Günter's derivatives* are defined as follows (cf. [Gu1], [KGBB1], [Du1]);

$$\mathscr{D}_j := \partial_j - \nu_j(\mathscr{X})\partial_{\boldsymbol{\nu}} = \partial_{\boldsymbol{d}^j}\,. \tag{0.10}$$

Here $\partial_{\boldsymbol{\nu}} := \sum_{j=1}^n \nu_j \partial_j$ denotes the normal derivative. For each $1 \leq j \leq n$, the first-order differential operator $\mathscr{D}_j = \partial_{\boldsymbol{d}^j}$ is the directional derivative along the tangential vector $\boldsymbol{d}^j := \pi_{\mathscr{S}}\,\boldsymbol{e}^j$, the projection of $\boldsymbol{e}^j$ on the space of tangent vector fields to $\mathscr{S}$. Here

$$\pi_{\mathscr{S}} : \mathbb{R}^n \to \mathscr{V}(\mathscr{S}), \quad \pi_{\mathscr{S}}(t) = I - \boldsymbol{\nu}(t)\boldsymbol{\nu}^\top(t) = \left[\delta_{jk} - \nu_j(t)\nu_k(t)\right]_{n\times n}, \quad t \in \mathscr{S} \tag{0.11}$$

defines the canonical orthogonal projection $\pi^2_{\mathscr{S}} = \pi_{\mathscr{S}}$ onto the space of tangent vector fields $\mathscr{V}(\mathscr{S})$ and $(\boldsymbol{\nu}, \pi_{\mathscr{S}} v) = 0$ for all $v \in \mathbb{R}^n$.

For tangential vector fields $\boldsymbol{V} \in \mathscr{V}(\mathscr{S})$ and $\boldsymbol{U} \in \mathscr{V}(\mathscr{S})$ we have representations

$$\boldsymbol{V} = \sum_{j=1}^n V^0_j \boldsymbol{e}^j = \sum_{j=1}^n V^0_j \boldsymbol{d}^j, \qquad \boldsymbol{U} = \sum_{j=1}^n U^0_j \boldsymbol{e}^j = \sum_{j=1}^n U^0_j \boldsymbol{d}^j. \tag{0.12}$$

The *surface gradient* $\nabla_{\mathscr{S}}\varphi$ and the *surface divergence* $\operatorname{div}_{\mathscr{S}}\boldsymbol{U}$ are defined as follows

$$\nabla_{\mathscr{S}}\varphi := (\mathscr{D}_1\varphi, \dots, \mathscr{D}_n\varphi)^\top, \qquad \operatorname{div}_{\mathscr{S}}\boldsymbol{U} ; = \sum_{j=1}^n \mathscr{D}_j U^0_j \tag{0.13}$$

(cf. (0.2)) while for the *derivative of a vector field* $\boldsymbol{V}$ along $\boldsymbol{U}$ and the corresponding *covariant derivative* we have the formulae

$$\partial_{\boldsymbol{U}}\boldsymbol{V}=\sum_{j=1}^{n}U_j^0\mathscr{D}_j\boldsymbol{V},\qquad \partial_{\boldsymbol{U}}^{\mathscr{S}}\boldsymbol{V}=\sum_{j=1}^{n}U_j^0\mathscr{D}_j^{\mathscr{S}}\boldsymbol{V} \tag{0.14}$$

(cf. (0.4)). Here $\mathscr{D}_j^{\mathscr{S}}:\mathscr{V}(\mathscr{S})\to\mathscr{V}(\mathscr{S})$ is the *covariant Günter's derivative*

$$\mathscr{D}_j^{\mathscr{S}}\boldsymbol{V}:=\pi_{\mathscr{S}}\mathscr{D}_j\boldsymbol{V}=\mathscr{D}_j\boldsymbol{V}-\langle\boldsymbol{\nu},\mathscr{D}_j\boldsymbol{V}\rangle\boldsymbol{\nu},\qquad j=1,\dots,n. \tag{0.15}$$

The Lamé operator $\mathscr{L}_{\mathscr{S}}$ on $\mathscr{S}$ is the natural operator associated with the Euler-Lagrange equations for a variational integral. The starting point is the total free (elastic) energy

$$\mathscr{E}[\boldsymbol{U}]:=\int_{\mathscr{S}}E(y,\mathscr{D}^{\mathscr{S}}\boldsymbol{U}(y))\,dS,\quad \mathscr{D}^{\mathscr{S}}\boldsymbol{U}:=\left[(\mathscr{D}_j^{\mathscr{S}}\boldsymbol{U})_k^0\right]_{n\times n},\quad \boldsymbol{U}\in\mathscr{V}(\mathscr{S}), \tag{0.16}$$

ignoring at the moment the displacement boundary conditions (Koiter's model). Equilibria states correspond to minimizers of the above variational integral (see [NH1, § 5.2]). The kernel $E=(\mathfrak{S}_{\mathscr{S}},\mathrm{Def}_{\mathscr{S}})$ depends bi-linearly on the stress $\mathfrak{S}_{\mathscr{S}}=\left[\mathfrak{S}^{jk}\right]_{n\times n}$ and the deformation $\mathscr{D}^{\mathscr{S}}$ tensors. The following form of the important deformation (strain) tensor was identified in [DMM1]

$$\mathrm{Def}_{\mathscr{S}}(\boldsymbol{U})=\left[\mathfrak{D}_{jk}(\boldsymbol{U})\right]_{n\times n},\quad \boldsymbol{U}=\sum_{j=1}^{n}U_j^0\boldsymbol{d}^j\in\mathscr{V}(\mathscr{S}),\quad j,k=1,\dots,n, \tag{0.17}$$

$$\mathfrak{D}_{jk}(\boldsymbol{U}):=\frac{1}{2}\left[(\mathscr{D}_j^{\mathscr{S}}\boldsymbol{U})_k^0+(\mathscr{D}_k^{\mathscr{S}}\boldsymbol{U})_j^0\right]=\frac{1}{2}\left[\mathscr{D}_kU_j^0+\mathscr{D}_jU_k^0+\sum_{m=1}^{n}U_m^0\mathscr{D}_m\big(\nu_j\nu_k\big)\right],$$

where $(\mathscr{D}_j^{\mathscr{S}}\boldsymbol{U})_k^0:=\langle\mathscr{D}_j^{\mathscr{S}}\boldsymbol{U},\boldsymbol{e}^k\rangle$. Hooke's law states that $\mathfrak{S}_{\mathscr{S}}=\mathbb{T}\,\mathrm{Def}_{\mathscr{S}}$ for some linear fourth-order tensor $\mathbb{T}:=\left[c_{jk\ell m}\right]_{n\times n\times n\times n}$, which is positive definite:

$$\langle\mathbb{T}\zeta,\zeta\rangle:=\sum_{i,j,k,\ell=1}^{n}c_{ijk\ell}\zeta_{ij}\overline{\zeta}_{k\ell}\geq C_0\sum_{i,j=1}^{n}|\zeta_{i,j}|^2:=C_0|\zeta|^2 \tag{0.18}$$

for all symmetric tensors $\zeta_{ij}=\zeta_{ji}\in\mathbb{C}$, $\zeta:=\left[\zeta_{ij}\right]_{n\times n}$. Moreover, $\mathbb{T}$ has the following symmetry properties:

$$c_{ijk\ell}=c_{ij\ell k}=c_{k\ell ij}\qquad\qquad \forall\, i,j,k,\ell\,. \tag{0.19}$$

The following form of the Lamé operator for a linear anisotropic elastic medium was identified in [DMM1]:

$$\mathscr{L}_{\mathscr{S}}=\mathrm{Def}_{\mathscr{S}}^{*}\mathbb{T}\,\mathrm{Def}_{\mathscr{S}}=\Big[\sum_{\ell m=1}^{n}c_{jk\ell m}\mathscr{D}_j^{\mathscr{S}}\mathscr{D}_\ell^{\mathscr{S}}\Big]_{n\times n},\qquad \boldsymbol{U}\in\mathscr{V}(\mathscr{S}), \tag{0.20}$$

The adjoint operator to the deformation tensor

$$\mathrm{Def}^*_{\mathscr{S}}\mathfrak{U} := \frac{1}{2}\sum_{j=1}^{n}\left\{(\mathscr{D}^{\mathscr{S}}_j)^*\left[\mathfrak{U}^{jk}+\mathfrak{U}^{kj}\right]\right\}_{k=1}^{n} \qquad \text{for} \quad \mathfrak{U}=\|\mathfrak{U}^{jk}\|_{n\times n} \tag{0.21}$$

maps tensor functions to vector functions.

For an isotropic medium

$$c_{jklm} = \lambda\delta_{jk}\delta_{lm} + \mu\left[\delta_{jl}\delta_{km}+\delta_{jm}\delta_{kl}\right] \tag{0.22}$$

and the Lamé operator acquires a simpler form

$$\begin{aligned}\mathscr{L}_{\mathscr{S}}\boldsymbol{U} &= -\lambda\nabla_{\mathscr{S}}\,\mathrm{div}_{\mathscr{S}}\boldsymbol{U} + 2\mu\,\mathrm{Def}^*_{\mathscr{S}}\mathrm{Def}_{\mathscr{S}}\boldsymbol{U}\\ &= -\mu\pi_{\mathscr{S}}\Delta_{\mathscr{S}}\boldsymbol{U} - (\lambda+\mu)\,\nabla_{\mathscr{S}}\mathrm{div}_{\mathscr{S}}\boldsymbol{U} - \mu\,\mathscr{H}^0_{\mathscr{S}}\mathscr{W}_{\mathscr{S}}\boldsymbol{U}\,, \quad \boldsymbol{U}\in\mathscr{V}(\mathscr{S})\end{aligned} \tag{0.23}$$

(cf. (0.11) for the projection $\pi_{\mathscr{S}}$). $\lambda,\mu\in\mathbb{R}$ are the Lamé moduli, whereas

$$\mathscr{H}^0_{\mathscr{S}} = -\mathrm{div}_{\mathscr{S}}\boldsymbol{\nu} := -\sum_{j=1}^{n}\mathscr{D}_j\nu_j = \mathrm{Tr}\,\mathscr{W}_{\mathscr{S}}\,, \qquad \mathscr{W}_{\mathscr{S}} = -\left[\mathscr{D}_j\nu_k\right]_{n\times n}. \tag{0.24}$$

Note, that $\mathscr{H}_{\mathscr{S}} := (n-1)^{-1}\mathscr{H}^0_{\mathscr{S}}$ represents the *mean curvature* of the surface $\mathscr{S}$; $\mathscr{W}_{\mathscr{S}}$ is the *Weingarten curvature tensor* of $\mathscr{S}$; Eigenvalues of $\mathscr{W}_{\mathscr{S}}$, except one which is 0, represent all principal curvatures of the surface $\mathscr{S}$.

Note, that Günter's derivatives were already applied in [MM1] to minimal surfaces and in [Gu1], [KGBB1] to the problems of 3D elasticity.

We believe that our results should be useful in numerical and engineering applications (cf. [AN1], [Be1], [Ce1], [Co1], [DaL1], [BGS1], [Sm1]). Having in mind applications, equations in Cartesian coordinates are simpler for approximation and numerical treatment.

The paper is organized as follows. § 1 is auxiliary. In § 2 we prove generalized Lions' Lemma for the Bessel potential spaces $\mathbb{H}^s_p(\mathscr{S})$ on a hypersurface with and without boundary. The result is applied to the proof of important Korn's inequality for Killing's vector fields.

In § 3 we investigate Killing's vector fields, which constitute the kernel of the Lamé operator and represent analogues of rigid motions in $\mathbb{R}^n$. The most important result there states that the class of Killing's vector fields has the unique continuation property from the boundary: if such a field vanishes on a set of positive measure on the boundary of a hypersurface with boundary, it vanishes on this hypersurface identically. The result is applied to prove further Korn's inequality "without boundary condition" and to the investigation of basic BVPs for the Lamé equation.

In § 4 we prove the ellipticity of the Lamé operator, which follows also from the Gårding's inequality

$$(\mathscr{L}_{\mathscr{S}}\boldsymbol{U},\boldsymbol{U})_{\mathscr{S}} \geq C_1\|\boldsymbol{U}|\mathbb{H}^1(\mathscr{S})\|^2 - C_0\|\boldsymbol{U}|\mathbb{L}_2(\mathscr{S})\|^2.$$

For a hypersurface without boundary $\mathscr{S}$ the kernel Ker $\mathscr{L}_{\mathscr{S}}$ coincides with the space of Killing's vector fields. Moreover, the operator $\mathscr{L}_{\mathscr{S}} + \mathscr{B}I \ : \ \mathbb{H}^s_p(\mathscr{S}) \to \mathbb{H}^{s-2}_p(\mathscr{S})$ is invertible if $\mathscr{S}$ has no boundary and is smooth, $1 < p < \infty$, $s \in \mathbb{R}$ and $\mathscr{B} \neq 0$ is a non-negative function.

In §§ 5–7 we investigate the Dirichlet and the Neumann boundary value problems for the Lamé operator on a hypersurface with boundary $\mathscr{C}$ under a minimal requirements on the surface. Namely, we require that the immersion Θ in (0.1) (or the implicit function $\Psi_{\mathscr{S}}$ in (0.8)), representing the surface $\mathscr{C}$, has the bounded second derivative $\Theta \in (\mathbb{H}^2_\infty)^n$ ($\Psi_{\mathscr{S}} \in \mathbb{H}^2_\infty$, respectively). The Dirichlet problem

$$\begin{cases} \mathscr{L}_{\mathscr{S}} \boldsymbol{U} = \boldsymbol{F} & \text{in} \quad \mathscr{C}, \\ \boldsymbol{U}\big|_\Gamma = \boldsymbol{G} & \text{on} \quad \Gamma := \partial\mathscr{S}, \end{cases} \qquad \boldsymbol{F} \in \widetilde{\mathbb{H}}^{-1}(\mathscr{C}), \quad \boldsymbol{G} \in \mathbb{H}^{1/2}(\Gamma), \tag{0.25}$$

where $\boldsymbol{U} = \sum_{j=1}^n U_j^0 \boldsymbol{d}^j \in \mathscr{V}(\mathscr{C}) \cap \mathbb{H}^1(\mathscr{C})^n$ is the (tangential) generalized displacement vector field of the elastic hypersurface $\mathscr{S}$, is reduced to an equivalent Dirichlet BVP with vanishing boundary data $\boldsymbol{G} = 0$, which, in its turn, is equivalent to the invertibility of the operator

$$\mathscr{L}_{\mathscr{C}} \ : \ \widetilde{\mathbb{H}}^{-1}(\mathscr{C}) \to \mathbb{H}^{-1}(\mathscr{C}).$$

The invertibility is derived from Gårding's inequality proved there. For the investigation of the Neumann BVP we apply the Lax-Milgramm Lemma, based on the coerciveness of the corresponding sesquilinear form.

1. Sobolev spaces and Bessel potential operators

Proposition 1.1. (*cf.* [DMM1]). *The surface divergence* $\operatorname{div}_{\mathscr{S}}$ *and the surface gradient* $\nabla_{\mathscr{S}}$ *(cf.* (0.13)) *are dual operators* $(\nabla_{\mathscr{S}}\varphi, \boldsymbol{U})_{\mathscr{S}} := -(\varphi, \operatorname{div}_{\mathscr{S}} \boldsymbol{U})_{\mathscr{S}}$ *with respect to the usual scalar product of (square integrable) vector functions on the surface* $\mathscr{S}$

$$(\boldsymbol{U}, \boldsymbol{V})_{\mathscr{S}} = \int_{\mathscr{S}} \langle \boldsymbol{U}(t), \overline{\boldsymbol{V}(t)} \rangle \, dS \qquad \forall \boldsymbol{U}, \boldsymbol{V} \in \mathscr{V}(\mathscr{S}). \tag{1.1}$$

The Laplace-Beltrami operator $\Delta_{\mathscr{S}} := \operatorname{div}_{\mathscr{S}} \nabla_{\mathscr{S}}$ *on* $\mathscr{S}$ *writes*

$$\boldsymbol{\Delta}_{\mathscr{S}} \psi = \operatorname{div}_{\mathscr{S}} \nabla_{\mathscr{S}} \psi = \sum_{j=1}^n \mathscr{D}_j^2 \psi \qquad \forall \psi \in C^2(\mathscr{S}). \tag{1.2}$$

We remind that the surface gradient $\nabla_{\mathscr{S}}$ maps scalar functions to the tangential vector fields

$$\nabla_{\mathscr{S}} \ : \ C^1(\mathscr{S}) \to \mathscr{V}(\mathscr{S}) \subset C(\mathscr{S}, \mathbb{C}^n) \tag{1.3}$$

and the scalar product with the normal vector vanishes $\langle \boldsymbol{\nu}(\mathscr{X}), \nabla_{\mathscr{S}}\varphi(\mathscr{X}) \rangle \equiv 0$ for all $\varphi \in C^1(\mathscr{S})$ and all $\mathscr{X} \in \mathscr{S}$.

Tangential derivatives can be applied to a definition of Sobolev spaces $\mathbb{H}^m_p(\mathscr{S})$, $m \in \mathbb{N}^0$, $1 \le p < \infty$ on an ℓ-smooth surface $\mathscr{S}$ if $m \le \ell$:

$$\mathbb{H}^m_p(\mathscr{S}) := \{\varphi \in D'(\mathscr{S}) \ : \ \mathscr{D}^\alpha_{\mathscr{S}}\varphi \in \mathbb{L}_p(\mathscr{S}) \, , \ \forall\, \alpha \in \mathbb{N}^n_0 \, , \ |\alpha| \le m\} \, , \tag{1.4}$$

$$\mathscr{D}^\alpha_{\mathscr{S}} := \mathscr{D}^{\alpha_1}_1 \cdots \mathscr{D}^{\alpha_n}_n \, .$$

The derivative of $\varphi \in D'(\mathscr{S})$ in (1.4) is understood, as usual, in the distributional sense

$$(\mathscr{D}_j\varphi, \psi)_{\mathscr{S}} := (\varphi, \mathscr{D}^*_j\psi)_{\mathscr{S}},$$

where $\mathscr{D}^*_j$ is the formal dual operator to $\mathscr{D}_j$ (cf. [DMM1]):

$$\mathscr{D}^*_j\varphi = -\mathscr{D}_j\varphi - \nu_j\mathscr{H}^0_{\mathscr{S}}\varphi \, , \qquad \varphi \in C^1(\mathscr{S}) \, . \tag{1.5}$$

The space $\mathbb{H}^1_p(\mathscr{S})$ is well defined if $\mathscr{S}$ is a Lipschitz hypersurface.

Equivalently, $\mathbb{H}^m_p(\mathscr{S})$ is the closure of the space $C^\ell(\mathscr{S})$ (or of $C^\infty(\mathscr{S})$ if $\mathscr{S}$ is infinitely smooth $\ell = \infty$) with respect to the norm

$$\|\varphi \,|\, \mathbb{H}^m_p(\mathscr{S})\,\| := \left[\sum_{|\alpha|\le m} \|\mathscr{D}^\alpha_{\mathscr{S}}\varphi \,|\, \mathbb{L}_p(\mathscr{S})\|_p\right]^{1/p} \, . \tag{1.6}$$

Moreover, $\mathbb{H}^m_2(\mathscr{S})$ is a Hilbert space with the scalar product

$$(\varphi, \psi)^{(m)}_{\mathscr{S}} := \sum_{|\alpha|\le m} \oint_{\mathscr{S}} \mathscr{D}^\alpha_{\mathscr{S}}\varphi)(\mathscr{x})\overline{\mathscr{D}^\alpha_{\mathscr{S}}\psi(\mathscr{x})}dS \, . \tag{1.7}$$

As usual, $\mathbb{H}^{-m}_2(\mathscr{S})$ with an integer $m \in \mathbb{N}$ denotes the space of distributions of the negative order $-m$ which is dual to the Sobolev space $\mathbb{H}^m_2(\mathscr{S})$.

We write, as customary, $\mathbb{H}^m(\mathscr{S})$ instead of $\mathbb{H}^m_2(\mathscr{S})$.

To accomplish the definition of the Banach spaces $\mathbb{H}^m_p(\mathscr{S})$ we need to prove the following.

Lemma 1.2. *For $\varphi \in C^1(\mathscr{S})$ the surface gradient vanishes $\nabla_{\mathscr{S}}\varphi \equiv 0$ if and only if $\varphi(\mathscr{x}) \equiv$ const.*

Proof. We only have to show that $\nabla_{\mathscr{S}}\varphi \equiv 0$ implies $\varphi(\mathscr{x}) \equiv$ const. The inverse implication is trivial. Let

$$\Omega^\varepsilon_{\mathscr{S}} := \mathscr{S} \times [-\varepsilon, \varepsilon] := \{\mathscr{x} + t\boldsymbol{\nu} \ : \ \mathscr{x} \in \mathscr{S}, \quad -\varepsilon < t < \varepsilon\}$$

be the *tubular neighborhood* of the surface of the thickness 2ε, with the middle surface $\mathscr{S}$. Taking ε sufficiently small, we can assume that the domain $\Omega^\varepsilon_{\mathscr{S}}$ has no self-intersections. Any function $\varphi \in C^1(\mathscr{S})$ is extended as a constant along the normal vector: $\widetilde{\varphi}(x,t) := \varphi(x)$, $x, t \in \Omega^\varepsilon_{\mathscr{S}}$. Then the normal derivatives are applicable and vanish: $\partial_{\boldsymbol{\nu}}\widetilde{\varphi} = 0$. Therefore the coordinate derivatives also are applicable and $\partial_j\widetilde{\varphi}(\mathscr{x},t) = \mathscr{D}_j\varphi(\mathscr{x}) \equiv 0$ for all $j = 1, \ldots, n$ and all $(\mathscr{x}, t) \in \Omega^\varepsilon_{\mathscr{S}}$. But this implies $\widetilde{\varphi}(\mathscr{x}) \equiv$ const and, restricted to the surface, $\varphi(\mathscr{x}) = \gamma_{\mathscr{S}}\widetilde{\varphi}(\mathscr{x}) =$ const. □

To the equivalence of the norm in (1.6) with the usual one defined by a partition of unity we only remark that among the "pull back" operators of n covariant derivatives there always can be selected locally $n-1$ linearly independent linear differential operators of order 1 of the variable $x \in \mathbb{R}^{n-1}$, which can equivalently be replaced by the coordinate derivatives $\partial_1, \dots, \partial_{n-1}$.

Lemma 1.3. *Let $\varphi \in \mathbb{H}^2(\Omega^\varepsilon_{\mathscr{S}}) \cap C^1(\mathscr{S})$ and $\gamma_{\mathscr{S}} \nabla \varphi$, $\gamma_{\mathscr{S}} \partial_{\boldsymbol{\nu}} \varphi$ denote the traces on $\mathscr{S}$ of the spatial gradient and of the normal derivative, while $\nabla_{\mathscr{S}} \varphi$ denote the surface gradient. Then*

$$\|\gamma_{\mathscr{S}} \nabla \varphi \,|\, \mathbb{L}_2(\mathscr{S})\|_2^2 = \|\nabla_{\mathscr{S}} \varphi \,|\, \mathbb{L}_2(\mathscr{S})\|_2^2 + \|\gamma_{\mathscr{S}} \partial_{\boldsymbol{\nu}} \varphi \,|\, \mathbb{L}_2(\mathscr{S})\|_2^2 . \tag{1.8}$$

Proof. Indeed,

$$\begin{aligned}
\|\nabla_{\mathscr{S}} \varphi \,|\, \mathbb{L}_2(\mathscr{S})\|^2 &= \sum_{j=1}^n \oint_{\mathscr{S}} \mathscr{D}_j \varphi(\mathscr{X}) \overline{\mathscr{D}_j \varphi(\mathscr{X})}\, dS \\
&= \sum_{j=1}^n \oint_{\mathscr{S}} (\partial_j \varphi(\mathscr{X}) - \nu_j(\mathscr{X}) \partial_{\boldsymbol{\nu}(x)} \varphi(\mathscr{X})) \overline{(\partial_j \varphi(\mathscr{X}) - \nu_j(\mathscr{X}) \partial_{\boldsymbol{\nu}(x)} \varphi(\mathscr{X}))}\, dS \\
&= \sum_{j=1}^n \Bigg[\oint_{\mathscr{S}} (\partial_j \varphi(\mathscr{X}) \overline{\partial_j \varphi(\mathscr{X})}\, dS - \oint_{\mathscr{S}} \nu_j(\mathscr{X}) \partial_j \varphi(\mathscr{X}) \overline{\partial_{\boldsymbol{\nu}(x)} \varphi(\mathscr{X})}\, dS \\
&\quad - \oint_{\mathscr{S}} \partial_{\boldsymbol{\nu}(x)} \varphi(\mathscr{X}) \overline{\nu_j(\mathscr{X}) \partial_j \varphi(\mathscr{X})}\, dS + \nu_j^2(\mathscr{X}) \oint_{\mathscr{S}} \partial_{\boldsymbol{\nu}(x)} \varphi(\mathscr{X}) \overline{\partial_{\boldsymbol{\nu}(x)} \varphi(\mathscr{X})}\, dS \Bigg] \\
&= \|\gamma_{\mathscr{S}} \nabla \varphi \,|\, \mathbb{L}_2(\mathscr{S})\|^2 - \|\gamma_{\mathscr{S}} \partial_{\boldsymbol{\nu}} \varphi \,|\, \mathbb{L}_2(\mathscr{S})\|^2
\end{aligned}$$

and (1.8) follows. □

Lemma 1.4. *The operator*

$$\boldsymbol{\Delta}_{\mathscr{S},\mu} := \mu I - \boldsymbol{\Delta}_{\mathscr{S}} \,:\, \mathbb{H}^1(\mathscr{S}) \to \mathbb{H}^{-1}(\mathscr{S}), \qquad \mu = \text{const} > 0 \tag{1.9}$$

is positive definite, elliptic and invertible. For arbitrary $s \in \mathbb{R}$ the power $\boldsymbol{\Delta}^s_{\mathscr{S},\mu}$ is a self-adjoint positive definite pseudodifferential operator with a trivial kernel $\operatorname{Ker} \boldsymbol{\Delta}^r_{\mathscr{S},\mu} = \{0\}$ *in the Sobolev space $\mathbb{W}^m_p(\mathscr{S}) = \mathbb{H}^m_p(\mathscr{S})$ for all $m = 1, \dots, m$ and all $1 < p < \infty$.*

Proof. The positive definiteness (also implying self-adjointness, ellipticity and invertibility) of $\boldsymbol{\Delta}_{\mathscr{S},\mu}$ follows from Proposition 1.1

$$((\mu I - \boldsymbol{\Delta}_{\mathscr{S}})\varphi, \varphi)_{\mathscr{S}} = \mu \|\varphi | \mathbb{L}_2(\mathscr{S})\|^2 + \|\nabla_{\mathscr{S}} \varphi | \mathbb{L}_2(\mathscr{S})\| \geq C \|\varphi | \mathbb{H}^1(\mathscr{S})\|^2$$

with $C := \min\{1, \mu\} > 0$. Then the powers $\boldsymbol{\Delta}^s_{\mathscr{S},\mu}$, $s \in \mathbb{R}$ exist and are pseudodifferential operators (cf., e.g., [Sh1]). We quote [DNS1] (also see [Ag1, Du2, Ka1] and [DNS2] for a most general result) that an elliptic pseudodifferential operators on a manifold without boundary has the same kernel and cokernel in the spaces $\mathbb{H}^m_p(\mathscr{S})$ for all $m = 1, \dots, \ell$ and all $1 < p < \infty$. □

Now we are able to define the Bessel potential space $\mathbb{H}^s_p(\mathscr{S})$ for arbitrary $s \in \mathbb{R}$ and $1 < p < \infty$:

$$\mathbb{H}^s_p(\mathscr{S}) := \left\{\varphi \; : \; \|\varphi \,|\, \mathbb{H}^s_p(\mathscr{S})\,\| := \|\boldsymbol{\Delta}^{s/2}_{\mathscr{S},1}\varphi \,|\, \mathbb{L}_p(\mathscr{S})\,\| < \infty\right\}. \tag{1.10}$$

The Sobolev spaces with negative indices $\mathbb{H}^{-s}_p(\mathscr{S})$, $s < 0$, $1 < p < \infty$ are dual to $\mathbb{H}^s_{p'}(\mathscr{S})$, $p' := \dfrac{p}{p-1}$, with respect to the sesquilinear form $(\varphi,\psi)_{\mathscr{S}}$ (cf. (1.1)) extended by continuity to duality between pairs $\varphi \in \mathbb{H}^s_{p'}(\mathscr{S})$ and $\psi \in \mathbb{H}^{-s}_p(\mathscr{S})$.

The embeddings $\mathbb{H}^s_p(\mathscr{S}) \subset \mathbb{L}_p(\mathscr{S}) \subset \mathbb{H}^{-s}_p(\mathscr{S})$, for $s > 0$, are continuous, even compact, and for integer-valued parameter $s = m$ the space $\mathbb{H}^{-m}_p(\mathscr{S})$ is the convex linear hull of distributional derivatives of $\mathbb{L}_p(\mathscr{S})$-functions:

$$\mathbb{H}^{-m}_p(\mathscr{S}) := \mathscr{L}\left\{\mathscr{D}^\alpha\varphi \; : \; \varphi \in \mathbb{L}_p(\mathscr{S}) \text{ for all } \mathscr{D}^\alpha = \mathscr{D}^{\alpha_1}_1 \cdots \mathscr{D}^{\alpha_n}_n\,, \quad |\alpha| \le m\right\}.$$

If $\mathscr{C}$ is a subsurface with Lipschitz boundary $\Gamma = \partial\mathscr{C} \neq \emptyset$, $\widetilde{\mathbb{H}}^s_p(\mathscr{C})$ denotes the space of functions $\varphi \in \mathbb{H}^s_p(\mathscr{S})$, supported in $\overline{\mathscr{C}}$, $\mathscr{C} \subset \mathscr{S}$, where $\mathscr{S}$ is a smooth surface without boundary which extends the surface $\mathscr{C}$. Let $\mathscr{C}^+ := \mathscr{C}$ and $\mathscr{C}^- := \mathscr{C}^c = \mathscr{S} \setminus \overline{\mathscr{C}}$ denote the complemented subsurface with boundary $\mathscr{S} = \mathscr{C}^+ \cup \overline{\mathscr{C}^-}$; the notation $\mathbb{H}^s_p(\mathscr{C})$ is used for the factor space $\mathbb{H}^s_p(\mathscr{S})/\widetilde{\mathbb{H}}^s_p(\mathscr{C}^-)$; the space $\mathbb{H}^s_p(\mathscr{C})$ can also be viewed as the space of restrictions $r_{\mathscr{C}}\varphi := \varphi\big|_{\mathscr{C}}$ of all functions $\varphi \in \mathbb{H}^s_p(\mathscr{S})$ to the subsurface $\mathscr{C} = \mathscr{C}^+$.

We refer to [Tr1] and [DS1] for details about similar spaces.

2. Lions' Lemma and Korn's inequalities

The following generalizes essentially J.L. Lions' Lemma (cf. [DaL1, p.111], [Ta1], [AG1, Proposition 2.10], [Ci3, § 1.7], [Mc1]).

Lemma 2.1. *Let $\mathscr{S}$ be a 2-smooth hypersurface without boundary in $\mathbb{R}^n$. Then the inclusions $\varphi \in \mathbb{H}^{-1}_p(\mathscr{S})$, $\mathscr{D}_j\varphi \in \mathbb{H}^{-1}_p(\mathscr{S})$, for all $j = 1,\dots,n$ imply $\varphi \in \mathbb{L}_p(\mathscr{S})$.*

Moreover, the assertion remains valid for a hypersurface $\mathscr{C}$ with the Lipschitz boundary $\Gamma := \partial\mathscr{C}$ and the spaces $\mathbb{H}^{-1}_p(\mathscr{C})$ and $\widetilde{\mathbb{H}}^{-1}_p(\mathscr{C})$.

Proof. First we assume that $\mathscr{S}$ is a hypersurface without boundary. The proof is based on the following facts (cf. [Hr1, Sh1, Ta2, Tr1]):

A. The "lifting operators" (the Bessel potential operator) $\Lambda^{\pm 1}_{\mathscr{S}}(\mathscr{X}, D) := \boldsymbol{\Delta}^{\pm 1/2}_{\mathscr{S},1}$ (cf. Lemma 1.4 and (1.10)), are invertible $\Lambda^{\pm 1}_{\mathscr{S}}(\mathscr{X}, D)\Lambda^{\mp 1}_{\mathscr{S}}(\mathscr{X}, D) = I$, mapping isometrically the spaces

$$\begin{aligned} \Lambda^{-1}_{\mathscr{S}}(\mathscr{X}, D) \; &: \; \mathbb{H}^{m-1}_p(\mathscr{S}) \to \mathbb{H}^m_p(\mathscr{S}), \\ \Lambda_{\mathscr{S}}(\mathscr{X}, D) \; &: \; \mathbb{H}^m_p(\mathscr{S}) \to \mathbb{H}^{m-1}_p(\mathscr{S}) \end{aligned} \tag{2.1}$$

for arbitrary $m = 0, \pm 1, \dots$ and are pseudodifferential operators of order ± 1, respectively.

B. The commutant

$$[\mathscr{D}_j, \Lambda^{-1}_{\mathscr{S}}(\mathscr{X}, D)] := \mathscr{D}_j \Lambda^{-1}_{\mathscr{S}}(\mathscr{X}, D) - \Lambda^{-1}_{\mathscr{S}}(\mathscr{X}, D)\mathscr{D}_j \tag{2.2}$$

with the pseudodifferential operator $\mathscr{D}_j$ has order -1 and maps continuously the spaces

$$[\mathscr{D}_j, \Lambda^{-1}_{\mathscr{S}}(\mathscr{X}, D)] \; : \; \mathbb{H}^{-1}_p(\mathscr{S}) \to \mathbb{L}_p(\mathscr{S}).$$

The assertion (**B**) is a well-known property of pseudodifferential operators and can be found in many sources [Hr1, Sh1, Ta1, Tr1].

Let $\varphi \in \mathbb{H}^{-1}_p(\mathscr{S})$, $\mathscr{D}_j\varphi \in \mathbb{H}^{-1}_p(\mathscr{S})$, for all $j = 1, \dots, n$. Then, due to (2.1), $\psi := \Lambda^{-1}_{\mathscr{S}}(\mathscr{X}, D)\varphi \in \mathbb{L}_p(\mathscr{S})$ and, due to (2.2), $\mathscr{D}_j\psi = [\mathscr{D}_j, \Lambda^{-1}_{\mathscr{S}}(\mathscr{X}, D)]\varphi + \Lambda^{-1}_{\mathscr{S}}(\mathscr{X}, D)\mathscr{D}_j\varphi \in \mathbb{L}_p(\mathscr{S})$ for all $j = 1, \dots, n$. By the definition of the space $\mathbb{H}^1_p(\mathscr{S}) = \mathbb{H}^1_p(\mathscr{S})$ in (1.6) we conclude that $\psi \in \mathbb{H}^1_p(\mathscr{S})$. Due to (2.1) we get finally $\varphi = \Lambda_{\mathscr{S}}(\mathscr{X}, D)\psi \in \mathbb{L}_p(\mathscr{S})$.

If $\mathscr{C}$ has non-empty Lipschitz boundary $\Gamma \neq \emptyset$, there exist pseudodifferential operators

$$\begin{aligned} \Lambda^{-1}_{-}(\mathscr{X}, D) \; &: \; \mathbb{H}^{-1}_p(\mathscr{C}) \to \mathbb{L}_p(\mathscr{C}), \\ \Lambda^{-1}_{+}(\mathscr{X}, D) \; &: \; \widetilde{\mathbb{H}}^{-1}_p(\mathscr{C}) \to \widetilde{\mathbb{L}}_p(\mathscr{C}), \end{aligned} \tag{2.3}$$

of order -1, arranging isomorphisms between the indicated spaces, and their inverses are $\Lambda_{\pm}(\mathscr{X}, D)$, respectively (cf. [DS1]).

Moreover, the commutants $[\mathscr{D}_j, \Lambda^{-1}_{\pm}(\mathscr{X}, D)] := \mathscr{D}_j\Lambda^{-1}_{\pm}(\mathscr{X}, D) - \Lambda^{-1}_{\pm}(\mathscr{X}, D)\mathscr{D}_j$ have order -1, i.e., map continuously the spaces $[\mathscr{D}_j, \Lambda^{-1}_{-}(\mathscr{X}, D)] \; : \; \mathbb{H}^{-1}_p(\mathscr{C}) \to \mathbb{L}_p(\mathscr{C})$ and $[\mathscr{D}_j, \Lambda^{-1}_{+}(\mathscr{X}, D)] \; : \; \widetilde{\mathbb{H}}^{-1}_p(\mathscr{C}) \to \mathbb{L}_p(\mathscr{C})$.

By using the formulated assertions the proof is completed as in the case of a hypersurface without boundary $\mathscr{S}$. □

The foregoing Lemma 2.1 has the following generalization for the Bessel potential spaces $\widetilde{\mathbb{H}}^s_p(\mathscr{S})$.

Lemma 2.2. *If $\mathscr{S}$ has no boundary, is sufficiently smooth, $1 < p < \infty$, $s \in \mathbb{R}$, $m = 1, 2, \dots$ and*

$$\varphi \in \mathbb{H}^{s-m}_p(\mathscr{S}), \quad \mathscr{D}^\alpha\varphi = \mathscr{D}^{\alpha_1}_1 \cdots \mathscr{D}^{\alpha_n}_n\varphi \in \mathbb{H}^{s-m}_p(\mathscr{S}) \quad \textit{for all} \quad |\alpha| \leq m,$$

then $\varphi \in \mathbb{H}^s_p(\mathscr{S})$.

Moreover, the assertion remains valid for a hypersurface $\mathscr{C}$ with the Lipschitz boundary $\Gamma := \partial\mathscr{C}$ and the spaces $\mathbb{H}^s_p(\mathscr{C})$ and $\widetilde{\mathbb{H}}^s_p(\mathscr{C})$.

Proof. Assume first $\mathscr{S}$ has no boundary. The proof is based on similar facts as in the foregoing case:

A. The "lifting operator" (the Bessel potential operator) $\Lambda^r_{\mathscr{S}}(\mathscr{X}, D) := \mathbf{\Delta}^{r/2}_{\mathscr{S},1}$ (cf. Lemma 1.4 and (1.10)) maps isometrically the spaces

$$\Lambda^r_{\mathscr{S}}(\mathscr{X}, D) \; : \; \mathbb{H}^s_p(\mathscr{S}) \to \mathbb{H}^{s-r}_p(\mathscr{S}), \quad r \in \mathbb{R} \tag{2.4}$$

has the inverse $\Lambda^{-r}_{\mathscr{S}}(\mathscr{X}, D)$ and is a pseudodifferential operator of order r.

B. The commutant

$$[\mathscr{D}^\alpha, \Lambda^r_{\mathscr{S}}(\mathscr{X}, D)] := \mathscr{D}^\alpha \Lambda^r_{\mathscr{S}}(\mathscr{X}, D) - \Lambda^r_{\mathscr{S}}(\mathscr{X}, D)\mathscr{D}^\alpha \tag{2.5}$$

is a pseudodifferential operator of order $|\alpha| + r - 1$ and maps continuously the spaces

$$[\mathscr{D}^\alpha, \Lambda^r_{\mathscr{S}}(\mathscr{X}, D)] : \mathbb{H}^\gamma_p(\mathscr{S}) \to \mathbb{H}^{\gamma-|\alpha|-r+1}_p(\mathscr{S}), \qquad \forall\, \gamma \in \mathbb{R}.$$

Assume that $m = 1$. Then $\varphi \in \mathbb{H}^{s-1}_p(\mathscr{S})$ and, due to (2.4), (2.5), it follows that $\psi := \Lambda^{s-1}_{\mathscr{S}}(\mathscr{X}, D)\varphi \in \mathbb{L}_p(\mathscr{S})$, $\mathscr{D}_j\psi = [\mathscr{D}_j, \Lambda^{s-1}_{\mathscr{S}}(\mathscr{X}, D)]\varphi + \Lambda^{s-1}_{\mathscr{S}}(\mathscr{X}, D)\mathscr{D}_j\varphi \in \mathbb{L}_p(\mathscr{S})$ for all $j = 1, \dots, n$. By the definition of the space $\mathbb{H}^1_p(\mathscr{S}) = \mathbb{H}^1_p(\mathscr{S})$ in (1.6) the inclusion $\psi \in \mathbb{H}^1_p(\mathscr{S})$ follows. Due to (2.4) we get finally $\varphi = \Lambda^{1-s}_{\mathscr{S}}(\mathscr{X}, D)\psi \in \mathbb{H}^s_p(\mathscr{S})$.

Now assume: $m = 2, 3, \dots$ and the assertion is valid for $m - 1$. Then, due to the hypothesis, $\psi_j := \mathscr{D}_j\varphi \in \mathbb{H}^{s-m}_p(\mathscr{S})$ for $j = 1, \dots, n$. Moreover, due to the same hypothesis,

$$\mathscr{D}^\alpha\psi_j := \mathscr{D}^\alpha\mathscr{D}_j\varphi \in \mathbb{H}^{s-m}_p(\mathscr{S}) \quad \text{for all} \quad |\alpha| \le m - 1 \quad \text{and all} \quad j = 1, \dots, n.$$

Hence, the induction hypothesis implies that $\psi_j := \mathscr{D}_j\varphi \in \mathbb{H}^{s-1}_p(\mathscr{S})$ for $j = 1, \dots, n$. Now it follows from the already considered case $m = 1$ that $\varphi \in \mathbb{H}^s_p(\mathscr{S})$.

If $\mathscr{C}$ has the non-empty Lipschitz boundary $\Gamma \neq \emptyset$, there exist pseudodifferential operators

$$\Lambda^r_-(\mathscr{X}, D) : \mathbb{H}^s_p(\mathscr{C}) \to \mathbb{H}^{s-r}_p(\mathscr{C}), \qquad \Lambda^r_+(\mathscr{X}, D) : \widetilde{\mathbb{H}}^s_p(\mathscr{C}) \to \widetilde{\mathbb{H}}^{s-r}_p(\mathscr{C}), \tag{2.6}$$

arranging isomorphisms between the indicated spaces, and their inverses are $\Lambda^{-r}_-(\mathscr{X}, D)$, $\Lambda^{-r}_+(\mathscr{X}, D)$ (cf. [DS1]).

Moreover, the pseudodifferential operators $\Lambda^{-r}_\pm(\mathscr{X}, D)$ have order $-r$ and the commutants $[\mathscr{D}^\alpha, \Lambda^{-r}_\pm(\mathscr{X}, D)] := \mathscr{D}^\alpha\Lambda^{-1}_\pm(\mathscr{X}, D) - \Lambda^{-r}_\pm(\mathscr{X}, D)\mathscr{D}^\alpha$ have order $|\alpha| - r - 1$, i.e., mapping continuously the spaces $[\mathscr{D}^\alpha, \Lambda^{-r}_-(\mathscr{X}, D)] : \mathbb{H}^\gamma_p(\mathscr{C}) \to \mathbb{H}^{\gamma+r+1-|\alpha|}_p(\mathscr{C})$ and $[\mathscr{D}^\alpha, \Lambda^{-r}_+(\mathscr{X}, D)] : \widetilde{\mathbb{H}}^\gamma_p(\mathscr{C}) \to \widetilde{\mathbb{H}}^{\gamma+r+1-|\alpha|}_p(\mathscr{C})$.

By using the formulated assertions the proof is completed as in the foregoing cases. □

Theorem 2.3. (Korn's I inequality "without boundary condition"). *Let $\mathscr{S} \subset \mathbb{R}^n$ be a Lipshitz hypersurface without boundary,* $\mathrm{Def}_{\mathscr{S}}(\boldsymbol{U}) := [\mathfrak{D}_{jk}(\boldsymbol{U})]_{n\times n}$ *be the deformation tensor* (*cf.* (0.17)) *and*

$$\big\|\mathrm{Def}_{\mathscr{S}}(\boldsymbol{U})\big|\mathbb{L}_p(\mathscr{S})\big\| := \left[\sum_{j,k=1}^n \big\|\mathfrak{D}_{jk}\boldsymbol{U}\big|\mathbb{L}_p(\mathscr{S})\big\|^p\right]^{1/p}, \qquad \boldsymbol{U} \in \mathbb{H}^1_p(\mathscr{S}) \tag{2.7}$$

for $1 < p < \infty$. *Then*

$$\big\|\boldsymbol{U}\big|\mathbb{H}^1_p(\mathscr{S})\big\| \le M \left[\big\|\boldsymbol{U}\big|\mathbb{L}_p(\mathscr{S})\big\|^p + \big\|\mathrm{Def}_{\mathscr{S}}(\boldsymbol{U})\big|\mathbb{L}_p(\mathscr{S})\big\|^p\right]^{1/p} \tag{2.8}$$

for some constant $M > 0$ or, equivalently, the mapping

$$\boldsymbol{U} \mapsto \left[\left\|\boldsymbol{U}\middle|\mathbb{L}_p(\mathscr{S})\right\|^p + \left\|\mathrm{Def}_{\mathscr{S}}(\boldsymbol{U})\middle|\mathbb{L}_p(\mathscr{S})\right\|^p\right]^{1/p}$$

is an equivalent norm on the space $\mathbb{H}^1_p(\mathscr{S})$.

Proof. Consider the space

$$\widehat{\mathbb{H}}^1_p(\mathscr{S}) := \left\{\boldsymbol{U} = \left(U_1,\dots,U_n\right)^\top \ : \ U_j\,,\ \mathfrak{D}_{jk}(\boldsymbol{U}) \in \mathbb{L}_p(\mathscr{S}) \ \text{ for all } j,k = 1,\dots,n\right\} \tag{2.9}$$

endowed with the norm (cf. (2.8)):

$$\left\|\boldsymbol{U}\middle|\widehat{\mathbb{H}}^1_p(\mathscr{S})\right\| := \left[\left\|\boldsymbol{U}\middle|\mathbb{L}_p(\mathscr{S})\right\|^p + \left\|\mathrm{Def}_{\mathscr{S}}(\boldsymbol{U})\middle|\mathbb{L}_p(\mathscr{S})\right\|^p\right]^{1/p}. \tag{2.10}$$

The derivatives here are understood in the distributional sense

$$(\mathfrak{D}_{jk}(\boldsymbol{U}),\psi)_{\mathscr{S}} := \frac{1}{2}(U_k,\mathscr{D}^*_j\psi)_{\mathscr{S}} + \frac{1}{2}(U_j,\mathscr{D}^*_k\psi)_{\mathscr{S}} \qquad \forall\,\psi \in C^1(\mathscr{S})$$

(cf. (1.5) for the formal dual operator $\mathscr{D}^*_j$).

It is sufficient to prove that the spaces $\mathbb{H}^1_p(\mathscr{S})$ and $\widehat{\mathbb{H}}^1_p(\mathscr{S})$ are identical. The inclusion $\mathbb{H}^1_p(\mathscr{S}) \subset \widehat{\mathbb{H}}^1_p(\mathscr{S})$ is trivial and we only check the inverse inclusion $\widehat{\mathbb{H}}^1_p(\mathscr{S}) \subset \mathbb{H}^1_p(\mathscr{S})$.

To this end take $\boldsymbol{U} \in \widehat{\mathbb{H}}^1_p(\mathscr{S})$ and note that the inclusions $\boldsymbol{U} \in \mathbb{L}_p(\mathscr{S})$, $\mathrm{Def}_{\mathscr{S}}(\boldsymbol{U}) \in \mathbb{L}_p(\mathscr{S})$ (i.e., $\mathfrak{D}_{jk}\boldsymbol{U} \in \mathbb{L}_p(\mathscr{S})$ for all $j,k = 1,\dots,n$) imply

$$\widetilde{\mathfrak{D}}_{jk}(\boldsymbol{U}) = \frac{1}{2}\Big[\mathscr{D}_k U_j + \mathscr{D}_j U_k\Big] = \mathfrak{D}_{jk}(\boldsymbol{U}) - \frac{1}{2}\sum_{r=1}^{n} \partial_r\big(\nu_j\nu_k\big)U_r \in \mathbb{L}_p(\mathscr{S}) \tag{2.11}$$

for all $j,k = 1,\dots,n$. Then (cf. [DMM1, Proposition 4.4.iv] for the commutator $[\mathscr{D}_j,\mathscr{D}_k]$):

$$\mathscr{D}_j U_k \in \mathbb{H}^{-1}_p(\mathscr{S}), \quad \big[\mathscr{D}_j,\mathscr{D}_k\big]U_m = \sum_{r=1}^{n}\big[\nu_j\mathscr{D}_k\nu_r - \nu_k\mathscr{D}_j\nu_r\big]\mathscr{D}_r U_m \in \mathbb{H}^{-1}_p(\mathscr{S}),$$

$$\mathscr{D}_k\mathscr{D}_j U_m = \mathscr{D}_j\widetilde{\mathfrak{D}}_{km}(\boldsymbol{U}) + \mathscr{D}_k\widetilde{\mathfrak{D}}_{jm}(\boldsymbol{U}) - \mathscr{D}_m\widetilde{\mathfrak{D}}_{jk}(\boldsymbol{U}) - \frac{1}{2}\big[\mathscr{D}_j,\mathscr{D}_k\big]U_m$$

$$-\frac{1}{2}\big[\mathscr{D}_j,\mathscr{D}_m\big]U_k - \frac{1}{2}\big[\mathscr{D}_k,\mathscr{D}_m\big]U_j \in \mathbb{H}^{-1}_p(\mathscr{S}) \quad \text{for} \quad j,k,m = 1,\dots,n\,,$$

Due to Lemma 2.1 of J.L. Lions this implies $\mathscr{D}_j U_m \in \mathbb{L}_p(\mathscr{S})$ for all $j,m = 1,\dots,n$ and the claimed result $\boldsymbol{U} \in \mathbb{H}^1_p(\mathscr{S})$ follows. □

Remark 2.4. The foregoing Theorem 2.3 is proved by P. Ciarlet in [Ci3] for the case $p = 2$, $m = 1$, manifold without boundary, for curvilinear coordinates and covariant derivatives.

A remarkable consequence of Korn's inequality (2.8) is that the space

$$\mathbb{H}^1_p(\mathscr{S}) := \left\{\boldsymbol{U} = \left(U_1,\dots,U_n\right)^\top \ : \ U_j\,,\ \mathscr{D}_k U_j \in \mathbb{L}_p(\mathscr{S}) \ \text{ for all } j,k = 1,\dots,n\right\}$$

(cf. (1.6)) and the space $\widehat{\mathbb{H}}^1_p(\mathscr{S})$ (cf. (2.9)) are isomorphic (i.e., can be identified), although only $\frac{n(n+1)}{2} < n^2$ linear combinations of the n^2 derivatives $\mathscr{D}_j U_k$, $j,k = 1,\dots,n$ participate in the definition of the space $\widehat{\mathbb{H}}^1_p(\mathscr{S})$.

3. Killing's vector fields and the unique continuation from the boundary

Definition 3.1. Let $\mathscr{S}$ be a hypersurface in the Euclidean space $\mathbb{R}^n$. The space $\mathscr{R}(\mathscr{S})$ of solutions to the deformation equations

$$\begin{aligned}\mathfrak{D}_{jk}(\boldsymbol{U}) &:= \frac{1}{2}\big[(\mathscr{D}^{\mathscr{S}}_j \boldsymbol{U})^0_k + (\mathscr{D}^{\mathscr{S}}_k \boldsymbol{U})^0_j\big]\\ &= \frac{1}{2}\Big[\mathscr{D}_k U^0_j + \mathscr{D}_j U^0_k + \sum_{m=1}^n U^0_m \mathscr{D}_m\big(\nu_j\nu_k\big)\Big] = 0\,,\end{aligned} \tag{3.1}$$

$$\boldsymbol{U} = \sum_{j=1}^n U^0_j \boldsymbol{d}^j \in \mathscr{V}(\mathscr{S}), \quad j,k = 1,\dots,n$$

(cf. (0.17)) is called the space of *Killing's vector fields.*

Killing's vector fields on a domain in the Euclidean space $\Omega \subset \mathbb{R}^n$ are known as the *rigid motions* and we start with this simplest class.

The space of rigid motions $\mathscr{R}(\Omega)$ extends naturally to the entire $\mathbb{R}^n$ and consists of linear vector functions

$$\boldsymbol{V}(x) = a + \mathscr{B}x\,, \qquad \mathscr{B} = \big[b_{jk}\big]_{n\times n}, \qquad a \in \mathbb{R}^n\,, \quad x \in \mathbb{R}^n\,, \tag{3.2}$$

where the matrix $\mathscr{B}$ is skew symmetric

$$\mathscr{B} := \begin{bmatrix} 0 & b_{12} & b_{13} & \cdots & b_{1(n-2)} & b_{1(n-1)}\\ -b_{12} & 0 & b_{21} & \cdots & b_{1(n-3)} & b_{2(n-2)}\\ \cdots & \cdots & \cdots & \cdots & \cdots & \cdots\\ -b_{1(n-2)} & -b_{2(n-3)} & -b_{3(n-4)} & \cdots & 0 & b_{(n-1)1}\\ -b_{1(n-1)} & -b_{2(n-2)} & -b_{3(n-3)} & \cdots & -b_{(n-1)1} & 0 \end{bmatrix} = -\mathscr{B}^\top \tag{3.3}$$

with real-valued entries $b_{jk} \in \mathbb{R}$. For $n = 3,4,\dots$ the space $\mathscr{R}(\mathbb{R}^n)$ is finite-dimensional and $\dim \mathscr{R}(\mathbb{R}^n) = n + \frac{n(n-1)}{2} = \frac{n(n+1)}{2}$.

Note that for $n = 3$ the vector field $\boldsymbol{V} \in \mathscr{R}(\Omega)$, $\Omega \subset \mathbb{R}^3$, is the classical rigid displacement

$$\begin{aligned}&\boldsymbol{V}(x) = a + \mathscr{B}x = a + b \wedge x\,,\\ &b := (b_1, b_2, b_3)^\top \in \mathbb{R}^3\,, \quad x \in \Omega\,,\end{aligned} \qquad \mathscr{B} := \begin{bmatrix} 0 & -b_3 & b_2\\ b_3 & 0 & -b_1\\ -b_2 & b_1 & 0\end{bmatrix}. \tag{3.4}$$

Definition 3.2. We call a subset $\mathscr{M} \subset \mathbb{R}^n$ *essentially m-dimensional* and write $\operatorname{ess\,dim} \mathscr{M} = m$, if there exist $m+1$ points $\mathscr{x}^0, \mathscr{x}^1, \cdots, \mathscr{x}^m \in \mathscr{M}$ such that the vectors $\big\{\mathscr{x}^j - \mathscr{x}^0\big\}_{j=1}^m$ are linearly independent.

Note, that any m-dimensional subset $\mathscr{M} \subset \mathbb{R}^m$ is essentially m-dimensional, because contains m linearly independent vectors. Moreover, any collection of $m+1$ points in $\mathbb{R}^m$ (a 0-dimensional subset!) is essentially m-dimensional, provided these points does not belong to any $m-1$-dimensional hyperplane.

Proposition 3.3. *Let*

$$\mathrm{Def}(\boldsymbol{U}) := \left[\mathfrak{D}^0_{jk}(\boldsymbol{U})\right]_{n\times n}, \tag{3.5}$$

$$\mathfrak{D}^0_{jk}(\boldsymbol{U}) = \frac{1}{2}\left[\partial_k U^0_j + \partial_j U^0_k\right], \qquad \boldsymbol{U} = \sum_{j=1}^{n} U^0_j \boldsymbol{e}^j$$

be the deformation tensor in Cartesian coordinates.

The linear space $\mathscr{R}(\mathbb{R}^n)$ of rigid motions (of Killing's vector fields) in $\mathbb{R}^n$ consists of vector fields $\boldsymbol{K} = (K^0_1, \dots, K^0_n)^\top$ which are solutions to the system

$$2\mathfrak{D}^0_{jk}(\boldsymbol{K})(x) = \partial_k K^0_j(x) + \partial_j K^0_k(x) = 0 \qquad x \in \mathscr{S} \quad \text{for all} \quad j,k = 1,\dots,n. \tag{3.6}$$

If a rigid motion vanishes on an essentially $(n-1)$-dimensional subset $\boldsymbol{K}(\mathscr{x}) = 0$ for all $\mathscr{x} \in \mathscr{M}$, $\operatorname{ess\,dim}\mathscr{M} = n-1$, or at infinity $\boldsymbol{K}(x) = \mathscr{o}(1)$ as $|x| \to \infty$, then $\boldsymbol{K}$ vanishes identically $\boldsymbol{K}(x) \equiv 0$ on $\mathbb{R}^n$.

Proof. The proof can be retrieved from many sources. We quote only two of them [Ci2, KGBB1]. □

Remark 3.4. For the deformation tensor in Cartesian coordinates $\mathrm{Def}(\boldsymbol{U})$ (cf. (3.5)) in a domain $\Omega \subset \mathbb{R}^n$ Korn's inequality

$$\left\|\boldsymbol{U}\big|\mathbb{H}^1_p(\Omega)\right\| \le M\left[\left\|\boldsymbol{U}\big|\mathbb{L}_p(\Omega)\right\|^P + \left\|\mathrm{Def}(\boldsymbol{U})\big|\mathbb{L}_p(\Omega)\right\|^p\right]^{1/p}, \quad 1 < p < \infty \tag{3.7}$$

with some constant $M > 0$ is well known and is proved, e.g., in [Ci2] (cf. (2.7) for a similar norm).

In contrast to the rigid motions in $\mathbb{R}^n$ nobody can describe Killing's vector fields on hypersurfaces explicitly so far. The next Theorem 3.5 underlines importance of Killing's vector fields for the Lamé equation on hypersurfaces. Later we investigate properties of Killing's vector fields to prepare tools for investigations of boundary value problems for the Lamé equation.

Theorem 3.5. *Let $\mathscr{S}$ be an ℓ-smooth hypersurface without boundary in $\mathbb{R}^n$ and $\ell \ge 2$. The Lamé operator $\mathscr{L}_\mathscr{S}$ for an isotropic media (cf. (0.23))*

$$\mathscr{L}_\mathscr{S} \,:\, \mathbb{H}^{s+1}_p(\mathscr{S}) \to \mathbb{H}^{s-1}_p(\mathscr{S}) \tag{3.8}$$

is self-adjoint $\mathscr{L}^*_\mathscr{S} = \mathscr{L}_\mathscr{S}$, elliptic, Fredholm and $\operatorname{Ind}\mathscr{L}_\mathscr{S} = 0$ for all $1 < p < \infty$ and all $s \in \mathbb{R}$, provided that $|s| \le \ell$.

The kernel of the operator $\operatorname{Ker}\mathscr{L}_\mathscr{S} \subset \mathbb{H}^s_p(\mathscr{S})$ is independent of the parameters p and s, coincides with the space of Killing's vector fields

$$\operatorname{Ker}\mathscr{L}_\mathscr{S} = \{\boldsymbol{U} \in \mathscr{V}(\mathscr{S}) \,:\, \mathscr{L}_\mathscr{S}\boldsymbol{U} = 0\} = \mathscr{R}(\mathscr{S}) \tag{3.9}$$

and, therefore, is finite-dimensional $\dim\mathscr{R}(\mathscr{S}) = \dim\operatorname{Ker}\mathscr{L}_\mathscr{S} < \infty$.

If $\mathscr{S}$ is C^∞ smooth, then the Killing's vector fields are smooth as well $\mathscr{R}(\mathscr{S}) \subset C^\infty(\mathscr{S})$.

$\mathscr{L}_{\mathscr{S}}$ is non-negative on the space $\mathbb{H}^1(\mathscr{S})$ and positive definite on the orthogonal complement $\mathbb{H}^1_{\mathscr{R}}(\mathscr{S})$ to the kernel $\mathscr{R}(\mathscr{S}) = \mathrm{Ker}\,\mathscr{L}_{\mathscr{S}}$

$$(\mathscr{L}_{\mathscr{S}}\boldsymbol{U},\boldsymbol{U})_{\mathscr{S}} \geq 0 \quad \text{for all} \quad \boldsymbol{U} \in \mathbb{H}^1(\mathscr{S}), \tag{3.10}$$

$$(\mathscr{L}_{\mathscr{S}}\boldsymbol{U},\boldsymbol{U})_{\mathscr{S}} \geq C\big\|\boldsymbol{U}\big|\mathbb{H}^1(\mathscr{S})\big\|^2 \quad \text{for all} \quad \boldsymbol{U} \in \mathbb{H}^1_{\mathscr{R}}(\mathscr{S}), \quad C > 0. \tag{3.11}$$

Moreover, the following Gårding's inequality

$$(\mathscr{L}_{\mathscr{S}}\boldsymbol{U},\boldsymbol{U})_{\mathscr{S}} \geq C_1\|\boldsymbol{U}|\mathbb{H}^1(\mathscr{S})\|^2 - C_0\|\boldsymbol{U}|\mathbb{H}^{-r}(\mathscr{S})\|^2 \tag{3.12}$$

holds for all $\boldsymbol{U} \in \mathbb{H}^1(\mathscr{S})$, with arbitrary $0 < r \leq \ell$ and some positive constants $C_0 > 0$, $C_1 > 0$.

The *proof* will be given later, in § 4. Here we draw the following consequence.

Corollary 3.6. *Let $\mathscr{S} \subset \mathbb{R}^n$ be a Lipschitz hypersurface without boundary,*

$$\mathrm{Def}_{\mathscr{S}}(\boldsymbol{U}) := [\mathfrak{D}_{jk}(\boldsymbol{U})]_{n\times n}$$

be the deformation tensor (cf. (0.17)) and the norm

$$\big\|\mathrm{Def}_{\mathscr{S}}(\boldsymbol{U})\big|\mathbb{L}_2(\mathscr{S})\big\|$$

be defined by (2.7).

Then the following Korn's inequality

$$\big\|\mathrm{Def}_{\mathscr{S}}(\boldsymbol{U})\big|\mathbb{L}_2(\mathscr{S})\big\| \geq c\big\|\boldsymbol{U}\big|\mathbb{H}^1(\mathscr{S})\big\| \qquad \forall\,\boldsymbol{U} \in \mathbb{H}^1_{\mathscr{R}}(\mathscr{S}) \tag{3.13}$$

holds for some constant $c > 0$ or, equivalently, the mapping

$$\boldsymbol{U} \mapsto \big\|\mathrm{Def}_{\mathscr{S}}(\boldsymbol{U})\big|\mathbb{L}_2(\mathscr{S})\big\|$$

is an equivalent norm on the orthogonal complement $\mathbb{H}^1_{\mathscr{R}}(\mathscr{S})$ to the space of Killing's vector fields.

Proof. Due to Korn's inequality (2.8) for $p = 2$

$$\big\|\boldsymbol{U}\big|\mathbb{L}_2(\mathscr{S})\big\|^2 \geq M_1\Big[\big\|\boldsymbol{U}\big|\mathbb{H}^1(\mathscr{S})\big\|^2 - \big\|\boldsymbol{U}\big|\mathbb{H}^1(\mathscr{S})\big\|\big\|\mathrm{Def}_{\mathscr{S}}(\boldsymbol{U})\big|\mathbb{L}_2(\mathscr{S})\big\|^2\Big]$$

the mapping $\mathrm{Def}_{\mathscr{S}} : \mathbb{H}^1_{\mathscr{R}}(\mathscr{S}) \to \mathbb{L}_2(\mathscr{S})$ is Fredholm and has index 0. The inequality (3.13) follows since the mapping is injective (has an empty kernel). □

Let us recall some results related to the uniqueness of solutions to arbitrary elliptic equation.

Definition 3.7. Let Ω be an open subset with the Lipschitz boundary $\partial\Omega \neq \emptyset$ either on a Lipschitz hypersurface $\mathscr{S} \subset \mathbb{R}^n$ or in the Euclidean space $\mathbb{R}^{n-1}$.

A class of functions $\mathscr{U}(\Omega)$ defined in a domain Ω in $\mathbb{R}^n$, is said to have the *strong unique continuation property*, if every $u \in \mathscr{U}(\Omega)$ in this class which vanishes to infinite order at one point must vanish identically.

If a surface $\mathscr{S}$ is C^∞-smooth, any elliptic operator on $\mathscr{S}$ has the strong unique continuation property due to Holmgren's theorem. But we can have more.

Lemma 3.8. *Let $\mathscr{S}$ be a C^2-smooth hypersurface in $\mathbb{R}^n$. The class of solutions to a second-order elliptic equation $\mathbb{A}(\mathscr{X},\mathscr{D})u = 0$, with Lipschitz continuous top-order coefficients on a surface $\mathscr{S}$ has the strong unique continuation property.*

In particular, if the solution $u(\mathscr{X}) = 0$ vanishes in any open subset of $\mathscr{S}$ it vanishes identically on entire $\mathscr{S}$.

Proof. The result was proved in [AKS1] for a domain $\Omega \subset \mathbb{R}^n$ by the method of "Carleman estimates" (also see [Hr1, Volume 3, Theorem 17.2.6]). Another proof, involving monotonicity of the frequency function was discovered by N. Garofalo and F. Lin (see [GL1, GL2]). A differential equation $\mathbb{A}(\mathscr{X},\mathscr{D})u(\mathscr{X}) = 0$ with Lipschitz continuous top-order coefficients on a C^2-smooth surface $\mathscr{S}$ is locally equivalent to a differential equation with Lipschitz continuous top-order coefficients on a domain $\Omega \subset \mathbb{R}^{n-1}$. Therefore a solution $u(\mathscr{X})$ has the strong unique continuation property locally (on each coordinate chart) on $\mathscr{S}$.

Since $\mathscr{S}$ is covered by a finite number of local coordinate charts which intersect on open neighborhoods, a solution $u(\mathscr{X})$ has the strong unique continuation property globally on $\mathscr{S}$. □

Remark 3.9. If the top-order coefficients of a second-order elliptic equation $\mathbb{A}(\mathscr{X},\mathscr{D})u = 0$ in open subsets $\Omega \subset \mathbb{R}^n$, $n \geq 3$, are merely Hölder continuous, with exponent less than 1, examples due to A. Plis [Pl1] and K. Miller [Mi1] show that a solution $u(x)$ does not have the strong unique continuation property.

Lemma 3.10. *Let $\mathscr{C}$ be a C^2-smooth hypersurface in $\mathbb{R}^n$ with the Lipschitz boundary $\Gamma := \partial\mathscr{C}$ and $\gamma \subset \Gamma$ be an open part of the boundary Γ. Let $\mathbb{A}(\mathscr{X},\mathscr{D})$ be a second-order elliptic system with Lipschitz continuous top-order matrix coefficients on a surface $\mathscr{S}$.*

The Cauchy problem

$$\begin{cases} \mathbb{A}(\mathscr{X},\mathscr{D})u = 0 & \text{on } \mathscr{C}, \quad u \in \mathbb{H}^1(\Omega), \\ u(\mathfrak{s}) = 0 & \text{for all } \quad \mathfrak{s} \in \gamma, \\ (\partial_{\boldsymbol{V}} u)(\mathfrak{s}) = 0 & \text{for all } \quad \mathfrak{s} \in \gamma, \end{cases} \tag{3.14}$$

where $\boldsymbol{V}$ is a non-tangent vector to Γ, but tangent to $\mathscr{S}$, has only a trivial solution $u(\mathscr{X}) = 0$ on entire $\mathscr{S}$.

Proof. With a local diffeomorphism the Cauchy problem (3.14) is transformed into a similar problem on a domain $\Omega \subset \mathbb{R}^{n-1}$ with the Cauchy data vanishing on some open subset of the boundary $\gamma \subset \Gamma := \partial\Omega$.

Let us, for simplicity, use the same notation $\gamma \subset \Gamma = \partial\Omega$, the non-tangent vector $\boldsymbol{V}$ to γ, the function u and the differential operator $\mathbb{A}(x,\mathscr{D})$ for the transformed Cauchy problem in the transformed domain Ω. Moreover, we will suppose that γ is a part of the hypersurface $x_1 = 0$ (otherwise we can transform the domain Ω again). We also use new variables $t = x_1$ and $x := (x_2, \dots, x_{n-1})$. Then $(0, x) \in \gamma$ while $(t, x) \in \Omega$ for all small $0 < t < \varepsilon$ and some $x \in \Omega'$.

Thus, the natural basis element $\boldsymbol{e}^1$ (cf. (0.6)) is orthogonal to γ and, therefore, $\boldsymbol{e}^1 = c_1(x)\boldsymbol{V}(0, x) + c_2(x)\boldsymbol{g}^j(x)$ for some unit tangential vector $\boldsymbol{g}^j(x)$ to γ for all $x \in \Omega'$ and some scalar functions $c_1(x)$, $c_2(x)$. Then, due to the third line in (3.14),

$$(\partial_t u)(0, x) = \partial_{\boldsymbol{e}^j} u(0, x) = c_1(x)\partial_{\boldsymbol{V}} u(0, x) + c_2(x)\partial \boldsymbol{g}^j u(0, x) = 0$$

because any derivative along tangential vector to γ vanishes $\partial \boldsymbol{g}^j u(0, x) = 0$ due to the second line in (3.14).

The second-order equation $\mathbb{A}(t, x; \mathscr{D})$ can be written in the form

$$\mathbb{A}(t, x, D)u = \mathbb{A}(t, x; \boldsymbol{e}^1)\partial_t^2 u + \mathbb{A}_1(t, x; D)\partial_t u + \mathbb{A}_2(t, x; D), \qquad D := -i\partial_x,$$

where $\mathbb{A}_1(t, x; \boldsymbol{e}^1)$ is the (invertible) matrix function, $\mathbb{A}_1(t, x; D)$ and $\mathbb{A}_2(t, x; D)$ are differential operators of orders 1 and 2 respectively, compiled of derivatives ∂_x, $x \in \Omega'$. Therefore, if $\mathbb{A}_j^0(t, x; D) := \mathbb{A}^{-1}(t, x; \boldsymbol{e}^1)\mathbb{A}_j(t, x; D)$, $j = 1, 2$, the Cauchy problem (3.14) transforms into

$$\begin{cases} \partial_t^2 u(t, x) + \mathbb{A}_1^0(t, x; D)\partial_t u(t, x) + \mathbb{A}_2^0(t, x; D)u(t, x) = 0 & \text{on } (t, x) \in \Omega_\varepsilon, \\ u(0, x) = 0 & \text{for all } x \in \Omega', \\ (\partial_t u)(0, x) = 0 & \text{for all } x \in \Omega', \end{cases} \tag{3.15}$$

where $\Omega_\varepsilon := (0, \varepsilon) \times \Omega' \subset \Omega$, $u \in \mathbb{H}^1(\Omega_\varepsilon)$ and $\gamma := \{(0, x) : x \in \Omega'\}$.

Now let us recall the inequality (see [Miz1, § 4.3, Theorem 4.3, § 6.14], [Sch1, § 4-7, Lemma 4-21]): There is a constant C which depends on ε and $\mathbb{A}(t, x; D)$ only and such that the inequality

$$\int_{\Omega_\varepsilon} e^{-\lambda t}|v(t, x)|^2 dt\, dx \leq C \int_{\Omega_\varepsilon} e^{-\lambda t}|(\mathbb{A}(t, x; D)v)(t, x)|^2 dt\, dx, \tag{3.16}$$

holds for $\mathbb{A}(t, x; D)v \in \mathbb{L}_2(\Omega_\varepsilon)$, $v \in C^\infty(\Omega_\varepsilon)$; moreover, $v(t, x)$ should vanish near $t = \varepsilon$ and should have vanishing Cauchy data $v(0, x) = (\partial_t v)(0, x) = 0$ for all $x \in \Omega'$.

Let $\rho \in C^2(0, \varepsilon)$ be a cut-off function: $\rho(t) = 1$ for $0 \leq t < \varepsilon/2$ and $\rho(t) = 0$ for $3\varepsilon/4 \leq t < \varepsilon$. Then $v := \rho u \in \mathbb{H}^1(\Omega_\varepsilon)$ and since $\mathbb{A}(t, x; D)u = 0$ on Ω_ε, we get

$$\begin{aligned} \mathbb{A}(t, x; D)(\rho u) &= \rho\mathbb{A}(t, x; D)u + (\partial_t^2 \rho)u + (\partial_t \rho)\partial_t u + (\partial_t \rho)\mathbb{A}_1^0(t, x; D)u \\ &= (\partial_t^2 \rho)u + (\partial_t \rho)\partial_t u + (\partial_t \rho)\mathbb{A}_1^0(t, x; D)u. \end{aligned}$$

We have asserted $u \in \mathbb{H}^1(\Omega_\varepsilon)$, $\rho \in C^2$ and this implies $(\partial_t^2 \rho)u \in \mathbb{H}^1(\Omega_\varepsilon) \subset \mathbb{L}_2(\Omega_\varepsilon)$, $(\partial_t \rho)\partial_t u \in \mathbb{L}_2(\Omega_\varepsilon)$. Note, that $(\partial_t \rho(t)$ vanishes for $0 < t < \varepsilon/2$. Therefore $(\partial_t \rho)\mathbb{A}_1^0(t, x; D)u$ vanishes in a neighborhood of the boundary $\gamma \subset \Gamma$. Due to a priori regularity result (cf. [LM1, Ch. 2, § 3.2, § 3.3]), a solution to an elliptic equation in (3.15) has additional regularity $u \in \mathbb{H}^2(\Omega_\varepsilon^0)$ for arbitrary Ω_ε^0 properly imbedded into Ω_ε. This implies $(\partial_t \rho)\mathbb{A}_1^0(t, x; D)u \in \mathbb{L}_2(\Omega_\varepsilon)$ and we conclude

$$\mathbb{A}(t, x; D)(\rho u) \in \mathbb{L}_2(\Omega_\varepsilon). \tag{3.17}$$

Introducing $v = \rho u$ into the inequality (3.16) we get

$$\int_{\Omega'}\int_0^{\varepsilon/4} e^{-\lambda t}|\rho(t)u(t,x)|^2 dt\, dx \leq \int_{\Omega_\varepsilon} e^{-\lambda t}|\rho(t)u(t,x)|^2 dt\, dx$$
$$\leq C\int_{\Omega'}\int_{\varepsilon/2}^{3\varepsilon/4} e^{-\lambda t}|(\mathbb{A}(t,x;D))\rho(t)u(t,x)|^2 dt\, dx.$$

This implies for $\lambda > 0$

$$\int_{\Omega'}\int_0^{\varepsilon/4} |\rho(t)u(t,x)|^2 dt\, dx \leq e^{-\lambda\varepsilon/4}\int_{\Omega_\varepsilon} |(\mathbb{A}(t,x;D))\rho(t)u(t,x)|^2\, dt\, dx \leq C_1 e^{-\lambda\varepsilon/4}.$$

where, due to (3.14), $C_1 > 0$ is a finite constant. By sending $\lambda \to \infty$ we get the desired result $u(t,x) = 0$ for all $0 \leq t \leq \varepsilon/4$ and all $x \in \Omega'$. Since $u(x)$ vanishes in a subset of the domain Ω, bordering γ, due to Lemma 3.8 the solution vanishes on entire Ω (on entire $\mathscr{C}$). □

Due to our specific interest (see next Lemma 3.12) and many applications, for example to control theory, the following boundary unique continuation property is of special interest.

Definition 3.11. Let $\mathscr{S}$ be a Lipschitz hypersurface in $\mathbb{R}^n$ and $\mathscr{C} \subset \mathscr{S}$ be a subsurface with Lipschitz boundary $\Gamma = \partial\mathscr{C}$.

We say that a class of functions $\mathscr{U}(\Omega)$ has the *strong unique continuation property from the boundary* if a vector function $\boldsymbol{U} \in \mathscr{U}(\Omega)$ which vanishes $\boldsymbol{U}(\mathfrak{s}) = 0$, $\forall\, \mathfrak{s} \in \gamma$ on an open subset of the boundary $\gamma \subset \Gamma$, vanishes on the entire $\mathscr{C}$.

Lemma 3.12. *Let $\mathscr{S}$ be a C^2-smooth hypersurface in $\mathbb{R}^n$ and $\mathscr{C} \subset \mathscr{S}$ be a C^2-smooth subsurface with boundary.*

The set of Killing's vector fields $\mathscr{R}(\mathscr{S})$ on the surface $\mathscr{C}$ with boundary has the strong unique continuation property from the boundary.

Proof. Let $\gamma \subset \Gamma := \partial\mathscr{C}$, $\operatorname{mes}\gamma > 0$ and $\boldsymbol{U}(\mathfrak{s}) = 0$ for all $\mathfrak{s} \in \gamma \subset \Gamma := \partial\mathscr{C}$. Then (cf. (3.1))

$$\begin{cases} (\mathscr{D}_j U_k^0)(\mathfrak{s}) + (\mathscr{D}_k U_j^0)(\mathfrak{s}) = -\sum\limits_{m=1}^{n} U_m^0(\mathfrak{s})\mathscr{D}_m\big(\nu_j(\mathfrak{s})\nu_k(\mathfrak{s})\big) = 0, \\ U_k^0(\mathfrak{s}) = 0 \qquad \forall\, \mathfrak{s} \in \gamma, \quad j,k = 1,\dots,n. \end{cases} \tag{3.18}$$

Among tangent vector fields generating the Günter's derivatives $\{\boldsymbol{d}^j(\mathfrak{s})\}_{j=1}^{n-1}$ only $n-1$ are linearly independent. One of vectors might collapse at a point $\boldsymbol{d}^j(\mathfrak{s}) = 0$ if the corresponding basis vector $\boldsymbol{e}^j$ is orthogonal to the surface at $\mathfrak{s} \in \mathscr{S}$, while others might be tangential to the subsurface Γ, except at least one, say $\boldsymbol{d}^n(\mathfrak{s})$, which is non-tangential to γ. Then from (3.18) follows

$$2(\mathscr{D}_n U_n^0)(\mathfrak{s}) = 0 \quad \text{and implies} \quad (\mathscr{D}_j U_n^0)(\mathfrak{s}) = 0 \tag{3.19}$$
$$\text{for all} \quad \mathfrak{s} \in \gamma \quad \text{and all} \quad j = 1,\dots,n.$$

Indeed, the vector $\boldsymbol{d}^j$, $1 \leq j = 1 \leq n-1$ is a linear combination $\boldsymbol{d}^j(\mathfrak{s}) = c_1(\mathfrak{s})\boldsymbol{d}^n(\mathfrak{s}) + c_2(\mathfrak{s})\boldsymbol{\tau}^j(\mathfrak{s})$ of the non-tangential vector $\boldsymbol{d}^n(\mathfrak{s})$ and of the projection $\boldsymbol{\tau}^j(\mathfrak{s}) := \pi_\gamma \boldsymbol{d}^j(\mathfrak{s})$ of $\boldsymbol{d}^j(\mathfrak{s})$ to the subsurface γ at the point $\mathfrak{s} \in \gamma$. Since U_n^0 vanishes identically on γ, the derivative $(\partial_{\boldsymbol{\tau}^j} U_n^0)(\mathfrak{s}) = 0$ vanishes as well and (3.19) follows:

$$(\mathscr{D}_j U_n^0)(\mathfrak{s}) = c_1(\mathfrak{s})(\partial_{\boldsymbol{d}^n} U_n^0)(\mathfrak{s}) + c_2(\mathfrak{s})(\partial_{\boldsymbol{\tau}^j} U_n^0)(\mathfrak{s}) = c_1(\mathfrak{s})(\mathscr{D}_n U_n^0)(\mathfrak{s}) = 0 \quad \forall\, \mathfrak{s} \in \gamma.$$

Equalities (3.18) and (3.19) imply

$$(\mathscr{D}_n U_j^0(\mathfrak{s}) = -(\mathscr{D}_j U_n^0)(\mathfrak{s}) = 0 \qquad \forall\, \mathfrak{s} \in \gamma \quad \text{and all} \quad j = 1, \dots, n. \tag{3.20}$$

Thus, we have the following Cauchy problem

$$\begin{cases} \mathscr{L}_{\mathscr{C}}(\mathscr{X}, \mathscr{D})\boldsymbol{U}(\mathscr{X}) = 0 & \text{on} \quad \mathscr{C}, \\ \boldsymbol{U}(\mathfrak{s}) = 0 & \text{for all} \quad \mathfrak{s} \in \gamma, \\ (\mathscr{D}_n \boldsymbol{U})(\mathfrak{s}) = (\partial_{\boldsymbol{d}^n} \boldsymbol{U})(\mathfrak{s}) = 0 & \text{for all} \quad \mathfrak{s} \in \gamma, \end{cases} \tag{3.21}$$

where $\boldsymbol{d}^n$ is a vector field non-tangential to Γ. Due to Lemma 3.10, $\boldsymbol{U}(\mathscr{X}) = 0$ for all $\mathscr{X} \in \mathscr{C}$. □

Corollary 3.13. (Korn's I inequality "with boundary condition"). *Let $\mathscr{C} \subset \mathbb{R}^n$ be a C^ℓ-smooth hypersurface with the Lipschitz boundary $\Gamma := \partial\mathscr{C} \neq \emptyset$ and $\ell \geq 2$, $|s| \leq \ell$. Then*

$$\left\| \boldsymbol{U} \middle| \mathbb{H}_p^s(\mathscr{C}) \right\| \leq M \left\| \mathrm{Def}_{\mathscr{C}}(\boldsymbol{U}) \middle| \mathbb{H}_p^{s-1}(\mathscr{C}) \right\| \qquad \forall\, \boldsymbol{U} \in \widetilde{\mathbb{H}}_p^s(\mathscr{C})$$

for some constant $M > 0$. In other words: the mapping

$$\boldsymbol{U} \mapsto \left\| \mathrm{Def}_{\mathscr{C}}(\boldsymbol{U}) \middle| \mathbb{H}_p^{s-1}(\mathscr{C}) \right\| \tag{3.22}$$

is an equivalent norm on the space $\widetilde{\mathbb{H}}_p^s(\mathscr{C})$.

Proof. If the claimed inequality (3.22) is false, there exists a sequence $\boldsymbol{U}^j \in \widetilde{\mathbb{H}}_p^s(\mathscr{C})$, $j = 1, 2, \dots$ such that

$$\left\| \boldsymbol{U}^j \middle| \mathbb{H}_p^s(\mathscr{C}) \right\| = 1 \quad \forall j = 1, 2, \dots \quad \lim_{j \to \infty} \left\| \mathrm{Def}_{\mathscr{C}}(\boldsymbol{U}^j) \middle| \mathbb{H}_p^{s-1}(\mathscr{C}) \right\| = 0.$$

Due to the compact embedding $\widetilde{\mathbb{H}}_p^s(\mathscr{C}) \subset \mathbb{H}_p^s(\mathscr{C}) \subset \mathbb{H}_p^{s-1}(\mathscr{C})$, a convergent subsequence $\boldsymbol{U}^{j_1}, \boldsymbol{U}^{j_2}, \dots$ in $\mathbb{H}_p^{s-1}(\mathscr{C})$ can be selected. Let $\boldsymbol{U}^0 = \lim_{k \to \infty} \boldsymbol{U}^{j_k}$. Then

$$\left\| \mathrm{Def}_{\mathscr{C}}(\boldsymbol{U}^0) \middle| \mathbb{H}_p^{s-1}(\mathscr{C}) \right\| = \lim_{k \to \infty} \left\| \mathrm{Def}_{\mathscr{C}}(\boldsymbol{U}^{j_k}) \middle| \mathbb{H}_p^{s-1}(\mathscr{C}) \right\| = 0$$

and $\boldsymbol{U}^0$ is a Killing's vector field. Since $\boldsymbol{U}(x) = 0$ on Γ, due to Lemma 3.12 $\boldsymbol{U}^0(x) = 0$ for all $x \in \mathscr{C}$ which contradicts to $\left\| \boldsymbol{U}^0 \middle| \mathbb{H}_p^s(\mathscr{C}) \right\| = \lim_{k \to \infty} \left\| \boldsymbol{U}^{j_k} \middle| \mathbb{H}_p^s(\mathscr{C}) \right\| = 1$. □

4. A local fundamental solution to the Lamé equation

Proof of Theorem 3.5. Let us check the ellipticity of $\mathscr{L}_{\mathscr{S}}$. The operator $\mathscr{L}_{\mathscr{S}}$ maps the tangential spaces and the principal symbol is defined on the cotangent space. The cotangent space is orthogonal to the normal vector and, therefore,

$$\mathscr{L}_{\mathscr{S}}(\mathscr{X},\xi)\eta=\mu|\xi|^2(1-\boldsymbol{\nu}\boldsymbol{\nu}^\top)\eta+(\lambda+\mu)\xi\xi^\top\eta=\mu|\xi|^2\eta+(\lambda+\mu)\xi\xi^\top\eta,\quad \forall\xi,\,\eta\perp\boldsymbol{\nu}.$$

Thus, while considering the principal symbol $\mathscr{L}_{\mathscr{S}}(\mathscr{X},\xi)$ we can ignore the projection $\pi_{\mathscr{S}}$. With this assumption, the principal symbol of $\mathscr{L}_{\mathscr{S}}$ reads

$$\mathscr{L}_{\mathscr{S}}(\mathscr{X},\xi)=\mu|\xi|^2+(\lambda+\mu)\xi\xi^\top \qquad \text{for}\quad (\mathscr{X},\xi)\in\mathbb{T}^*(\mathscr{S}). \tag{4.1}$$

The matrix $\mathscr{L}_{\mathscr{S}}(\mathscr{X},\xi)$ has eigenvalue $(\lambda+2\mu)|\xi|^2$ (the corresponding eigenvector is ξ) and $\mu|\xi|^2$ which has multiplicity $n-1$ (the corresponding eigenvectors θ^j are orthogonal to ξ: $\xi^\top\theta^j=\langle\xi,\theta^j\rangle=0$, $j=1,\dots,n-1$). Then

$$\det\mathscr{L}_{\mathscr{S}}(\mathscr{X},\xi)=(\lambda+2\mu)|\xi|^2\left[\mu|\xi|^2\right]^{n-1}=\mu^{n-1}(\lambda+2\mu)>0$$
$$\text{for}\quad(\mathscr{X},\xi)\in\mathbb{T}^*(\mathscr{S}),\quad |\xi|=1$$

and the ellipticity is proved.

The ellipticity of the differential operator $\mathscr{L}_{\mathscr{S}}=\mathscr{L}_{\mathscr{S}}(\mathscr{X},\mathscr{D})$ in (3.8) on a manifold without boundary $\mathscr{S}$, proved above, implies Fredholm property for all $1<p<\infty$ and all $s\in\mathbb{R}$. Indeed, $\mathscr{L}_{\mathscr{S}}(\mathscr{X},\mathscr{D})$ has a parametrix $\boldsymbol{R}_{\mathscr{S}}(\mathscr{X},\mathscr{D})$, which is a pseudodifferential operator (ΨDO) with the symbol $R_{\mathscr{S}}(\mathscr{X},\xi):=\chi(\xi)\mathscr{L}_{\mathscr{S}}^{-1}(\mathscr{X},\xi)$, where $\mathscr{L}_{\mathscr{S}}^{-1}(\mathscr{X},\xi)$ is the inverse symbol and $\chi\in C^\infty(\mathbb{R}^n)$ is a smooth function, $\chi(\xi)=1$ for $|\xi|>2$ and $\chi(\xi)=0$ for $|\xi|<1$. ΨDO $\boldsymbol{R}_{\mathscr{S}}(\mathscr{X},\mathscr{D})$ is a bounded operator between the spaces

$$\boldsymbol{R}_{\mathscr{S}}(\mathscr{X},\mathscr{D})\ :\ \mathbb{H}^{s-2}_p(\mathscr{S})\to\mathbb{H}^s_p(\mathscr{S}),\qquad \text{for all}\quad 1<p<\infty,\quad s\in\mathbb{R},$$

because the symbol $R_{\mathscr{S}}(\mathscr{X},\xi)=\mathscr{L}_{\mathscr{S}}^{-1}(\mathscr{X},\xi)$ belongs to the Hörmander class $S^{-2}(\mathscr{S},\mathbb{R}^n)$

$$\left|\mathscr{D}^\alpha\partial^\beta_\xi R_{\mathscr{S}}(\mathscr{X},\xi)\right|\le C_{\alpha,\beta}|\xi|^{-2-|\beta|}$$

for all multi-indices $\alpha,\beta\in\mathbb{Z}^n_+$ (cf. [Hr1, Sh1, Ta2] for details).

The Fredholm property for the case $p=2$ and $s=1$ follows from Gårding's inequality (3.12) as well (cf. [HW1, Thorem 5.3.10] and [Mc1, Thorem 2.33]).

The Fredholm property implies the finite-dimensional kernel

$$\dim\operatorname{Ker}\mathscr{L}_{\mathscr{S}}(\mathscr{X},\mathscr{D})<\infty.$$

To prove that the index is trivial $\operatorname{Ind}\mathscr{L}_{\mathscr{S}}(\mathscr{X},\mathscr{D})=0$ for all $1<p<\infty$, $s\in\mathbb{R}$ we apply Gårding's inequality (3.12) and homotopy. For this purpose first note that the symbol $\mathscr{L}_{\mathscr{S}}(\mathscr{X},\xi)$ is positive definite (cf. (4.1))

$$\langle\mathscr{L}_{\mathscr{S}}(\mathscr{X},\xi)\eta,\eta\rangle=\mu|\xi|^2|\eta|^2+(\lambda+\mu)\langle\xi\xi^\top\eta,\eta\rangle=\mu|\xi|^2|\eta|^2+(\lambda+\mu)\sum_{j=1^n}(\xi_j\eta_j)^2$$
$$\ge\mu|\xi|^2|\eta|^2\qquad\forall\,\mathscr{X}\in\mathscr{S},\quad\forall\xi,\eta\in\mathbb{R}^n. \tag{4.2}$$

Further recall that the Bessel potential operator $\Lambda^2_{\mathscr{S}}(\mathscr{X}, D) : \mathbb{H}^s_p(\mathscr{S}) \to \mathbb{H}^{s-2}_p(\mathscr{S})$ (cf. (2.4)) lifting the Bessel potential spaces, has positive definite symbol

$$\langle \Lambda^2_{\mathscr{S}}(\mathscr{X}, \xi)\eta, \eta\rangle \geq C|\xi|^2|\eta|^2 \qquad \forall\, \mathscr{X} \in \mathscr{S}, \quad \forall \xi, \eta \in \mathbb{R}^n \tag{4.3}$$

(cf. [DS1]). Now consider the symbols $\boldsymbol{B}_\tau(\mathscr{X}, \xi) = (1-\tau)\mathscr{L}_{\mathscr{S}}(\mathscr{X}, \xi) + \tau\Lambda^2_{\mathscr{S}}(\mathscr{X}, \xi)$ and the corresponding ΨDO

$$\boldsymbol{B}_\tau(\mathscr{X}, \mathscr{D}) = (1-\tau)\mathscr{L}_{\mathscr{S}}(\mathscr{X}, \mathscr{D}) + \tau\Lambda^2_{\mathscr{S}}(\mathscr{X}, \mathscr{D}) \; : \; \mathbb{H}^s_p(\mathscr{S}) \to \mathbb{H}^{s-2}_p(\mathscr{S}). \tag{4.4}$$

Obviously, $\boldsymbol{B}_\tau(\mathscr{X}, \mathscr{D})$ is a continuous (with respect to $0 \leq \tau \leq 1$) homotopy connecting the operator $\boldsymbol{B}_0(\mathscr{X}, \mathscr{D}) = \mathscr{L}_{\mathscr{S}}(\mathscr{X}, \mathscr{D})$ with $\boldsymbol{B}_1(\mathscr{X}, \mathscr{D}) = \Lambda^2_{\mathscr{S}}(\mathscr{X}, \mathscr{D})$. Since the symbol $B_\tau(\mathscr{X}, \xi)$ is positive definite

$$\langle B_\tau(\mathscr{X}, \xi)\eta, \eta\rangle \geq [(1-\tau)\mu + \tau C]|\xi|^2|\eta|^2 \quad \forall \xi, \eta \in \mathbb{R}^n$$

(cf. (4.2) and (4.3)), it is elliptic and the operator $\boldsymbol{B}_\tau(\mathscr{X}, \mathscr{D})$ is then Fredholm for all $0 \leq \tau \leq 1$. Then $\operatorname{Ind}\mathscr{L}_{\mathscr{S}}(\mathscr{X}, \mathscr{D}) = \operatorname{Ind}\boldsymbol{B}_0(\mathscr{X}, \mathscr{D}) = \operatorname{Ind}\boldsymbol{B}_1(\mathscr{X}, \mathscr{D}) = \operatorname{Ind}\Lambda^2_{\mathscr{S}}(\mathscr{X}, \mathscr{D}) = 0$, since the operator $\Lambda^2_{\mathscr{S}}(\mathscr{X}, \mathscr{D})$ is invertible.

From the representation (0.23) follows that the bilinear form $(\mathscr{L}_{\mathscr{S}}\boldsymbol{U}, \boldsymbol{U})_{\mathscr{S}}$ is non-negative

$$\begin{aligned}(\mathscr{L}_{\mathscr{S}}\boldsymbol{U}, \boldsymbol{U})_{\mathscr{S}} &= \lambda(\operatorname{div}^*_{\mathscr{S}} \operatorname{div}_{\mathscr{S}}\boldsymbol{U}, \boldsymbol{U})_{\mathscr{S}} + 2\mu(\operatorname{Def}^*_{\mathscr{S}}\operatorname{Def}_{\mathscr{S}}\boldsymbol{U}, \boldsymbol{U})_{\mathscr{S}} \\ &= \lambda\big\|\operatorname{div}_{\mathscr{S}}\boldsymbol{U}\big|\mathbb{L}_2(\mathscr{S})\big\|^2 + 2\mu\big\|\operatorname{Def}_{\mathscr{S}}\boldsymbol{U}\big|\mathbb{L}_2(\mathscr{S})\big\|^2 \geq 0 \end{aligned} \tag{4.5}$$

(cf. (3.10)) and only vanishes if $\boldsymbol{U}$ is a Killing's vector field $\operatorname{Def}_{\mathscr{S}}\boldsymbol{U} = 0$. Indeed, $\mathfrak{D}_{jj}\boldsymbol{U} = (\mathscr{D}^{\mathscr{S}}_j\boldsymbol{U})^0_j = 0$, $j = 1, \dots, n$, if $\operatorname{Def}_{\mathscr{S}}\boldsymbol{U} = 0$ and, due to (0.13),

$$\operatorname{div}_{\mathscr{S}}\boldsymbol{U} = \sum_{j=1}^n \mathscr{D}_j U^0_j = \sum_{j=1}^n \mathscr{D}_j U^0_j + \frac{1}{2}\sum_{j=1}^n \partial_{\boldsymbol{U}}(\nu_j)^2 = \sum_{j=1}^n (\mathscr{D}^{\mathscr{S}}_j\boldsymbol{U})^0_j = 0 \qquad \forall \boldsymbol{U} \in \mathscr{R}(\mathscr{S}) \tag{4.6}$$

since $|\boldsymbol{\nu}(\mathscr{X})| \equiv 1$. Thence, due to (4.5), $\mathscr{R}(\mathscr{S}) \subset \operatorname{Ker}\mathscr{L}_{\mathscr{S}}$. The inverse inclusion follows also from (4.5) because $\operatorname{Def}_{\mathscr{S}}(\boldsymbol{U}) = 0$ if $\mathscr{L}_{\mathscr{S}}(\mathscr{X}, \mathscr{D})\boldsymbol{U} = 0$. This accomplishes the proof of (3.9).

The estimate (3.11) is a direct consequence of (3.10) and of (3.9): Since the operator $\mathscr{L}_{\mathscr{S}}$ is Fredholm, self-adjoint and $\operatorname{Ker}\mathscr{L}_{\mathscr{S}} = \mathscr{R}(\mathscr{S})$, then also $\operatorname{Coker}\mathscr{L}_{\mathscr{S}} = \mathscr{R}(\mathscr{S})$ and, therefore, the mapping

$$\mathscr{L}_{\mathscr{S}} \; : \; \mathbb{H}^1_{\mathscr{R}}(\mathscr{S}) \longrightarrow \mathbb{H}^{-1}_{\mathscr{R}}(\mathscr{S})$$

is one-to-one, i.e., is invertible. The established invertibility implies the claimed inequality (3.11).

A priori regularity property of solutions to partial differential equations (cf. [Ta2, Hr1]) states that the ellipticity of $\mathscr{L}_{\mathscr{S}}(\mathscr{X}, \mathscr{D})$ provides $C^\ell(\mathscr{S})$-smoothness of any solution $\boldsymbol{K}$ to the homogeneous equation $\mathscr{L}_{\mathscr{S}}(\mathscr{X}, \mathscr{D})\boldsymbol{K} = 0$ (the hypersurface $\mathscr{S}$ is C^ℓ-smooth). Due to the embeddings $\mathbb{H}^r_q(\mathscr{S}) \subset \mathbb{H}^s_p(\mathscr{S})$, $s \leq r$, $p \leq q$, then the kernel $\operatorname{Ker}\mathscr{L}_{\mathscr{S}}(\mathscr{X}, \mathscr{D})$ is independent of the space $\mathbb{H}^s_p(\mathscr{S})$ provided that the spaces are well defined, which is the case if $|s| \le \ell$ (cf. [Ag1, Du2, DNS2, Ka1] for similar assertions).

In particular, the Killing's vector fields $\mathscr{R}(\mathscr{S}) = \operatorname{Ker} \mathscr{L}_{\mathscr{S}}(\mathscr{X}, \mathscr{D})$ are smooth $\mathscr{R}(\mathscr{S}) \subset C^\infty(\mathscr{S})$ provided that the hypersurface $\mathscr{S}$ is C^∞-smooth.

Let $\{\boldsymbol{K}_j\}_{j=1}^m$ be an orthogonal basis $(\boldsymbol{K}_j, \boldsymbol{K}_k)_{\mathscr{S}} = \delta_{jk}$ in the finite-dimensional space of Killing's vector fields $\mathscr{R}(\mathscr{S})$. Let

$$\boldsymbol{T}\boldsymbol{U}(\mathscr{X}) := \sum_{j=1}^m (\boldsymbol{K}_j, \boldsymbol{U})_{\mathscr{S}} \, \boldsymbol{K}_j(\mathscr{X}), \qquad \mathscr{X} \in \mathscr{S}. \tag{4.7}$$

Due to the proved part $\{\boldsymbol{K}_j\}_{j=1}^m \subset C^\ell(\mathscr{S})$ and the operator $\boldsymbol{T}$ is smoothing $\boldsymbol{T} : \mathbb{H}^{-r}(\mathscr{S}) \to \mathbb{H}^r(\mathscr{S})$ (is infinitely smoothing if $\ell = \infty$). Then, the operator

$$\mathscr{L}_{\mathscr{S}} + \boldsymbol{T} : \mathbb{H}^1(\mathscr{S}) \to \mathbb{H}^{-1}(\mathscr{S})$$

is invertible and non-negative

$$(\mathscr{L}_{\mathscr{S}} + \boldsymbol{T})\boldsymbol{U}, \boldsymbol{U})_{\mathscr{S}} = (\mathscr{L}_{\mathscr{S}}\boldsymbol{U}, \boldsymbol{U})_{\mathscr{S}} + \sum_{j=1}^m (\boldsymbol{K}_j, \boldsymbol{U})_{\mathscr{S}}^2 \ge 0$$

(cf. (4.5)). This implies that $\mathscr{L}_{\mathscr{S}} + \boldsymbol{T}$ is positive definite

$$(\mathscr{L}_{\mathscr{S}}\boldsymbol{U}, \boldsymbol{U})_{\mathscr{S}} + \boldsymbol{T})\boldsymbol{U}, \boldsymbol{U})_\Gamma \ge C_1 \big\|\boldsymbol{U}\big|\mathbb{H}^1(\mathscr{S})\big\|^2$$

and we write

$$\begin{aligned}(\mathscr{L}_{\mathscr{S}}\boldsymbol{U}, \boldsymbol{U})_{\mathscr{S}} &:= ((\mathscr{L}_{\mathscr{S}} + \boldsymbol{T})\boldsymbol{U}, \boldsymbol{U})_{\mathscr{S}} + (\boldsymbol{T}\boldsymbol{U}, \boldsymbol{U})_{\mathscr{S}} \\ &\ge C_1 \big\|\boldsymbol{U}\big|\mathbb{H}^1(\mathscr{S})\big\|^2 + (\boldsymbol{T}\boldsymbol{U}, \boldsymbol{U})_{\mathscr{S}} \\ &\ge C_1 \big\|\boldsymbol{U}\big|\mathbb{H}^1(\mathscr{S})\big\|^2 - C_2 \big\|\boldsymbol{U}\big|\mathbb{H}^{-r}(\mathscr{S})\big\|^2,\end{aligned}$$

which proves (3.12). □

Remark 4.1. Gårding's inequality (3.12), but in a weaker form $r = 0$, is a direct consequence of the inequality (4.5) and Korn's inequality (2.8) for $p = 2$.

Theorem 4.2. *Let $\mathscr{S}$ be a ℓ-smooth hypersurface without boundary, $\ell \ge 2$ and $\mathscr{B} \in C^\ell(\mathbb{R}^n)$ be a real-valued and non-negative $\mathscr{B} \ge 0$ function with non-trivial support* $\operatorname{mes}\operatorname{supp}\mathscr{B} \ne 0$.

The perturbed operator

$$\mathscr{L}_{\mathscr{S}}(\mathscr{X}, \mathscr{D}) + \mathscr{B}I : \mathbb{H}_p^{\theta+1}(\mathscr{S}) \to \mathbb{H}_p^{\theta-1}(\mathscr{S}) \tag{4.8}$$

is invertible for all $|\theta| \le \ell - 1$ and all $1 < p < \infty$.

Proof. The principal symbol of the operator $\mathscr{L}_{\mathscr{S}}(\mathscr{X}, \mathscr{D}) + \mathscr{B}I$ in (4.8) ignores lower-order terms and coincides with $\mathscr{L}_{\mathscr{S}}(\mathscr{X}, \xi)$ and is elliptic (cf. Theorem 3.5). Therefore on the hypersurface without boundary $\mathscr{S}$ the operator $\mathscr{L}_{\mathscr{S}}(\mathscr{X}, \mathscr{D}) + \mathscr{B}I$ in (4.8) is Fredholm for all $\theta = 0, 1, \dots$ (cf. Theorem 3.5). On the other hand, if $(\mathscr{L}_{\mathscr{S}}(\mathscr{X}, \mathscr{D}) + \mathscr{B}I)\boldsymbol{U} = 0$, then

$$0 = ((\mathscr{L}_{\mathscr{S}}(\mathscr{X}, \mathscr{D}) + \mathscr{B})\boldsymbol{U}, \boldsymbol{U})_{\mathscr{S}} = (\mathscr{L}_{\mathscr{S}}(\mathscr{X}, \mathscr{D})\boldsymbol{U}, \boldsymbol{U})_{\mathscr{S}} + (\mathscr{B}\boldsymbol{U}, \boldsymbol{U})_{\mathscr{S}}$$

and (3.10) implies that $(\mathscr{B}\boldsymbol{U},\boldsymbol{U})_{\mathscr{S}}=0$. Since $\mathscr{B}\geq 0$, the obtained equality implies $\boldsymbol{U}=0$ for all $\mathscr{X}\in\operatorname{supp}\mathscr{B}$ and, due to the strong unique continuation property $\boldsymbol{U}=0$ (cf. Lemma 3.8).

Thus, the operator

$$\mathscr{L}_{\mathscr{S}}(\mathscr{X},\mathscr{D})+\mathscr{B}I\,:\,\mathbb{H}^1(\mathscr{S})\to\mathbb{H}^{-1}(\mathscr{S}) \tag{4.9}$$

has the trivial kernel $\operatorname{Ker}(\mathscr{L}_{\mathscr{S}}(\mathscr{X},\mathscr{D})+\mathscr{B}I)=\{0\}$. Since $\mathscr{L}_{\mathscr{S}}(\mathscr{X},\mathscr{D})+\mathscr{B}I$ is formally self-adjoint (cf. Theorem 3.5), the same is true for the dual operator and $\operatorname{Coker}(\mathscr{L}_{\mathscr{S}}(\mathscr{X},\mathscr{D})+\mathscr{B}I)=\{0\}$. The invertibility of the Fredholm operator $\mathscr{L}_{\mathscr{S}}(\mathscr{X},\mathscr{D})+\mathscr{B}I$ in (4.9) for $p=2$ and $\theta=0$ follows.

The invertibility of $\mathscr{L}_{\mathscr{S}}(\mathscr{X},\mathscr{D})+\mathscr{B}I$ in (4.9) for arbitrary p and θ is a consequence of the ellipticity of $\mathscr{L}_{\mathscr{S}}(\mathscr{X},\mathscr{D})+\mathscr{B}I$ (cf. a similar arguments in the proof of Theorem 3.5). □

Corollary 4.3. *Let $\mathscr{S}$ be a C^∞-smooth hypersurface in $\mathbb{R}^n$ and $\mathscr{C}\subset\mathscr{S}$ be a proper subsurface $\mathscr{S}\setminus\overline{\mathscr{C}}\neq\emptyset$. Then $\mathscr{L}_{\mathscr{S}}(\mathscr{X},\mathscr{D})$ has a fundamental solution on $\mathscr{S}$, which we call a local fundamental solution on $\mathscr{C}$, viewed as the Schwartz kernel of the inverse operator to $\mathscr{L}_{\mathscr{S}}(\mathscr{X},\mathscr{D})+\mathscr{B}I$, where $\operatorname{supp}\mathscr{B}\subset\mathscr{S}\setminus\overline{\mathscr{C}}$.*

Proof. The Schwartz kernel $\mathscr{K}_{\mathscr{S}}(\mathscr{X},\tau)$ of the inverse operator to $\mathscr{L}_{\mathscr{S}}(\mathscr{X},\mathscr{D})+\mathscr{B}I$, satisfies the equality

$$\mathscr{L}_{\mathscr{S}}(\mathscr{X},\mathscr{D})\mathscr{K}_{\mathscr{S}}=\delta(\mathscr{X})I,\qquad \mathscr{X}\in\mathscr{C}$$

since $\mathscr{B}(\mathscr{X})=0$ for $\mathscr{X}\in\mathscr{C}$, and can be viewed as a local fundamental solution of $\mathscr{L}_{\mathscr{S}}(\mathscr{X},\mathscr{D})$ on $\mathscr{C}$. □

Remark 4.4. The operator $\mathscr{L}_{\mathscr{S}}(\mathscr{X},\mathscr{D})$ itself has a fundamental solution on the entire hypersurface $\mathscr{S}$ if and only if the space of Killing's vector fields on $\mathscr{S}$ is trivial $\mathscr{R}(\mathscr{S})=\{0\}$. The situation is essentially different from the case of the Euclidean space $\mathbb{R}^n$, where the condition at infinity

$$\boldsymbol{U}(\mathscr{X})=\mathscr{O}(1)\qquad\text{as}\quad |\mathscr{X}|\to\infty$$

eliminates the kernel of any linear partial differential operator with constant coefficients and the fundamental solution (the inverse operator) exists.

A compact hypersurface with certain symmetry might possess non-trivial Killing's vector fields.

5. BVPs for the Lamé equation and Green's formulae

Throughout the present section, if not stated otherwise, $\mathscr{S}$ is a C^2-smooth surface, $\mathscr{C}\subset\mathscr{S}$ denotes a C^2-smooth subsurface with the Lipschitz boundary $\partial\mathscr{C}=\Gamma\neq\emptyset$ and $r_{\mathscr{C}}$ is the restriction to the surface $\mathscr{C}$. Under the operation $r_{\mathscr{C}}\mathscr{L}_{\mathscr{C}}(\mathscr{X},\mathscr{D})\boldsymbol{U}$ on a function (distribution) $\boldsymbol{U}\in\mathbb{H}^s_p(\mathscr{C})$ is meant that the operator $r_{\mathscr{C}}\mathscr{L}_{\mathscr{C}}(\mathscr{X},\mathscr{D})$ acts on a vector function $\boldsymbol{U}$ extended to a function $\widetilde{\boldsymbol{U}}\in\mathbb{H}^s_p(\mathscr{S})$ on the entire surface, $r_{\mathscr{C}}\widetilde{\boldsymbol{U}}=\boldsymbol{U}$. Since $\mathscr{L}_{\mathscr{C}}(\mathscr{X},\mathscr{D})$ is a local (differential) operator, the result is, after restriction, independent of the extension. Moreover, $\mathscr{L}_{\mathscr{C}}(\mathscr{X},\mathscr{D})$ does not extends supports of vector functions: if $\operatorname{supp}\boldsymbol{U}\subset\mathscr{C}$ then $\operatorname{supp}\mathscr{L}_{\mathscr{C}}(\mathscr{X},\mathscr{D})\boldsymbol{U}\subset\mathscr{C}$. Therefore we will drop the restriction operator $r_{\mathscr{C}}$ and write $(\mathscr{L}_{\mathscr{C}}(\mathscr{X},\mathscr{D})\boldsymbol{U})(\mathscr{X})$ for all $\mathscr{X}\in\mathscr{C}$.

We can not relax the constraint on a surface $\mathscr{C}$ (we remind that the underlying surface is C^2-smooth), because in the definition of equation

$$\mathscr{L}_{\mathscr{C}}(\mathscr{X},\mathscr{D})\boldsymbol{U}=\boldsymbol{F},\qquad \boldsymbol{U}\in\mathbb{H}^1(\mathscr{C}),\quad \boldsymbol{F}\in\widetilde{\mathbb{H}}^{-1}(\mathscr{C}),\tag{5.1}$$

is participating the gradient $\nabla_{\mathscr{S}}\boldsymbol{\nu}=[\mathscr{D}_j\nu_k]_{n\times n}$ of the unit normal vector field $\boldsymbol{\nu}$ (see (0.20)–(0.24)). $\boldsymbol{\nu}(\mathscr{X})$ is defined almost everywhere on $\mathscr{C}$ is just C^1-smooth. We can actually require that $\mathscr{S}$ is $\mathbb{H}^2_\infty$ (i.e., corresponding parameterizations of the surface have, instead of continuous, bounded second derivatives).

Equation (5.1) is actually understood in a weak sense:

$$(\mathscr{L}_{\mathscr{C}}(\mathscr{X},\mathscr{D})\boldsymbol{U},\boldsymbol{V})_{\mathscr{C}}:=(\mathbb{T}\operatorname{Def}_{\mathscr{C}}\boldsymbol{U},\operatorname{Def}_{\mathscr{C}}\boldsymbol{V})_{\mathscr{C}}=(\boldsymbol{F},\boldsymbol{V})_{\mathscr{C}},\tag{5.2}$$

$$\forall\boldsymbol{U}\in\mathbb{H}^1(\mathscr{C}),\boldsymbol{V}\in\widetilde{\mathbb{H}}^1(\mathscr{C}).$$

In particular, for the Lamé operator in isotropic media we have

$$(\mathscr{L}_{\mathscr{C}}(\mathscr{X},\mathscr{D})\boldsymbol{U},\boldsymbol{V})_{\mathscr{C}}:=\lambda(\nabla_{\mathscr{C}}\boldsymbol{U},\nabla_{\mathscr{C}}\boldsymbol{V})_{\mathscr{C}}+\mu(\operatorname{Def}_{\mathscr{C}}\boldsymbol{U},\operatorname{Def}_{\mathscr{C}}\boldsymbol{V})_{\mathscr{C}}=(\boldsymbol{F},\boldsymbol{V})_{\mathscr{C}},\tag{5.3}$$

$$\forall\boldsymbol{V}\in\widetilde{\mathbb{H}}^1_2(\mathscr{S})$$

(cf. (0.23)).

Let $\boldsymbol{\nu}_\Gamma=\left(\nu_\Gamma^1,\dots,\nu_\Gamma^n\right)^\top$ be the tangential to $\mathscr{C}$ and outer unit normal vector field to Γ.

If a tangential vector field $\boldsymbol{U}\in\mathbb{H}^1_p(\mathscr{C})\cap\mathscr{V}(\mathscr{C})$ denotes the displacement, the natural boundary value problems for $\mathscr{L}_{\mathscr{C}}$ are the following:

I. The Dirichlet problem when the displacement is prescribed on the boundary

$$\begin{cases}(\mathscr{L}_{\mathscr{C}}(\mathscr{X},\mathscr{D})\boldsymbol{U})(\mathscr{X})=\boldsymbol{F}(\mathscr{X}), & \mathscr{X}\in\mathscr{C},\\ \boldsymbol{U}^+(\tau)=\boldsymbol{G}(\tau), & \tau\in\Gamma,\end{cases}\tag{5.4}$$
$$\boldsymbol{F}\in\widetilde{\mathbb{H}}^{-1}(\mathscr{C}),\quad \boldsymbol{G}\in\mathbb{H}^{1/2}(\Gamma),\quad \boldsymbol{U}\in\mathbb{H}^1(\mathscr{C});$$

the first (basic) equation in the domain is understood in a weak sense (see (5.2), (5.3)) and

$$\gamma_D^+\boldsymbol{U}:=\boldsymbol{U}^+\tag{5.5}$$

is the Dirichlet trace operator on the boundary.

II. The Neumann problem when the traction is prescribed on the boundary:

$$\begin{cases}(\mathscr{L}_{\mathscr{C}}(\mathscr{X},\mathscr{D})\boldsymbol{U})(\mathscr{X})=\boldsymbol{F}(\mathscr{X}), & \mathscr{X}\in\mathscr{C},\\ (\mathfrak{T}_{\mathscr{C}}(\boldsymbol{\nu}_\Gamma,\mathscr{D})\,\boldsymbol{U})^+(\tau)=\boldsymbol{H}(\tau), & \tau\in\Gamma,\end{cases}\tag{5.6}$$
$$\boldsymbol{F}\in\widetilde{\mathbb{H}}^{-1}(\mathscr{C}),\quad \boldsymbol{H}\in\mathbb{H}^{-1/2}(\Gamma),\quad \boldsymbol{U}\in\mathbb{H}^1(\mathscr{C});$$

here

$$\gamma_N^+ \boldsymbol{U} := (\mathfrak{T}_{\mathscr{C}}(\boldsymbol{\nu}_\Gamma, \mathscr{D})\, \boldsymbol{U})^+, \tag{5.7}$$

$$\mathfrak{T}_{\mathscr{C}}(\boldsymbol{\nu}_\Gamma, \mathscr{D})\boldsymbol{U} := -\lambda(\mathrm{div}_{\mathscr{C}}\boldsymbol{U})\boldsymbol{\nu}_\Gamma - 2\mu \sum_{j=1}^{n} \left\{(\nu_\Gamma^j + \mathscr{H}_{\mathscr{C}}^0 \nu_j)\mathfrak{D}_{jk}(\boldsymbol{U})\right\}_{k=1}^{n} \tag{5.8}$$

$$= -\mu \mathscr{D}_{\boldsymbol{\nu}_\Gamma}\boldsymbol{U} - (\lambda+\mu)(\mathrm{div}_{\mathscr{C}}\boldsymbol{U})\boldsymbol{\nu}_\Gamma \tag{5.9}$$

is the Neumann trace operator on the boundary (the traction) with

$$\mathscr{D}_{\boldsymbol{\nu}_\Gamma}\varphi := \sum_{j=1}^{n} \nu_\Gamma^j \mathscr{D}_j \varphi. \qquad \varphi \in \mathbb{H}^1(\mathscr{C}). \tag{5.10}$$

In Lemma 7.3 below it will be shown that the trace $\gamma_N^+ \boldsymbol{U}$ exists provided that $\boldsymbol{U}$ is a solution to the basic (first) equation in (5.6).

A crucial role in the investigation of BVPs (5.4)–(5.10) belongs to the Green formula.

Theorem 5.1. *Let $\mathscr{B} \in C^1(\mathscr{S})$ and $\mathscr{C}^+ := \mathscr{C}$, $\mathscr{C}^- := \mathscr{C}^c = \mathscr{S} \setminus \overline{\mathscr{C}}$ denote the complemented surfaces with boundary $\mathscr{S} = \overline{\mathscr{C}^+ \cup \mathscr{C}^-}$.*

For a solution to the equation

$$\mathscr{L}_{\mathscr{C}}\boldsymbol{U} + \mathscr{B}\boldsymbol{U} = \boldsymbol{F}, \qquad \boldsymbol{F} \in \widetilde{\mathbb{H}}^{-1}(\mathscr{C}^\pm), \quad \boldsymbol{U} \in \mathbb{H}^1(\mathscr{C}^\pm) \tag{5.11}$$

(for $\mathscr{B} = 0$ cf. (5.1) and the basic equations in (5.4)–(5.10)) the following Green formula are valid

$$\begin{aligned}((\mathscr{L}_{\mathscr{C}} + \mathscr{B}I)\boldsymbol{U}, \boldsymbol{V})_{\mathscr{C}^\pm} &= \int_{\mathscr{C}^\pm} \langle (\mathscr{L}_{\mathscr{C}} + \mathscr{B}I)\boldsymbol{U}(\mathscr{X}), \boldsymbol{V}(\mathscr{X})\rangle\, dS \\ &= \pm \oint_\Gamma \langle \gamma_N^\pm \boldsymbol{U}(\tau), \gamma_D^\pm \boldsymbol{V}(\tau)\rangle\, d\mathfrak{s} \;+\; \mathscr{E}_\pm(\boldsymbol{U}, \boldsymbol{V}),\end{aligned} \tag{5.12}$$

$$\begin{aligned}&\mathscr{E}_\pm(\boldsymbol{U}, \boldsymbol{V}) \\ &:= \int_{\mathscr{C}^\pm} \Big[\lambda\langle \mathrm{div}_{\mathscr{C}}\, \boldsymbol{U}, \mathrm{div}_{\mathscr{C}}\, \boldsymbol{V}\rangle + 2\mu\, \langle \mathrm{Def}_{\mathscr{C}}\, \boldsymbol{U}, \mathrm{Def}_{\mathscr{C}}\, \boldsymbol{V}\rangle + \mathscr{B}\langle \boldsymbol{U}, \boldsymbol{V}\rangle\Big]\, dS\,, \\ &\mathscr{D}_{\nu_\Gamma} := \sum_{m=1}^{n} \nu_\Gamma^j \mathscr{D}_j, \quad \boldsymbol{U} = \sum_{j=1}^{n} U_j^0 \boldsymbol{d}^j, \quad \boldsymbol{V} = \sum_{j=1}^{n} V_j^0 \boldsymbol{d}^j \in \mathbb{H}^1(\mathscr{C}) \cap \mathscr{V}(\mathscr{C})\,,\end{aligned} \tag{5.13}$$

Here the index $^\pm$ denotes the traces on Γ from the surfaces $\mathscr{C}^\pm$ and the scalar product of matrices is defined as follows:

$$\langle M, N\rangle := \mathrm{Tr}\big[MN^\top\big], \qquad M = [M_{jk}]_{n\times n}, \quad N = [N_{jk}]_{n\times n}. \tag{5.14}$$

Proof. We apply the integration by parts formula

$$\int_{\mathscr{C}^\pm} \langle (\mathscr{D}_j \boldsymbol{U}), \boldsymbol{V}\rangle\, dS = \pm \oint_\Gamma \nu_\Gamma^j \langle \boldsymbol{U}^\pm, \boldsymbol{V}^\pm\rangle\, d\mathfrak{s} + \int_{\mathscr{C}^\pm} \langle \boldsymbol{U}, (\mathscr{D}_j^* \boldsymbol{V})\rangle\, dS\,, \tag{5.15}$$

$$\boldsymbol{U},\ \boldsymbol{V} \in \mathbb{H}^1(\mathscr{C}^\pm) \qquad j = 1, \ldots, n,$$

proved in [DMM1] (cf. (1.5) for the formal adjoint $\mathscr{D}_j^*$), and proceed as follows

$$((\mathscr{L}_{\mathscr{C}}+\mathscr{B}I)\boldsymbol{U},\boldsymbol{V})_{\mathscr{C}^\pm} = \int_{\mathscr{C}^\pm}\langle\lambda\nabla_{\mathscr{C}}\operatorname{div}_{\mathscr{C}}\boldsymbol{U}+2\mu\operatorname{Def}^*_{\mathscr{C}}\operatorname{Def}_{\mathscr{C}}\boldsymbol{U}+\mathscr{B}\boldsymbol{U},\boldsymbol{V}\rangle\,dS$$

$$=\int_{\mathscr{C}^\pm}\langle\lambda\nabla_{\mathscr{C}}\operatorname{div}_{\mathscr{C}}\boldsymbol{U}+2\mu\sum_{j=1}^{n}\left\{\mathscr{D}_j^*\mathfrak{D}_{jk}(\boldsymbol{U})\right\}_{k=1}^{n}+\mathscr{B}\boldsymbol{U},\boldsymbol{V}\rangle\,dS$$

$$=\int_{\mathscr{C}^\pm}\langle\lambda\nabla_{\mathscr{C}}\operatorname{div}_{\mathscr{C}}\boldsymbol{U}-2\mu\sum_{j=1}^{n}\left\{(\mathscr{D}_j+\nu_\Gamma^j\mathscr{H}^0_{\mathscr{C}})\mathfrak{D}_{jk}(\boldsymbol{U})\right\}_{k=1}^{n}+\mathscr{B}\boldsymbol{U},\boldsymbol{V}\rangle\,dS$$

$$=\pm\oint_{\Gamma}\langle\overset{\pm}{\gamma}_N\boldsymbol{U}(\tau),\overset{\pm}{\gamma}_D\boldsymbol{V}(\tau)\rangle\,d\mathfrak{s}\;+\;\mathscr{E}_\pm(\boldsymbol{U},\boldsymbol{V}),$$

where $\mathfrak{T}_{\mathscr{C}}(\boldsymbol{\nu}_\Gamma,\mathscr{D})$ is gives by formula (5.8). We have applied formulae (1.5), (0.21), (0.15) and the equalities

$$\langle\pi_{\mathscr{S}}\boldsymbol{U},\boldsymbol{V}\rangle=\langle\boldsymbol{U},\boldsymbol{V}\rangle,\qquad\langle(\mathscr{D}_j^{\mathscr{C}})^*\boldsymbol{U},\boldsymbol{V}\rangle=\langle\pi_{\mathscr{S}}\mathscr{D}_j^*\boldsymbol{U},\boldsymbol{V}\rangle=\langle\mathscr{D}_j^*\boldsymbol{U},\boldsymbol{V}\rangle\quad\forall\,\boldsymbol{V}\in\mathscr{V}.$$

To obtain another representation (5.9) of $\mathfrak{T}_{\mathscr{C}}(\boldsymbol{\nu}_\Gamma,\mathscr{D})$ we start by second representation of $\mathscr{L}_{\mathscr{C}}$ in (0.23) and proceed similarly. □

6. The Dirichlet BVP for the Lamé equation

Throughout this section $\mathscr{C}$ is a C^2-smooth hypersurface with the Lipschitz boundary $\Gamma=\partial\mathscr{C}$.

Theorem 6.1. *The Dirichlet problem* (5.4) *has a unique solution* $\boldsymbol{U}\in\mathbb{H}^1(\mathscr{C})$ *for arbitrary data* $\boldsymbol{F}\in\widetilde{\mathbb{H}}^{-1}(\mathscr{C})$ *and* $\boldsymbol{G}\in\mathbb{H}^{1/2}(\Gamma)$.

The proof will be exposed at the end of the section after we prove some auxiliary results.

Lemma 6.2. (Gårding's inequality "with boundary condition"). *The Lamé operator*

$$\mathscr{L}_{\mathscr{C}}(\mathscr{X},\mathscr{D})\;:\;\widetilde{\mathbb{H}}^1(\mathscr{C})\to\mathbb{H}^{-1}(\mathscr{C})\tag{6.1}$$

is positive definite: there exists some constant $C>0$ *such that*

$$(\mathscr{L}_{\mathscr{C}}(\mathscr{X},\mathscr{D})\boldsymbol{U},\boldsymbol{U})_{\mathscr{C}}\geq C\big\|\boldsymbol{U}\big|\mathbb{H}^1(\mathscr{C})\big\|^2\qquad\forall\,\boldsymbol{U}\in\widetilde{\mathbb{H}}^1(\mathscr{C}).\tag{6.2}$$

Proof. Due to (3.11) inequality (6.1) holds for all $\boldsymbol{U}\in\mathbb{H}^1_{\mathscr{R}}(\mathscr{S})$, i.e., for $\boldsymbol{U}\in\mathbb{H}^1(\mathscr{S})$ and $\boldsymbol{U}\notin\mathscr{R}(\mathscr{S})$. Since $\boldsymbol{U}\in\widetilde{\mathbb{H}}^1(\mathscr{C})$ due to the strong unique continuation from the boundary (cf. Lemma 3.12), all Killing's vector fields $\boldsymbol{K}\in\widetilde{\mathbb{H}}^1(\mathscr{C})$ are identically 0. Therefore, (3.11) holds for all $\boldsymbol{U}\in\widetilde{\mathbb{H}}^1(\mathscr{C})$. □

Corollary 6.3. *The Lamé operator* $\mathscr{L}_{\mathscr{C}}(\mathscr{X},\mathscr{D})$ *in* (6.1) *is invertible.*

Proof. From the inequality (6.2) follows that $\mathscr{L}_{\mathscr{C}}(\mathscr{X},\mathscr{D})$ is normally solvable (has the closed range) and the trivial kernel $\operatorname{Ker}\mathscr{L}_{\mathscr{C}}(\mathscr{X},\mathscr{D})=\{0\}$. Since $\mathscr{L}_{\mathscr{C}}(\mathscr{X},\mathscr{D})$ is self-adjoint, the co-kernel (the kernel of the adjoint operator) is trivial as well $\operatorname{Ker}\mathscr{L}^*_{\mathscr{C}}(\mathscr{X},\mathscr{D})=\operatorname{Ker}\mathscr{L}_{\mathscr{C}}(\mathscr{X},\mathscr{D})=\{0\}$. Therefore $\mathscr{L}_{\mathscr{C}}(\mathscr{X},\mathscr{D})$ is invertible. □

Definition 6.4. (see [LM1, Ch.2, § 1.4]). A partial differential operator

$$\mathbf{A}(x,\mathscr{D}):=\sum_{|\alpha|\le m}a_\alpha(x)\nabla^\alpha_{\mathscr{C}},\quad \nabla^\alpha_{\mathscr{C}}u=\mathscr{D}_1^{\alpha_1}\cdots\mathscr{D}_n^{\alpha_n},\quad a_\alpha\in C(\mathscr{C},C^{N\times N})\tag{6.3}$$

is called normal on Γ if

$$\inf|\det\mathscr{A}_0(\mathscr{X},\boldsymbol{\nu}(\mathscr{X}))|\ne 0,\quad \mathscr{X}\in\Gamma,\quad |\xi|=1,\tag{6.4}$$

where $\mathscr{A}_0(x,\xi)$ is the homogeneous principal symbol of $\mathbf{A}$

$$\mathscr{A}_0(x,\xi):=\sum_{|\alpha|=m}a_\alpha(x)(-i\xi)^\alpha,\quad x\in\overline{\mathscr{C}},\quad \xi\in\mathbb{R}^n.\tag{6.5}$$

Definition 6.5. A system $\{\mathbf{B}_j(\mathscr{X},D)\}_{j=0}^{k-1}$ of differential operators with matrix $N\times N$ coefficients is called a Dirichlet system of order k if all participating operators are normal on Γ (see Definition 6.4) and $\operatorname{ord}\mathbf{B}_j=j$, $j=0,1,\dots,k-1$.

Let us assume $\mathscr{C}$ is k-smooth and $m\le k$ $(m,k=1,2,\dots)$ and define the trace operator (cf. (5.10)):

$$\mathscr{R}_m u:=\{\gamma_\Gamma\mathbf{B}_1u,\dots,\gamma_\Gamma\mathbf{B}_mu\}^\top,\qquad u\in\mathbb{C}_0^k(\overline{\mathscr{C}}).\tag{6.6}$$

Proposition 6.6. *Let $\mathscr{C}$ be k-smooth, $1\le p\le\infty$, $m=1,2,\dots$, $m\le k$ and $m<s-1/p\notin\mathbb{N}_0$. The trace operator*

$$\mathscr{R}_m\ :\ \mathbb{H}^s_p(\overline{\mathscr{C}})\to\bigotimes_{j=0}^{m}\mathbb{W}_p^{s-1/p-j}(\Gamma),\tag{6.7}$$

where $\mathbb{W}^r_p(\overline{\mathscr{C}})=\mathbb{B}^r_{p,p}(\overline{\mathscr{C}})$ is the Sobolev-Slobodecki-Besov space (cf. [Tr1] for details) is a retraction, i.e., is continuous and has a continuous right inverse, called a coretraction

$$\begin{aligned}(\mathscr{R}_m)^{-1}\ &:\ \bigotimes_{j=0}^{m}\mathbb{W}_p^{s-1/p-j}(\mathscr{S})\to\mathbb{H}^s_p(\overline{\Omega})\\ \mathscr{R}_m(\mathscr{R}_m)^{-1}\Phi=\Phi,&\qquad\forall\Phi\in\bigotimes_{j=0}^{m}\mathbb{W}_p^{s-1/p-j}(\mathscr{S}).\end{aligned}\tag{6.8}$$

Proof. The result was proved in [Tr1, Theorem 2.7.2, Theorem 3.3.3] for a domain $\Omega\subset\mathbb{R}^{n-1}$ and the classical Dirichlet trace operator $\mathscr{R}_mu:=\{\gamma_\Gamma\partial_{\boldsymbol{\nu}}u,\dots,\gamma_\Gamma\partial^m_{\boldsymbol{\nu}}u\}^\top$. In [Du3] the theorem was proved for a domain $\Omega\subset\mathbb{R}^{n-1}$ and for arbitrary trace operator $\mathscr{R}_mu$.

A surface $\mathscr{C}=\cup_{j=1}^N\mathscr{C}_j$ is covered by a finite number of local coordinate charts $\varkappa_j\ :\ \Omega_j\to\mathscr{C}_j$, $\Omega_j\subset\mathbb{R}^{n-1}$. After transformation, the Dirichlet trace operator $\mathscr{R}_mu$ on a portion $\mathscr{C}_j$ of the surface transform into another Dirichlet trace operator on the coordinate domains Ω_j. Therefore, we prove the assertion locally on each coordinate chart $\mathscr{C}_j \subset \mathscr{C}$ and, by applying a partition of unity, extend it to the entire surface $\mathscr{C}$. □

Proof of Theorem 6.1. Let $\widetilde{\boldsymbol{G}} = (\mathscr{R}_0)^{-1}\boldsymbol{G} \in \mathbb{H}^1(\mathscr{C})$ be the continuation of the Dirichlet boundary data $\boldsymbol{G} \in \mathbb{H}^{1/2}(\Gamma)$ from BVP (5.4) into the surface $\mathscr{C}$ from the boundary Γ, found with the help of a coretraction from Proposition 6.6. Then the Dirichlet BVP

$$\begin{cases} (\mathscr{L}_{\mathscr{C}}(\mathscr{X},\mathscr{D})\widetilde{\boldsymbol{U}})(\mathscr{X}) = \boldsymbol{F}_0(\mathscr{X}), & \mathscr{X} \in \mathscr{C}, \\ \widetilde{\boldsymbol{U}}^+(\tau) = 0, & \tau \in \Gamma, \end{cases} \tag{6.9}$$

$$\boldsymbol{F}_0 := \boldsymbol{F} - \mathscr{L}_{\mathscr{C}}(\mathscr{X},\mathscr{D})\widetilde{\boldsymbol{G}} \in \widetilde{\mathbb{H}}^{-1}(\mathscr{C}),$$

is an equivalent reformulation of BVP (5.4) and the solutions are related by the equality $\widetilde{\boldsymbol{U}} := \boldsymbol{U} - \widetilde{\boldsymbol{G}}$. On the other hand, since

$$\widetilde{\mathbb{H}}^{-1}(\mathscr{C}) := \{\boldsymbol{U} \in \mathbb{H}^{-1}(\mathscr{C}) \ : \ \boldsymbol{U}^+ = 0\},$$

the solvability of BVP (6.9) is equivalent to the invertibility of the operator $\mathscr{L}_{\mathscr{C}}(\mathscr{X},\mathscr{D})$ in (6.1). Now the unique solvability of BVP (6.9) (and of the equivalent BVP (5.4)) follows from Corollary 6.3. □

7. The Neumann BVP for the Lamé equation

Throughout this section $\mathscr{C}$ is a C^2-smooth hypersurface with the Lipschitz boundary $\Gamma = \partial\mathscr{C}$.

Theorem 7.1. *The Neumann problem* (5.6) *has a solution* $\boldsymbol{U} \in \mathbb{H}^1(\mathscr{C})$ *only for those right-hand sides* $\boldsymbol{F} \in \widetilde{\mathbb{H}}^{-1}(\Gamma)$ *and* $\boldsymbol{H} \in \mathbb{H}^{-1/2}(\Gamma)$ *which satisfy the equality*

$$\int_{\mathscr{C}} \boldsymbol{F}(\mathscr{X})\boldsymbol{K}(\mathscr{X})dS = \oint_{\Gamma} \boldsymbol{H}(\tau)\gamma_D^+\boldsymbol{K}(\tau)d\mathfrak{s} \qquad \forall \boldsymbol{K} \in \mathscr{R}(\mathscr{C}). \tag{7.1}$$

If the condition (7.1) *holds, the Neumann problem has a general solution* $\boldsymbol{U} = \boldsymbol{U}^0 + \boldsymbol{K} \in \mathbb{H}^1(\mathscr{C})$, *where* $\boldsymbol{U}^0 \in \mathbb{H}^1(\mathscr{C})$ *is a particular solution and* $\boldsymbol{K} \in \mathscr{R}(\mathscr{C})$ *is a Killing's vector field.*

The proof will be exposed at the end of the section after we prove some auxiliary results. The proof is based on the celebrated Lax-Milgram lemma.

Lemma 7.2. (Lax-Milgram). *Let* $\mathfrak{B}$ *be a Banach space and* $A(\varphi,\psi)$ *be a bilinear* $A(\cdot,\cdot) : \mathfrak{B} \times \mathfrak{B} \to \mathbb{R}$, *positive definite form: the inequality*

$$A(\varphi,\varphi) \geq C\|\varphi|\mathfrak{B}\|^2 \tag{7.2}$$

holds for some constant $C > 0$ *and all* $\varphi \in \mathfrak{B}$. *Further let* $L(\cdot) : \mathfrak{B} \to \mathbb{R}$ *be a continuous linear form (a functional).*

A linear equation

$$A(\varphi,\psi) = L(\psi) \tag{7.3}$$

has a unique solution $\varphi \in \mathfrak{B}$ *for arbitrary* $\psi \in \mathfrak{B}$.

Proof. The proof can be retrieved from many sources (cf., e.g., [Ci3, § 6.3]), mostly for symmetric forms (we have dropped this requirement). For a non-symmetric form the proof can be found in the original paper [LaM1]. □

Lemma 7.3. *If* $\boldsymbol{U} \in \mathbb{H}^1_p(\mathscr{C})$, $1 < p < \infty$, *is a solution to the first equations in* (5.6), *then the Neumann trace on the boundary exists and* $\gamma_N^+ \boldsymbol{U} = (\mathfrak{T}_{\mathscr{C}}(\boldsymbol{\nu}_\Gamma, \mathscr{D})\ \boldsymbol{U})^+ \in \mathbb{H}^{-1/p}_p(\Gamma)$.

Proof. For $\mathscr{B} = 0$, $\mathscr{C}^+ = \mathscr{C}$ the Green's formulae (5.12), (5.13) become:

$$(\mathscr{L}_{\mathscr{C}} \boldsymbol{U}, \boldsymbol{V})_{\mathscr{C}} = (\gamma_N^+ \boldsymbol{U}, \gamma_D^+ \boldsymbol{V})_\Gamma + \mathscr{E}(\boldsymbol{U}, \boldsymbol{V}), \tag{7.4}$$

$$\mathscr{E}(\boldsymbol{U}, \boldsymbol{V}) = \lambda(\operatorname{div}_{\mathscr{C}} \boldsymbol{U}, \operatorname{div}_{\mathscr{C}} \boldsymbol{V})_{\mathscr{C}} + 2\mu\, (\operatorname{Def}_{\mathscr{C}} \boldsymbol{U}, \operatorname{Def}_{\mathscr{C}} \boldsymbol{V})_{\mathscr{C}} \tag{7.5}$$

Introducing the value $\mathscr{L}_{\mathscr{C}}(\mathscr{X}, \mathscr{D})\ \boldsymbol{U} = \boldsymbol{F}$ into the Green formula (7.4) we rewrite it

$$(\gamma_N^+ \boldsymbol{U}, \gamma_D^+ \boldsymbol{V})_\Gamma = (\boldsymbol{F}, \boldsymbol{V})_{\mathscr{C}} - \mathscr{E}(\boldsymbol{U}, \boldsymbol{V}),$$

where $\boldsymbol{V} \in \mathbb{H}^1_p(\mathscr{C})$ is arbitrary. The bilinear forms $(\boldsymbol{F}, \boldsymbol{V})_{\mathscr{C}}$ and $\mathscr{E}(\boldsymbol{U}, \boldsymbol{V})$ are continuous for $\boldsymbol{F} \in \widetilde{\mathbb{H}}^{-1}_p(\mathscr{C})$ and $\boldsymbol{U} \in \mathbb{H}^1_p(\mathscr{C})$, $\boldsymbol{V} \in \mathbb{H}^1_{p'}(\mathscr{C})$, $p' := p/(p-1)$; the bilinear form $\big((\mathfrak{T}_{\mathscr{C}}(\boldsymbol{\nu}_\Gamma, \mathscr{D})\boldsymbol{U})^+, \boldsymbol{V}^+\big)_\Gamma$ is well defined and, by a duality argument, $\big(\mathfrak{T}_{\mathscr{C}}(\boldsymbol{\nu}_\Gamma, \mathscr{D})\boldsymbol{U}\big)^+ \in \mathbb{H}^{-1/p}_p(\Gamma)$ since $\boldsymbol{V}^+ \in \mathbb{H}^{1-1/p'}_{p'}(\mathscr{C}) = \mathbb{H}^{1/p}_{p'}(\mathscr{C})$ is arbitrary. □

Lemma 7.4. *The condition* (7.1) *is necessary for the Neumann problem* (5.6) *to have a solution* $\boldsymbol{U} \in \mathbb{H}^1(\mathscr{C})$.

Proof. First note that for a Killing's vector field $\boldsymbol{K} \in \mathscr{R}(\mathscr{C})$,

$$\mathscr{L}_{\mathscr{C}}(\mathscr{X}, \mathscr{D})\boldsymbol{K} = 0 \quad \text{and} \qquad \gamma_N^+ \boldsymbol{K} = \big(\mathfrak{T}_{\mathscr{C}}(\boldsymbol{\nu}_\Gamma, \mathscr{D})\boldsymbol{K}\big)^+ = 0. \tag{7.6}$$

Indeed, if $\boldsymbol{K} \in \mathscr{R}(\mathscr{C})$ is naturally extended to $\widetilde{\boldsymbol{K}} \in \mathscr{R}(\mathscr{S})$, then $\mathscr{L}_{\mathscr{C}}(\mathscr{X}, \mathscr{D})\boldsymbol{K}(\mathscr{X}) = \mathscr{L}_{\mathscr{C}}(\mathscr{X}, \mathscr{D})\widetilde{\boldsymbol{K}}(\mathscr{X}) = 0$ for $\mathscr{X} \in \mathscr{C}$ (cf. (3.9)) and the first equality follows.

The second equality in (7.6) follows from (5.8) if we recall that $\operatorname{Def}_{\mathscr{C}}(\boldsymbol{U}) = 0$ (cf. Definition 3.1) and this also implies $\operatorname{div}_{\mathscr{C}} \boldsymbol{U} = 0$ (cf. (4.6)).

Since $\operatorname{div}_{\mathscr{S}} \boldsymbol{K} = \operatorname{Def}_{\mathscr{S}} \boldsymbol{K} = 0$ for $\boldsymbol{K} \in \mathscr{R}(\mathscr{C})$, we get

$$\begin{aligned} \mathscr{E}(\boldsymbol{K}, \boldsymbol{U}) &= \mathscr{E}(\boldsymbol{U}, \boldsymbol{K}) \\ &= \int_{\mathscr{C}} \Big[\lambda \langle \operatorname{div}_{\mathscr{C}} \boldsymbol{U}, \operatorname{div}_{\mathscr{C}} \boldsymbol{K}\rangle + 2\mu\, \langle \operatorname{Def}_{\mathscr{C}} \boldsymbol{U}, \operatorname{Def}_{\mathscr{C}} \boldsymbol{K}\rangle\Big]\, dS = 0 \\ &\qquad \text{for all} \quad \boldsymbol{U} \in \mathbb{H}^1(\mathscr{C}) \quad \text{and all} \quad \boldsymbol{K} \in \mathscr{R}(\mathscr{C}). \end{aligned} \tag{7.7}$$

Introducing into the Green formula (5.12) $\mathscr{B} = 0$, $\boldsymbol{F} = \mathscr{L}_{\mathscr{C}}(\mathscr{X}, \mathscr{D})\boldsymbol{U}$, $\boldsymbol{V} = \boldsymbol{K} \in \mathscr{R}(\mathscr{C})$ and the obtained equality, we get the claimed orthogonality condition (7.1). □

Lemma 7.5. *The bilinear form* (*cf.* (7.4) *and* (7.5))

$$\mathbb{A}_N(\boldsymbol{U}, \boldsymbol{V}) := (\mathscr{L}_{\mathscr{C}}(\mathscr{X}, \mathscr{D})\boldsymbol{U}, \boldsymbol{V})_{\mathscr{C}} - (\gamma_N^+ \boldsymbol{U}, \gamma_D^+ \boldsymbol{V})_\Gamma = \mathscr{E}(\boldsymbol{U}, \boldsymbol{V}) \tag{7.8}$$

is well defined, symmetric $\mathbb{A}_N(\boldsymbol{U},\boldsymbol{V}) = \mathbb{A}_N(\boldsymbol{V},\boldsymbol{U})$ *for all* $\boldsymbol{U}, \boldsymbol{V} \in \mathbb{H}^1(\mathscr{C})$ *and non-negative* $\mathbb{A}_N(\boldsymbol{U},\boldsymbol{U}) \geq 0$ *for* $\boldsymbol{U} \in \mathbb{H}^1(\mathscr{S})$ *(cf.* (5.13)*). Moreover, the form is positive definite*

$$\mathbb{A}_N(\boldsymbol{U},\boldsymbol{U}) \geq M_3 \big\|\boldsymbol{U}\big|\mathbb{H}^1(\mathscr{S})\big\|^2 \qquad \forall \boldsymbol{U} \in \mathbb{H}^1_{\mathscr{R}}(\mathscr{S}) \tag{7.9}$$

on the orthogonal complement $\mathbb{H}^1_{\mathscr{R}}(\mathscr{S})$ *to the finite-dimensional subspace of Killing's vector fields* $\mathscr{R}(\mathscr{C})$ *in the Hilbert-Sobolev space* $\mathbb{H}^1(\mathscr{C})$.

Proof. The estimate

$$\big|\mathbb{A}_N(\boldsymbol{U},\boldsymbol{V})\big| = \big|\mathscr{E}(\boldsymbol{U},\boldsymbol{V})\big| \leq \big\|\boldsymbol{U}\big|\mathbb{H}^1(\mathscr{S})\big\| \, \big\|\boldsymbol{V}\big|\mathbb{H}^1(\mathscr{S})\big\|$$

follows from the definition of the form $\mathscr{E}(\boldsymbol{U},\boldsymbol{V})$ in (5.13) and proves that $\mathbb{A}_N(\boldsymbol{U},\boldsymbol{V})$ is well defined. Moreover, the equality proves that the form is symmetric and non-negative

$$\mathbb{A}_N(\boldsymbol{U},\boldsymbol{V}) = \mathscr{E}(\boldsymbol{U},\boldsymbol{V}) = \mathscr{E}(\boldsymbol{V},\boldsymbol{U}) = \mathbb{A}_N(\boldsymbol{V},\boldsymbol{U}),$$

$$\mathbb{A}_N(\boldsymbol{U},\boldsymbol{U}) = \mathscr{E}(\boldsymbol{U},\boldsymbol{U}) \geq 0.$$

From (5.12) and (5.13) follows

$$\mathbb{A}_N(\boldsymbol{U},\boldsymbol{U}) = \mathscr{E}(\boldsymbol{U},\boldsymbol{U}) = \lambda\big\|\mathrm{div}_{\mathscr{C}}\boldsymbol{U}\big|\mathbb{L}_2(\mathscr{S})\big\|^2 + 2\mu\big\|\mathrm{Def}_{\mathscr{C}}\,\boldsymbol{U}\big|\mathbb{L}_2(\mathscr{S})\big\|^2$$

$$\geq 2\mu\big\|\mathrm{Def}_{\mathscr{C}}\,\boldsymbol{U}\big|\mathbb{L}_2(\mathscr{S})\big\|^2 \geq 2\mu\, c^2\big\|\boldsymbol{U}\big|\mathbb{H}^1(\mathscr{S})\big\|^2 \qquad \forall \boldsymbol{U} \in \mathbb{H}^1_{\mathscr{R}}(\mathscr{S}) \tag{7.10}$$

and accomplishes the proof. □

Proof of Theorem 7.1. The space of Killing's vector fields $\mathscr{R}(\mathscr{S})$ is finite-dimensional and consists of continuous vector-fields with bounded second derivatives (these fields are actually as smooth as the surface $\mathscr{C}$, i.e., are infinitely smooth if $\mathscr{S}$ is infinitely smooth; see Theorem 3.5). Let $\boldsymbol{K}_1, \dots, \boldsymbol{K}_m$ be the finite-dimensional orthonormal basis in $\mathscr{R}(\mathscr{C})$, $(\boldsymbol{K}_j, \boldsymbol{K}_r)_{\mathscr{C}} = \delta_{jr}$, $j, r = 1, \dots, m$. Consider the finite rank smoothing operator $\boldsymbol{T}\boldsymbol{U}$ introduced in (4.7). As we already know the operator $\boldsymbol{T}$ is symmetric and non-negative:

$$(\boldsymbol{T}\boldsymbol{U},\boldsymbol{V})_{\mathscr{C}} = (\boldsymbol{T}\boldsymbol{V},\boldsymbol{U})_{\mathscr{C}}. \quad (\boldsymbol{T}\boldsymbol{U},\boldsymbol{U})_{\mathscr{C}} = \sum_{j=1}^{m} (\boldsymbol{U},\boldsymbol{K}_j)^2_{\mathscr{C}} \geq 0 \tag{7.11}$$
$$\forall \boldsymbol{U}, \, \boldsymbol{V} \in \mathbb{H}^1(\mathscr{C}).$$

Consider the modified bilinear form

$$\mathbb{A}^{\#}_N(\boldsymbol{U},\boldsymbol{V}) := ((\mathscr{L}_{\mathscr{C}}(t,\mathscr{D}) + \boldsymbol{T})\boldsymbol{U},\boldsymbol{V})_{\mathscr{C}} - (\gamma^+_N \boldsymbol{U}, \gamma^+_D \boldsymbol{V})_\Gamma$$

$$= \mathscr{E}(\boldsymbol{U},\boldsymbol{V}) + (\boldsymbol{T}\boldsymbol{U},\boldsymbol{V})_{\mathscr{C}} \qquad \boldsymbol{U}, \, \boldsymbol{V} \in \mathbb{H}^1(\mathscr{C}) \tag{7.12}$$

(cf. (7.4)). The form is symmetric because both summands are

$$\mathbb{A}^{\#}_N(\boldsymbol{U},\boldsymbol{V}) = \mathscr{E}(\boldsymbol{U},\boldsymbol{V}) + (\boldsymbol{T}\boldsymbol{U},\boldsymbol{V})_{\mathscr{C}} = \mathscr{E}(\boldsymbol{V},\boldsymbol{U}) + (\boldsymbol{T}\boldsymbol{V},\boldsymbol{U})_{\mathscr{C}} = \mathbb{A}^{\#}_N(\boldsymbol{V},\boldsymbol{U})$$

(cf. Lemma 7.5 and the first equality in (7.11)).

Moreover, the corresponding quadratic form is strongly positive

$$\mathbb{A}^{\#}_N(\boldsymbol{U},\boldsymbol{U}) = \mathscr{E}(\boldsymbol{U},\boldsymbol{U}) + (\boldsymbol{T}\boldsymbol{U},\boldsymbol{U})_{\mathscr{C}} \geq C\big\|\boldsymbol{U}\big|\mathbb{H}^1(\mathscr{C})\big\|\tag{7.13}$$

for some $C > 0$. Indeed, $\mathbb{A}^{\#}_N(\boldsymbol{U},\boldsymbol{U}) = 0$ due to the positivity of the summands implies: $\mathscr{E}(\boldsymbol{U},\boldsymbol{U}) = 0$, and further $\boldsymbol{U} \in \mathscr{R}(\mathscr{C})$ (cf. Lemma 7.5), $(\boldsymbol{T}\boldsymbol{U},\boldsymbol{U})_{\mathscr{C}} = 0$ and further $(\boldsymbol{U},\boldsymbol{K}_j) = 0$ for all $j = 1,\dots,m$. Then $\boldsymbol{U} = \sum_{j=1}^m (\boldsymbol{U},\boldsymbol{K}_j)\boldsymbol{K}_j = 0$. A non-negative symmetric form with the property $\mathbb{A}^{\#}_N(\boldsymbol{U},\boldsymbol{U}) = 0$ if and only if $\boldsymbol{U} = 0$ is positive definite.

According to Lax-Milgram's Lemma 7.2 the equation

$$\mathbb{A}^{\#}_N(\boldsymbol{U},\boldsymbol{V}) = (\boldsymbol{F},\boldsymbol{V})_{\mathscr{C}} - (\boldsymbol{H},\boldsymbol{V}^+)_{\Gamma}\tag{7.14}$$

has a unique solution $\boldsymbol{U} \in \mathbb{H}^1(\mathscr{C})$ for all $\boldsymbol{V} \in \mathbb{H}^1(\mathscr{C})$. This solves the problem

$$\begin{cases} (\mathscr{L}_{\mathscr{C}}(t,\mathscr{D})\boldsymbol{U})(t) + \boldsymbol{T}\boldsymbol{U}(t) = \boldsymbol{F}(t), & t \in \mathscr{C}, \\ (\mathfrak{T}_{\mathscr{C}}(\boldsymbol{\nu}_{\Gamma},\mathscr{D})\,\boldsymbol{U})^+(\tau) = \boldsymbol{H}(\tau), & \tau \in \Gamma, \end{cases}\tag{7.15}$$

which is a modified Neumann's problem (5.6).

Now assume that the vector functions $\boldsymbol{F} \in \widetilde{\mathbb{H}}^{-1}(\mathscr{C})$ and $\boldsymbol{H} \in \mathbb{H}^{-1/2}(\Gamma)$ satisfy the orthogonality condition (7.1) from Theorem 7.1 and $\boldsymbol{U}^0 \in \mathbb{H}^1(\mathscr{C})$ be a solution of (7.15). Since

$$(\boldsymbol{T}\boldsymbol{U}^0,\boldsymbol{K}_k)_{\mathscr{C}} = (\boldsymbol{U}^0,\boldsymbol{K}_k)_{\mathscr{C}}, \quad \mathbb{A}_N(\boldsymbol{U}^0,\boldsymbol{K}_k) = \mathscr{E}(\boldsymbol{U}^0,\boldsymbol{K}_k) = 0 \qquad k = 1,2,\dots,m$$

(cf. (7.5)) from (7.14) we get

$$\begin{aligned} 0 = (\boldsymbol{F},\boldsymbol{K}_k)_{\mathscr{C}} - (\boldsymbol{H},\boldsymbol{K}_k)_{\Gamma} = \mathbb{A}^{\#}_N(\boldsymbol{U}^0,\boldsymbol{K}_k) &= \mathbb{A}_N(\boldsymbol{U}^0,\boldsymbol{K}_k) + (\boldsymbol{T}\boldsymbol{U}^0,\boldsymbol{K}_k)_{\mathscr{C}} \\ &= (\boldsymbol{U}^0,\boldsymbol{K}_k)_{\mathscr{C}} \qquad k = 1,2,\dots,m. \end{aligned}$$

Therefore, $\boldsymbol{T}\boldsymbol{U}^0 = \sum_{k=1}^m (\boldsymbol{U}^0,\boldsymbol{K}_k)_{\mathscr{C}}\boldsymbol{K}_k = 0$ and BVP (7.15), which is uniquely solvable, coincides with BVP (5.6) provided that the right-hand sides satisfy the orthogonality condition (7.1). Since the kernel of BVP (5.6) coincides with the space of Killing's vector fields $\mathscr{R}(\mathscr{C})$, a general solution of BVP (5.6) has the form $\boldsymbol{U} = \boldsymbol{U}^0 + \boldsymbol{K}$ with arbitrary $\boldsymbol{K} \in \mathscr{R}(\mathscr{C})$. □

Remark 7.6. If the surface is smooth, by invoking a local fundamental solution to the Lamé equation (cf. Corollary 4.3) and the potential method, it is possible to prove that BVPs (5.4), (5.6) and (7.17) have the same solvability properties if the constraints (5.4) and (5.6) are replaced by the following non-classical constraints

$$\boldsymbol{F} \in \widetilde{\mathbb{H}}^{s-2}_p(\mathscr{C}), \quad \boldsymbol{G} \in \mathbb{H}^{s-1/p}_p(\Gamma), \quad \boldsymbol{H} \in \mathbb{H}^{s-1/p-1}_p(\Gamma), \tag{7.16}$$
$$1 < p < \infty, \quad s \geq 1$$

and $\boldsymbol{U} \in \mathbb{H}^s_p(\mathscr{C})$ is unknown.

Moreover, by the potential method we can investigate *the mixed problem*: find the tangential displacement vector field $\boldsymbol{U} \in \mathbb{H}^s_p(\mathscr{C})$, prescribed on the part Γ_D of the boundary, while on the remainder part $\Gamma_N := \Gamma\setminus\Gamma_D$ is prescribed the traction:

$$\begin{cases} (\mathscr{L}_{\mathscr{C}}(\mathscr{X},\mathscr{D})\, \boldsymbol{U})(\mathscr{X}) = \boldsymbol{F}(\mathscr{X}), & \mathscr{X}\in\mathscr{C},\\ \boldsymbol{U}^+(\tau) = \boldsymbol{G}_0(\tau), & \tau\in\Gamma_D,\\ (\mathfrak{T}_{\mathscr{C}}(\boldsymbol{\nu}_\Gamma,\mathscr{D})\, \boldsymbol{U})^+(\tau) = \boldsymbol{H}_0(\tau), & \tau\in\Gamma_N\,. \end{cases} \tag{7.17}$$

The unique solvability of the mixed problem follows under the following conditions

$$\boldsymbol{F}\in\widetilde{\mathbb{H}}^{s-2}_p(\mathscr{C}), \quad \boldsymbol{G}_0\in\mathbb{H}^{s-1/p}_p(\Gamma_D), \quad \boldsymbol{H}_0\in\mathbb{H}^{s-1/p-1}_p(\Gamma_N), \tag{7.18}$$

$$1<p<\infty, \quad s\geq 1, \quad \frac{1}{p}-\frac{1}{2}<s<\frac{1}{p}+\frac{1}{2}.$$

References

[AMM1] P.M. Adler, A.E. Malevich, V. Mityushev, Macroscopic diffusion on rough surfaces. *Phys. Rev.* **E (3) 69** (2004), 1–10.

[Ag1] M. Agranovich. Elliptic singular integrodifferential operators, *Uspekhi Matematicheskikh Nauk* **20**, 5(125), 1965, 3–120 (Russian).

[AN1] H. Ammari, J.C. Nédélec, Generalized impedance boundary conditions for the Maxwell equations as singular perturbations problems, *Communications on Partial Differential Equations* **24** (1999), 821–849.

[AG1] C. Amrouche, V. Girault, Decomposition of vector spaces and applications to the Stokes problem in arbitrary dimension, *Czech. Math. J.* **44**, 1994, 109–140.

[AC1] L. Andersson, P.T. Chrusciel, Cauchy data for vacuum Einstein equations and obstructions to smoothness of null infinity. *Phys. Rev. Lett.* **70**, 1993, 2829–2832.

[Ar1] R. Aris, *Vectors, Tensors, and the Basic Equations of Fluid Mechanics*, Prentice-Hall, Englewood Cliffs, N.J., 1962.

[AKS1] N. Aronszajn, A. Krzywicki, I. Szarski, A unique continuation theorem for exterior differential forms on Riemannian manifolds, *Ark. Math.* **4**, 417–453, 1962.

[Be1] A. Bendali, Numerical analysis of the exterior boundary value problem for the time-harmonic Maxwell equations by a boundary finite element method. Part I: The continuous problem, *Mathematics of Computation* **43** (1984), 29–46.

[BGS1] B.D. Bonner, I.G. Graham, V.P. Smyshlyaev, The computation of conical diffraction coefficients in high-frequency acoustic wave scattering. *SIAM J. Numer. Anal.* **43**, No. 3 (2005), 1202–1230.

[Ce1] M. Cessenat, *Mathematical methods in electromagnetism. Linear theory and applications*, Series on Advances in Mathematics for Applied Sciences **41**, World Scientific Publishing Co., Inc., River Edge, NJ, 1996.

[Ci1] P.G. Ciarlet, *Introduction to Linear Shell Theory*, Series in Applied Mathematics, Vol. 1, Gauthier-Villars, Éditions Scientifiques et Médicales Elsevier, Paris, North-Holland, Amsterdam, 1998.

[Ci2] P.G. Ciarlet, *Mathematical Elasticity, Vol. I: Three-dimensional elasticity*, Studies in Mathematics and Applications, 20, Elsevier, North-Holland, Amsterdam, 1988.

[Ci3] P.G. Ciarlet, *Mathematical Elasticity, Vol. III: Theory of Shells*, Studies in Mathematics and Applications, 29, Elsevier, North-Holland, Amsterdam, 2000.

[Ci4] P.G. Ciarlet, Mathematical modelling of linearly elastic shells. *Acta Numer.* **10** (2001), 103–214

[Co1] M. Costabel, A coercive bilinear form for Maxwell's equations, *J. Math. Anal. Appl.* **157** (1991), 527–541.

[DaL1] R. Dautray, J.-L. Lions, *Mathematical Analysis and Numerical Methods for Science and Technology*, Springer-Verlag, Berlin, 1990.

[Du1] R. Duduchava, On multidimensional singular integral operators. I–II, *J. Oper. Theory* **11** (1984), 41–76, 199–214.

[Du2] R.Duduchava, On singular integral operators on piecewise smooth lines; in: *Research Notes in Mathematics* **8**, *"Functional Theoretic Methods in Differential Equations"*, Pitman, London, pp. 109–131, 1976.

[Du3] R. Duduchava, The Green formula and layer potentials, *Integral Equations and Operator Theory* **41**, 2 (2001), 127–178.

[DMM1] R.Duduchava, D.Mitrea, M.Mitrea, Differential operators and boundary value problems on surfaces. *Mathematische Nachrichten* **279**, No. 9–10 (2006), 996–1023.

[DNS1] R. Duduchava, D. Natroshvili, E. Shargorodsky, Boundary value problems of the mathematical theory of cracks, *Proceedings of I. Vekua Institute of Applied Mathematics,* Tbilisi State University **39**, 1990, 68–84.

[DNS2] R. Duduchava, D. Natroshvili, E. Shargorodsky, Basic boundary value problems of thermoelasticity for anisotropic bodies with cuts. I–II, *Georgian Mathematical Journal* **2**, 1995, 123–140, 259–276.

[DS1] R. Duduchava, F.-O. Speck, Pseudodifferential operators on compact manifolds with Lipschitz boundary. *Mathematische Nachrichten* **160**, 1993, 149–191.

[GL1] N. Garofalo, F. Lin, Monotonicity properties of variational integrals, A_p weights and unique continuation, *Indiana U. Math. J.* **35**, 245–267, 1986.

[GL2] N. Garofalo, F. Lin, Unique continuation for elliptic operators: A geometric variational approach, *Comm. Pure. Appl. Math.* **40**, 347–366, 1987.

[Go1] A.L. Goldenveízer, *Theory of elastic thin shells*, Translation from the Russian edited by G. Herrmann. International Series of Monographs on Aeronautics and Astronautics, published for the American Society of Mechanical Engineers by Pergamon Press, Oxford-London-New York-Paris, 1961.

[Gu1] N. Günter, *Potential Theory and its Application to the Basic Problems of Mathematical Physics*, Fizmatgiz, Moscow 1953 (Russian. Translation in French: Gauthier-Villars, Paris 1994).

[Ha1] W. Haack, *Elementare Differentialgeometrie* (German), Lehrbücher und Monographien aus dem Gebiete der exakten Wissenschaften, 20, Basel-Stuttgart: Birkhäuser Verlag, VIII, 1955.

[Hr1] L. Hörmander, *The Analysis of Linear Partial Differential Operators*, **v. I–IV**, Springer-Verlag, Heidelberg 1983.

[HW1] G.C. Hsiao, W.L. Wendland, *Boundary Integral Equations*, Applied Mathematical Sciences Vol. **164**, Springer, Berlin Heidelberg 2008, ISBN 978-3-540-15284-2.

[Ka1] R. Kapanadze, On some properties of singular integral operators in spaces with norms, *Proceedings of the Tbilisi State University* **129**, 1968, 263–277.

[Ko1] V.A. Kondrat'ev, Boundary value problems for elliptic equations in domains with conical or angular points, *Trudy Moskov. Mat. Obshestva* **16** (1967), 209–292 (Russian).

[Ko2] W.T. Koiter, On the foundations of the linear theory of thin elastic shells. I, II. *Nederl. Akad. Wetensch. Proc.* Ser. B **73** (1970), 169–182; ibid **73** 1970 183–195.

[KGBB1] V. Kupradze, T. Gegelia, M. Basheleishvili, T. Burchuladze, *Three-Dimensional Problems of the Mathematical Theory of Elasticity and Thermoelasticity*, North-Holland, Amsterdam 1979 (Russian edition: Nauka, Moscow 1976).

[LaM1] P.D. Lax, A.N. Milgram, Parabolic equations. *Contributions to the theory of partial differential equations.* Annals of Mathematics Studies **33**, 167–190. Princeton University Press, Princeton, N.J., 1954.

[LM1] J.L. Lions, E. Magenes, *Non-homogeneous Boundary Value Problems and Applications I*, Springer-Verlag, Heidelberg 1972.

[MM1] U. Massari, M. Miranda, *Minimal Surfaces of Codimension One*, North-Holland Mathematics Studies **91**. Notas de Matematica, 95. Amsterdam – New York – Oxford: North-Holland, 1984.

[Mc1] W. McLean, *Strongly elliptic systems and boundary integral equations.* Cambridge: Cambridge University Press **XIV**, 2000.

[Mi1] K. Miller, Nonunique continuation for uniformly parabolic and elliptic equations in self divergence form with Hölder continuous coefficients, *Bull. AMS* **79**, 350–354, 1973.

[Miz1] S. Mizohata, *The theory of partial differential equations.* London: Cambridge University Press. XI,490 p. (1973).

[NH1] J.-C. Nećas, J. Hlasváćek, *Mathematical Theory of Elastic and Elasto-Plastic Bodies: An Introduction*, Studies in Applied Mechanics, 3. Elsevier Scientific Publishing Co., Amsterdam 1980.

[Pl1] A. Plis, On nonuniqueness in the Cauchy problem for an elliptic second order differential equation, *Bull. Acad. Sci. Polon., Ser. Sci. Math. Astro. Phys.* **11**, 95–100, 1963.

[Sch1] M. Schechter, *Modern methods in partial differential equations. An introduction*, McGraw-Hill Inc., New York 1977.

[Sh1] M. Shubin, *Pseudodifferential Operators and Spectral Theory.* 2nd ed. (English, translated from Russian), Berlin: Springer 2001.

[Sm1] V.P. Smyshlyaev, Diffraction by conical surfaces at high frequencies, *Wave Motion* **12**, No.4, 329–339 (1990).

[Ta1] M. Taylor, *Analysis on Morrey spaces and applications to Navier-Stokes and other evolution equations*, Comm. Partial Differential Equations **17** (1992), 1407–1456.

[Ta2] M. Taylor, *Partial Differential Equations*, Vol. I–III, Springer-Verlag, 1996.

[TZ1] R. Temam, M. Ziane, *Navier-Stokes equations in thin spherical domains*, Optimization Methods in Partial Differential Equations, Contemporary Math., AMS **209** (1997), 281–314.

[TW1] R. Temam, S.H. Wang, *Inertial forms of Navier-Stokes equations on the sphere*, J. Funct. Anal. **117** (1993), 215–242.

[Tr1] H. Triebel, *Interpolation Theory, Function Spaces, Differential Operators*, 2nd edition, Johann Ambrosius Barth Verlag, Heidelberg-Leipzig 1995.

Roland Duduchava
A. Raznmadze Mathematical Institute
and I. Javakhishvili State University, Tbilisi
President of the Georgian Mathematical Union
M. Alexidze str. 1
Tbilisi 0193, Georgia
e-mail: dudu@rmi.acnet.ge
RolDud@gmail.com

Operator Theory:
Advances and Applications, Vol. 210, 79–106

On the Bergman Theory for Solenoidal and Irrotational Vector Fields, I: General Theory

José Oscar González-Cervantes, María Elena Luna-Elizarrarás and Michael Shapiro

To Nikolai Vasilevski on the occasion of his 60th birthday

Abstract. For solenoidal and irrotational vector fields as well as for quaternionic analysis of the Moisil-Théodoresco operator we introduce the notions of the Bergman space and the Bergman reproducing kernel; main properties of them are studied. Among other objects of our interest are: the analogues of the Bergman projections; the behavior of the Bergman theory for a given domain whenever the domain is transformed by a conformal map.

Mathematics Subject Classification (2000). 37C10, 32A25, 47B32, 30G35.

Keywords. Vector fields, quaternionic analysis; hyperholomorphic Bergman spaces; quaternionic Möbius transformations; categories and functors.

1. Introduction

1.1. The Bergman spaces and Bergman operators were first introduced for holomorphic functions of one complex variable where they proved to be of interest and importance both for the theory itself and for numerous applications. Of course, this has led to developments in different directions, in particular, for other classes of functions: for harmonic functions, for holomorphic functions of several complex variables, etc.

The present work deals with solenoidal and irrotational, or Laplacian, vector fields in $\mathbb{R}^3$. Although it is known that such vector fields generalize holomorphic functions in one complex variable, but the generalization is rather formal in the

The first author was partially supported by CONACYT and by Instituto Politécnico Nacional as Doctoral scholarship and PIFI scholarship recipient.

The second and the third author were partially supported by CONACYT projects as well as by Instituto Politécnico Nacional in the framework of COFAA and SIP programs.

sense that any holomorphic function can be considered as a solenoidal and irrotational vector field, and at same time the corresponding theories are quite different and many properties of holomorphic functions do not have their direct analogs for vector fields. So instead of working with the whole class of solenoidal and irrotational vector fields we have chosen a subclass of them for which it has proved to be possible to construct a theory preserving the main structural peculiarities of its antecedent in one complex variable. How to make the choice was suggested by the fact that solenoidal and irrotational vector fields can be identified with a subset of the set of null-solutions of the Moisil-Théodoresco operator which are called here hyperholomorphic quaternion-valued functions. The matter is that quaternionic analysis for the Moisil-Théodoresco operator does preserve a deep analogy with one-dimensional complex analysis and thus, the latter allows us to conjecture reasonably for the former.

Thus, as a matter of fact, we have developed the two Bergman-type theories: the one for the Laplacian vector fields which we consider as our primary goal, and the other for quaternionic analysis of the Moisil-Théodoresco operator which is a kind of an auxiliary tool here although of course it has its own undoubted interest and importance.

1.2. We preferred to divide the whole work in two parts, and this is the first of them. When talking about Bergman theory for a class of functions we are interested in finding out, first of all, what are the proper notions of the Bergman space and the Bergman reproducing kernel for the class of functions as well as which are their main properties. The other objects of our interest are: what is the Bergman projection of an $\mathcal{L}_2$-space onto the corresponding Bergman space; what is the behavior of the Bergman theory for a given domain whenever the domain is transformed by a conformal map? Section 2 describes the answers to all these questions for solenoidal and irrotational vector fields, but this is being done on the level of statements and explanations, without proofs which are postponed until the last section. The answers for the Moisil-Théodoresco quaternionic analysis the reader can find in Section 3 which contains both the statements and their complete proofs. The proofs of the vectorial statements of Section 2 are presented in Section 4 in the form of direct corollaries of the corresponding facts obtained for hyperholomorphic function theory which is being developed in Section 3.

1.3. As we mentioned already, the solenoidal and irrotational vector fields (the exact definitions are given in Subsection 2.1 below) describe many important physical phenomena. Besides, they are a particular case, for $\mathbb{R}^3$, of a more general notion of a system of conjugate harmonic functions, see, e.g., [18] but [14] as well, which is one of the main tools of harmonic analysis. What is more, as one may read in [18, p. 229], "Equations (1.6) [*which define a system of conjugate harmonic functions – the authors*] represent perhaps the most direct generalization of the Cauchy-Riemann equations". In spite of all this we are not aware of any work on the Bergman theory in this context which was one of the motivations for us.

Perhaps the most direct predecessors of the actual work are our papers [4] and [5] which deal with other classes of functions but employ again the relations with a version of quaternionic analysis, now for the Fueter operator.

2. Solenoidal and irrotational vector fields: main results

2.1. Vector fields are widely used in physics to model numerous phenomena. Given a domain $\Omega \subset \mathbb{R}^3$, we shall work with solenoidal and irrotational vector fields defined on Ω.

Recall that:

- A vector field $\vec{f} \in C^1(\Omega, \mathbb{R}^3)$ is called **solenoidal** if $\operatorname{div} \vec{f} = 0$. Some examples of solenoidal vector fields are: magnetic vector fields, the vector field of velocities of an incompressible fluid flow, etc.
- A vector field $\vec{f} \in C^1(\Omega, \mathbb{R}^3)$ is called **irrotational** if $\operatorname{rot} \vec{f} = \vec{0}$. An example of irrotational vector fields is every conservative vector field; in particular the vector field associated to the force of gravity is a conservative vector field, therefore it is also an irrotational vector field.

The $\mathbb{R}$-linear space of **solenoidal and irrotational vector fields**, or the **SI-vector fields**, in Ω is defined by

$$\vec{\mathfrak{M}}(\Omega) := \{\vec{f} \in C^1(\Omega, \mathbb{R}^3) \mid \operatorname{div} \vec{f} = 0, \ \operatorname{rot} \vec{f} = \vec{0} \ \text{ in } \Omega\}.$$

In this section we present the main results of the Bergman theory for SI-vector fields without proofs; the proofs will be presented in Section 4. We shall use the commonly accepted notations $\langle \cdot, \cdot \rangle_{\mathbb{R}^3}$ and $[\cdot, \cdot]_{\mathbb{R}^3}$ for the scalar (inner) and the vectorial (cross) products respectively.

2.2. The SI-Bergman space and the SI-Bergman kernel

Following the idea of S. Bergman himself one is tempted to use the whole space $\mathcal{L}_2$ in the definition of the Bergman space of solenoidal and irrotational vector fields associated to Ω. But it turns out that such a set is too large and loses its relations with the Bergman space in contrast to the situation of the usual complex Bergman space, see [3]. We are interested in obtaining the results which preserve not only the formal similarity with the complex analysis case but, which is most important for us, the underlying structures. In order to achieve this objective we "diminish", first of all, the space $\mathcal{L}_2(\Omega, \mathbb{R}^3)$, which is based on the following

Theorem. 2.2.1. *Let $\Omega \subset \mathbb{R}^3$ be a domain, there exist a unique scalar field $\mathcal{B}_{\Omega,0} \in C^1(\Omega \times \Omega, \mathbb{R})$ and a unique vector field $\vec{\mathcal{B}}_\Omega \in C^1(\Omega \times \Omega, \mathbb{R}^3)$, such that*

- $\mathcal{B}_{\Omega,0}(\vec{x}, \vec{v}) = \mathcal{B}_{\Omega,0}(\vec{v}, \vec{x}), \quad \vec{\mathcal{B}}_\Omega(\vec{x}, \vec{v}) = -\vec{\mathcal{B}}_\Omega(\vec{v}, \vec{x}), \quad (\vec{x}, \vec{v}) \in \Omega \times \Omega,$
- $\operatorname{div}_{\vec{x}} \vec{\mathcal{B}}_\Omega(\cdot, \vec{v}) = 0,$ *for a.e.* $\vec{v} \in \Omega,$
- $\operatorname{grad}_{\vec{x}} \mathcal{B}_{\Omega,0}(\cdot, \vec{v}) = -\operatorname{rot}_{\vec{x}} \vec{\mathcal{B}}_\Omega(\cdot, \vec{v}),$ *for a.e.* $\vec{v} \in \Omega,$
- $\mathcal{B}_{\Omega,0}(\cdot, \vec{v}) \in \mathcal{L}_2(\Omega, \mathbb{R}), \quad \vec{\mathcal{B}}_\Omega(\cdot, \vec{v}) \in \mathcal{L}_2(\Omega, \mathbb{R}^3), \quad \vec{v} \in \Omega.$

- *The space*

$$\hat{\mathcal{L}}_2(\Omega,\mathbb{R}^3) := \{\vec{f} \in \mathcal{L}_2(\Omega,\mathbb{R}^3) \mid \int_\Omega \left\langle \vec{\mathcal{B}}_{M,\Omega}(\vec{x},\vec{v}), \vec{f}(\vec{v}) \right\rangle_{\mathbb{R}^3} d\mu_{\vec{v}} = 0, \text{ for a.e. } \vec{x} \in \Omega\},$$

equipped with the norm inherited from $\mathcal{L}_2(\Omega,\mathbb{R}^3)$*, is a real Banach space and*

$$\vec{\mathfrak{M}}(\Omega) \cap \hat{\mathcal{L}}_2(\Omega,\mathbb{R}^3) = \vec{\mathfrak{M}}(\Omega) \cap \mathcal{L}_2(\Omega,\mathbb{R}^3).$$

Recall that the norm of $\vec{f} \in \mathcal{L}_2(\Omega,\mathbb{R}^3)$ is $\left(\int_\Omega \|\vec{f}\|^2_{\mathbb{R}^3} d\mu\right)^{\frac{1}{2}}$, where $d\mu$ is the three-dimensional volume measure, and $\|\cdot\|_{\mathbb{R}^3}$ is the Euclidean norm of $\mathbb{R}^3$.

We will see that the "appropriate" $\mathcal{L}_2$-space will be the space $\hat{\mathcal{L}}_2(\Omega,\mathbb{R}^3)$.

Definition. 2.2.2. The pair of fields $\left(\mathcal{B}_{\Omega,0}, \vec{\mathcal{B}}_\Omega\right)$ is called the **SI-Bergman kernel** associated to Ω.

Definition. 2.2.3. We define the **solenoidal and irrotational Bergman space**, or the **SI-Bergman space** associated to Ω by

$$\vec{\mathcal{A}}(\Omega) := \vec{\mathfrak{M}}(\Omega) \cap \hat{\mathcal{L}}_2(\Omega,\mathbb{R}^3).$$

Proposition. 2.2.4. *The space* $\vec{\mathcal{A}}(\Omega)$*, equipped with the norm inherited from* $\hat{\mathcal{L}}_2(\Omega,\mathbb{R}^3)$ *is a real Banach space. This norm will be denoted by* $\|\cdot\|_{\vec{\mathcal{A}}(\Omega)}$*. Besides,*

$$\int_\Omega \left(\mathcal{B}_{\Omega,0}(\vec{x},\vec{v})\vec{f}(\vec{v}) + \left[\vec{\mathcal{B}}_\Omega(\vec{x},\vec{v}), \vec{f}(\vec{v})\right]_{\mathbb{R}^3}\right) d\mu_{\vec{v}} = \vec{f}(\vec{x}),\ \vec{x} \in \Omega, \tag{1}$$

for all $\vec{f} \in \vec{\mathcal{A}}(\Omega)$.

What is more, given $\vec{f} \in \hat{\mathcal{L}}_2(\Omega,\mathbb{R}^3)$ *there holds:*

$$\vec{f}(\vec{x}) = \int_\Omega \left(\mathcal{B}_{\Omega,0}(\vec{x},\vec{v})\vec{f}(\vec{v}) + \left[\vec{\mathcal{B}}_\Omega(\vec{x},\vec{v}), \vec{f}(\vec{v})\right]_{\mathbb{R}^3}\right) d\mu_{\vec{v}} \iff \vec{f} \in \vec{\mathcal{A}}(\Omega). \tag{2}$$

2.2.4.1 Noticethat if we introduce the operation $\odot : \mathbb{R}^3 \times \left(\mathbb{R} \times \mathbb{R}^3\right) \longrightarrow \mathbb{R}^3$ as

$$\vec{a} \odot (r,\vec{b}) := \vec{a}r + \left[\vec{a},\vec{b}\right]_{\mathbb{R}^3}, \quad \forall \vec{a} \in \mathbb{R}^3 \quad \text{and} \quad \forall (r,\vec{b}) \in \left(\mathbb{R} \times \mathbb{R}^3\right), \tag{3}$$

then (1) and (2) respectively become

$$\int_\Omega \vec{f}(\vec{v}) \odot \left(\mathcal{B}_{\Omega,0}(\vec{x},\vec{v}), -\vec{\mathcal{B}}_\Omega(\vec{x},\vec{v})\right) d\mu_{\vec{v}} = \vec{f}(\vec{x}), \quad \forall \vec{f} \in \vec{\mathcal{A}}(\Omega), \tag{4}$$

and given $\vec{f} \in \hat{\mathcal{L}}_2(\Omega,\mathbb{R}^3)$ then

$$\vec{f}(\vec{x}) = \int_\Omega \vec{f}(\vec{v}) \odot \left(\mathcal{B}_{\Omega,0}(\vec{x},\vec{v}), -\vec{\mathcal{B}}_\Omega(\vec{x},\vec{v})\right) d\mu_{\vec{v}} \quad \iff \quad \vec{f} \in \vec{\mathcal{A}}(\Omega). \tag{5}$$

The reason for this will be clear from Section 3.

Note one more property of the SI-Bergman space.

Proposition. 2.2.5. *Let* $\vec{u} \in \Omega$ *and let* $\hat{\imath} \in \mathbb{R}^3$ *be a unit vector, define* $\vec{F}_{\vec{u},\hat{\imath}} := \{\vec{f} \in \vec{\mathcal{A}}(\Omega) \mid \vec{f}(\vec{u}) = \hat{\imath}\}$. *Then there exists a unique vector field* $\vec{f}_* \in \vec{F}_{\vec{u},\hat{\imath}}$, *which is a solution of the variational problem of finding* $\inf\{\|\vec{f}\|_{\vec{\mathcal{A}}(\Omega)} \mid \vec{f} \in \vec{F}_{\vec{u},\hat{\imath}}\}$; *i.e.,*

$$\|\vec{f}_*\|_{\vec{\mathcal{A}}(\Omega)} = \sigma := \inf\{\|\vec{f}\|_{\vec{\mathcal{A}}(\Omega)} \mid \vec{f} \in \vec{F}_{\vec{u},\hat{\imath}}\}.$$

Moreover,

$$\|\vec{f}_*\|^2_{\vec{\mathcal{A}}(\Omega)} > \frac{\displaystyle\int_\Omega |\mathcal{B}_{\Omega,0}(\vec{v},\vec{u})|^2 d\mu_{\vec{v}} + \|\vec{\mathcal{B}}_\Omega(\cdot,\vec{u})\|^2_{\mathcal{L}_2(\Omega,\mathbb{R}^3)}}{\left(\mathcal{B}_{\Omega,0}(\vec{u},\vec{u})\right)^2}. \tag{6}$$

A variational problem similar to the previous one has its antecedents in the theory of the classical complex Bergman spaces and in the theory of the quaternionic Bergman spaces where it has a unique solution which is the normalization of the corresponding Bergman kernel, see respectively [3] and [5]. In our case, the solution is quite different from the normalization of the SI-Bergman kernel.

2.3. SI-Bergman projection

Going along a deep analogy with the classical structure of the Bergman theory we introduce the operator $\mathbf{B}_\Omega$ defined on $\hat{\mathcal{L}}_2(\Omega,\mathbb{R}^3)$ by

$$\mathbf{B}_\Omega[\vec{f}](\vec{x}) := \int_\Omega \vec{f}(\vec{v}) \odot \left(\mathcal{B}_{\Omega,0}(\vec{x},\vec{v}), -\vec{\mathcal{B}}_\Omega(\vec{x},\vec{v})\right) d\mu_{\vec{v}}, \tag{7}$$

for all $\vec{f} \in \hat{\mathcal{L}}_2(\Omega,\mathbb{R}^3)$. It satisfies the following

Proposition. 2.3.1. *The operator* $\mathbf{B}_\Omega$ *is continuous,* $\mathbf{B}_\Omega(\hat{\mathcal{L}}_2(\Omega,\mathbb{R}^3)) = \vec{\mathcal{A}}(\Omega)$ *and* $\mathbf{B}^2_\Omega = \mathbf{B}_\Omega$.

The operator $\mathbf{B}_\Omega$ is called the **SI-Bergman projection** associated to Ω, and the above facts are quite analogous to properties of its counterparts in the theory of one complex variable and in the quaternionic theory which is owing to the "right choice" of the space $\hat{\mathcal{L}}_2$.

2.4. Decomposition of $\hat{\mathcal{L}}_2$

There is another advantage of working with $\hat{\mathcal{L}}_2(\Omega,\mathbb{R}^3)$ instead of $\mathcal{L}_2(\Omega,\mathbb{H})$, namely, this space has a decomposition into a direct sum of the two summands one of them being the SI-Bergman space and the other admits an explicit description. In order to show this fact, we denote by $\dot{W}^1_2(\Omega,\mathbb{R}^3)$ the $\mathbb{R}$-linear space of vector fields whose components belong to Sobolev space $\dot{W}^1_2(\Omega,\mathbb{R})$.

Proposition. 2.4.1. *The following decomposition holds:*

$$\hat{\mathcal{L}}_2(\Omega,\mathbb{R}^3) = \vec{\mathcal{A}}(\Omega) \oplus \left(\nabla \odot \left(\dot{W}^1_2(\Omega,\mathbb{R}) \times \dot{W}^1_2(\Omega,\mathbb{R}^3)\right)\right),$$

i.e., for any $\vec{f} \in \hat{\mathcal{L}}_2(\Omega, \mathbb{R}^3)$ *there exist unique elements* $\vec{g} \in \vec{\mathcal{A}}(\Omega)$ *and*

$$\vec{h} \in \nabla \odot \left(\dot{W}_2^1(\Omega, \mathbb{R}) \times \dot{W}_2^1(\Omega, \mathbb{R}^3) \right),$$

such that $\vec{f} = \vec{g} + \vec{h}$*, or equivalently, there exists* $(\ell_0, \vec{\ell}) \in \dot{W}_2^1(\Omega, \mathbb{R}) \times \dot{W}_2^1(\Omega, \mathbb{R}^3)$ *such that*

$$\vec{f} = \mathbf{B}_\Omega[\vec{f}] + (\text{grad } \ell_0 + \text{rot } \vec{\ell}).$$

In the theory of one complex variable the decomposition of $\mathcal{L}_2$ as a direct sum of two spaces, where one of them is the complex Bergman space is well known. One can see its quaternionic and Cliffordian counterparts respectively in [5] and [7]. Now, Proposition 2.4.1 is a purely vectorial fact deeply similar to this decomposition.

2.5. Möbius transformations in $\mathbb{R}^3$ in vectorial language

Liouville's theorem, see [2], establishes that in spaces of dimension greater than or equal to three, the only conformal mappings are: translations, rotations, inversions and dilations, as well as finite compositions of these. For this reason, the only conformal mappings in $\mathbb{R}^n$, $n > 2$, also bear historically the name of Möbius transformations.

The basic **Möbius transformations in** $\mathbb{R}^3$ and their explicit representations are the following:

Translation: $T_1(\vec{x}) = \vec{x} + \vec{a}, \quad \forall \vec{x} \in \mathbb{R}^3, \quad \text{with} \quad \vec{a} \in \mathbb{R}^3.$

Rotation: $T_2(\vec{x}) = r^2\vec{x} + 2r\,[\vec{a}, \vec{x}]_{\mathbb{R}^3} + \langle \vec{a}, \vec{x}\rangle_{\mathbb{R}^3}\,\vec{a} + [\vec{a}, [\vec{a}, \vec{x}]_{\mathbb{R}^3}]_{\mathbb{R}^3}, \quad \forall \vec{x} \in \mathbb{R}^3,$
with $r \in \mathbb{R}, \quad \vec{a} \in \mathbb{R}^3 \quad \text{such that} \quad r^2 + \|\vec{a}\|^2_{\mathbb{R}^3} = 1.$

Inversion: $T_3(\vec{x}) = -\dfrac{\vec{x}}{\|\vec{x}\|^2_{\mathbb{R}^3}}, \quad \forall \vec{x} \in \mathbb{R}^3 \setminus \{\vec{0}\}, \quad \text{and} \quad T_3(\vec{0}) := \infty.$

Dilation: $T_4(\vec{x}) = \lambda\vec{x}, \quad \forall \vec{x} \in \mathbb{R}^3, \quad$ with a positive real number λ.

The formulas for translations, dilations, and the inversion are obviously clear; the formula for rotations will be explained below, see equation (20).

Note that any Möbius transformations T, which is a finite composition of translations, dilations and rotations, can be represented by

$$T^*(\vec{x}) = r^2\vec{x} + 2r\,[\vec{a}, \vec{x}]_{\mathbb{R}^3} + \langle \vec{x}, \vec{a}\rangle_{\mathbb{R}^3}\,\vec{a} + [\vec{a}, [\vec{a}, \vec{x}]_{\mathbb{R}^3}]_{\mathbb{R}^3} + \vec{b}, \tag{8}$$

and if T has at least one inversion as a composition factor, then its representation is

$$T^{**}(\vec{x}) = -\frac{r^2\vec{x} + 2r\,[\vec{a}, \vec{x}]_{\mathbb{R}^3} + \langle \vec{x}, \vec{a}\rangle_{\mathbb{R}^3}\,\vec{a} + [\vec{a}, [\vec{a}, \vec{x}]_{\mathbb{R}^3}]_{\mathbb{R}^3} + \vec{c}}{\|r^2\vec{x} + 2r\,[\vec{a}, \vec{x}]_{\mathbb{R}^3} + \langle \vec{x}, \vec{a}\rangle_{\mathbb{R}^3}\,\vec{a} + [\vec{a}, [\vec{a}, \vec{x}]_{\mathbb{R}^3}]_{\mathbb{R}^3} + \vec{c}\|^2} + \vec{b}; \tag{9}$$

for both cases $r \in \mathbb{R}$ and $\vec{a}, \vec{b}, \vec{c} \in \mathbb{R}^3$ are such that $r^2 + \|\vec{a}\|^2_{\mathbb{R}^3} \neq 0$. In order to simplify the notation set $\gamma_{r,\vec{a}} := r^2 + \|\vec{a}\|^2_{\mathbb{R}^3}$.

2.6. $\mathbb{R}$-linear spaces of vector fields and Möbius transformations on $\mathbb{R}^3$

The aim of this subsection is to realize what happens with SI-vector fields whenever their domains are transformed by a Möbius transformation. In order to abbreviate some formulations we introduce the denotation: let $r \in \mathbb{R}$, $\vec{a} \in \mathbb{R}^3$ and let $\vec{g}$ be any vector field, then

$$P_{r,\vec{a}}[\vec{g}] := r^2 \vec{g} + 2r\,[\vec{g}, \vec{a}]_{\mathbb{R}^3} + \langle \vec{a}, \vec{g} \rangle_{\mathbb{R}^3}\, \vec{a} + [[\vec{a}, \vec{g}]_{\mathbb{R}^3}, \vec{a}]_{\mathbb{R}^3}\,.$$

The following proposition establishes how the operators div and rot interact with the Möbius transformations in $\mathbb{R}^3$.

Proposition. 2.6.1. *Let $\Omega, \Xi \subset \mathbb{R}^3$ be domains such that there exists a Möbius transformation T satisfying $T(\Xi) = \Omega$. Denote $\vec{y} = T(\vec{x})$.*

The divergence operator has the following properties:

- *If $T = T^*$, then*

$$\operatorname{div}_{\vec{x}} P_{r,\vec{a}}[\vec{f} \circ T] = (\gamma_{r,\vec{a}})^2\, (\operatorname{div}_{\vec{y}} \vec{f}) \circ T, \quad \forall \vec{f} \in C^1(\Omega, \mathbb{R}^3). \tag{10}$$

- *If $T = T^{**}$, then*

$$\begin{aligned} &\operatorname{div}_{\vec{x}} \left[-\frac{P_{r,-\vec{a}}[\vec{x}] + \vec{c}}{\|P_{r,-\vec{a}}[\vec{x}] + \vec{c}\|^3_{\mathbb{R}^3}}, P_{r,\vec{a}}[\vec{f} \circ T] \right]_{\mathbb{R}^3} \\ &\quad = \gamma_{r,\vec{a}} \|T - \vec{b}\|_{\mathbb{R}^3} \left\langle T - \vec{b}, P_{r,\vec{a}}[(\operatorname{rot}_{\vec{y}} \vec{f}) \circ T] \right\rangle_{\mathbb{R}^3}, \end{aligned} \tag{11}$$

for all $\vec{f} \in C^1(\Omega, \mathbb{R}^3)$.

The rotation operator satisfies:

- *If $T = T^*$, then*

$$\operatorname{rot}_{\vec{x}} P_{r,\vec{a}}[\vec{f} \circ T] = \gamma_{r,\vec{a}}\; P_{r,\vec{a}}[(\operatorname{rot}_{\vec{y}} \vec{f}) \circ T], \quad \forall \vec{f} \in C^1(\Omega, \mathbb{R}^3). \tag{12}$$

- *If $T = T^{**}$, then*

$$\begin{aligned} &- \operatorname{grad}_{\vec{x}} \left\langle -\frac{P_{r,-\vec{a}}[\vec{x}] + \vec{c}}{\|P_{r,-\vec{a}}[\vec{x}] + \vec{c}\|^3_{\mathbb{R}^3}}, P_{r,\vec{a}}[\vec{f} \circ T] \right\rangle_{\mathbb{R}^3} \\ &\quad + \operatorname{rot}_{\vec{x}} \left[-\frac{P_{r,-\vec{a}}[\vec{x}] + \vec{c}}{\|P_{r,-\vec{a}}[\vec{x}] + \vec{c}\|^3_{\mathbb{R}^3}}, P_{r,\vec{a}}[\vec{f} \circ T] \right]_{\mathbb{R}^3} \\ &\quad = -(\gamma_{r,\vec{a}})^2 \|T - \vec{b}\|_{\mathbb{R}^3} (T - \vec{b}) (\operatorname{div}_{\vec{y}} \vec{f}) \circ T \\ &\qquad - \gamma_{r,\vec{a}} \|T - \vec{b}\|_{\mathbb{R}^3} \left[T - \vec{b}, P_{r,\vec{a}}[(\operatorname{rot}_{\vec{y}} \vec{f}) \circ T] \right]_{\mathbb{R}^3}, \end{aligned} \tag{13}$$

for all $\vec{f} \in C^1(\Omega, \mathbb{R}^3)$.

The formulas (10)–(13) can be seen as analogs, in various directions, of the usual chain rule. Besides they allow already some conclusions about SI-vector fields.

Corollary. 2.6.2. *Under the same conditions, let $\vec{f}$ be a vector field defined on Ω. If $T = T^*$, then there holds:*

1. *The vector field $\vec{f}$ is solenoidal on Ω if and only if $P_{r,\vec{a}}[\vec{f} \circ T]$ is solenoidal on Ξ.*
2. *The vector field $\vec{f}$ is irrotational on Ω if and only if $P_{r,\vec{a}}[\vec{f} \circ T]$ is also irrotational on Ξ.*
3. *The vector field $\vec{f} \in \vec{\mathfrak{M}}(\Omega)$ if and only if $P_{r,\vec{a}}[\vec{f} \circ T] \in \vec{\mathfrak{M}}(\Omega)$.*

On the other hand, for $T = T^{**}$ we have:

4. *If $\vec{f}$ is an irrotational vector field in Ω then*

$$\left[-\frac{P_{r,-\vec{a}}[\vec{x}] + \vec{c}}{\|P_{r,-\vec{a}}[\vec{x}] + \vec{c}\|^3_{\mathbb{R}^3}}, P_{r,\vec{a}}[\vec{f} \circ T] \right]_{\mathbb{R}^3}$$

   is a solenoidal vector field in Ξ.

5. *If $\vec{f} \in \vec{\mathfrak{M}}(\Omega)$ then*

$$\left[-\frac{P_{r,-\vec{a}}[\vec{x}] + \vec{c}}{\|P_{r,-\vec{a}}[\vec{x}] + \vec{c}\|^3_{\mathbb{R}^3}}, P_{r,\vec{a}}[\vec{f} \circ T] \right]_{\mathbb{R}^3}$$

   is a solenoidal vector field in Ξ and satisfies

$$\begin{aligned} \operatorname{rot}_{\vec{x}} & \left[-\frac{P_{r,-\vec{a}}[\vec{x}] + \vec{c}}{\|P_{r,-\vec{a}}[\vec{x}] + \vec{c}\|^3_{\mathbb{R}^3}}, P_{r,\vec{a}}[\vec{f} \circ T] \right]_{\mathbb{R}^3} \\ & = \operatorname{grad}_{\vec{x}} \left\langle -\frac{P_{r,-\vec{a}}[\vec{x}] + \vec{c}}{\|P_{r,-\vec{a}}[\vec{x}] + \vec{c}\|^3_{\mathbb{R}^3}}, P_{r,\vec{a}}[\vec{f} \circ T] \right\rangle_{\mathbb{R}^3} . \end{aligned}$$

Observation. 2.6.3. Although the composition of solenoidal and irrotational vector fields with Möbius transformations does not act, in general, invariantly, nevertheless the facts 1.–3. of the previous corollary say that combining the composition and some additional operation generated by it we arrive again at a solenoidal and irrotational vector field. In other words, composing a SI-vector field $\vec{f}$ with a Möbius transformation T^*, we get an "almost" SI-vector field, which requires additionally the operation $P_{r,\vec{a}}$ that depends on T^* only.

2.6.4. Among the works related to the previous corollary there are [9], [20] and references therein. The first paper deals with the analytic vector fields, while the second with harmonic vector fields.

Corollary. 2.6.5. *Given a Möbius transformation $T = T^*$ and let*

$$V_T[\vec{f}] := P_{r,\vec{a}}[\vec{f} \circ T], \quad \forall \vec{f} \in C^1(\Omega, \mathbb{R}^3). \tag{14}$$

Then the operator V_T is an invertible real-linear operator, and its restriction $V_T\, |_{\vec{\mathfrak{M}}(\Omega)} \colon \vec{\mathfrak{M}}(\Omega) \to \vec{\mathfrak{M}}(\Xi)$ is an isomorphism of $\mathbb{R}$-linear spaces.

A similar construction for the space $\hat{\mathcal{L}}_2$ requires some adjustment of the operator V_T.

Proposition. 2.6.6. *Let $\Omega, \Xi \subset \mathbb{R}^3$ be domains such that there exists a Möbius transformation $T = T^*$ and $T(\Xi) = \Omega$. Let the operator $\mathcal{V}_T : \hat{\mathcal{L}}_2(\Omega, \mathbb{R}^3) \to \hat{\mathcal{L}}_2(\Xi, \mathbb{R}^3)$ be defined by*

$$\mathcal{V}_T[\vec{f}] := (\gamma_{r,\vec{a}})^{\frac{1}{2}} P_{r,\vec{a}}[\vec{f} \circ T], \quad \forall \vec{f} \in \hat{\mathcal{L}}_2(\Omega, \mathbb{R}^3), \tag{15}$$

then

$$\begin{aligned} \int_\Omega \left\langle \vec{f}, \vec{g} \right\rangle_{\mathbb{R}^3} d\mu &= \int_\Xi \left\langle \mathcal{V}_T[\vec{f}], \mathcal{V}_T[\vec{g}] \right\rangle_{\mathbb{R}^3} d\mu, \\ (\gamma_{r,\vec{a}})^{-1} P_{r,\vec{a}} \left[\int_\Omega \left[\vec{f}, \vec{g} \right]_{\mathbb{R}^3} d\mu \right] &= \int_\Xi \left[\mathcal{V}_T[\vec{f}], \mathcal{V}_T[\vec{g}] \right]_{\mathbb{R}^3} d\mu, \end{aligned} \tag{16}$$

for all $\vec{f}, \vec{g} \in \hat{\mathcal{L}}_2(\Omega, \mathbb{R}^3)$.

Corollary. 2.6.7. *The operator $\mathcal{V}_T \mid_{\vec{\mathcal{A}}(\Omega)} : \vec{\mathcal{A}}(\Omega) \longrightarrow \vec{\mathcal{A}}(\Xi)$ is an isometric isomorphism of real Banach spaces, which relates the SI-Bergman kernels of the spaces $\vec{\mathcal{A}}(\Omega)$ and $\vec{\mathcal{A}}(\Xi)$ as follows:*

$$\begin{aligned} \mathcal{B}_{M,\Xi,0}(\vec{x}, \vec{u}) &= (\gamma_{r,\vec{a}})^3 \mathcal{B}_{M,\Omega,0}(T(\vec{x}), T(\vec{u})), \\ \vec{\mathcal{B}}_{M,\Xi}(\vec{x}, \vec{u}) &= (\gamma_{r,\vec{a}})^2 P_{r,\vec{a}}[\vec{\mathcal{B}}_{M,\Omega}(T(\vec{x}), T(\vec{u}))], \end{aligned}$$

for all $(\vec{x}, \vec{u}) \in \Xi \times \Xi$.

Proposition 2.6.6 and its corollary tell us that the isometric isomorphism $\mathcal{V}_T$ between the spaces $\hat{\mathcal{L}}_2$ restricted onto the SI-Bergman spaces preserves the SI-vector fields, and in this way, it allows us to establish an isometric isomorphism between these SI-Bergman spaces which relates also their Bergman kernels.

Given a function space H, we denote by $\mathfrak{L}(H)$ the space of real-linear continuous operators from H to H.

Corollary. 2.6.8. *Let $\Omega, \Xi \subset \mathbb{R}^3$ be domains such that there exists a Möbius transformation $T = T^*$ and $T(\Xi) = \Omega$. Define the operator $\mathbf{J}_T$ on $\mathfrak{L}(\vec{\mathcal{A}}(\Omega))$ as follows:*

$$\mathbf{J}_T(Q)[\vec{f}] := (\gamma_{r,\vec{a}})^{-1} P_{r,\vec{a}}[\ Q[\ P_{r,-\vec{a}}[\vec{f}] \ \circ \ T^{-1}] \ \circ \ T], \quad \forall \vec{f} \in \vec{\mathcal{A}}(\Xi),$$

for all $Q \in \mathfrak{L}(\vec{\mathcal{A}}(\Omega))$. Moreover, the operator $\mathbf{J}_T : \mathfrak{L}(\vec{\mathcal{A}}(\Omega)) \longrightarrow \mathfrak{L}(\vec{\mathcal{A}}(\Xi))$ is an isomorphism of $\mathbb{R}$-linear algebras, which relates the SI-Bergman projections by $\mathbf{B}_\Xi = \mathbf{J}_T[\mathbf{B}_\Omega]$.

Note that the operator $\mathbf{J}_T$ was deduced from the construction of the operator $\mathcal{V}_T$. Note also that Proposition 2.6.6 and its corollaries are generalizations of important facts in complex and quaternionic analysis, see respectively [21] and [4].

3. The Bergman theory for Moisil-Théodoresco hyperholomorphy

3.1. Preliminaries

The skew-field of quaternions is denoted by $\mathbb{H}$, any quaternion can be represented by $a = a_0 + a_1 i + a_2 j + a_3 k$, where $a_n \in \mathbb{R}$ for $n = 0, 1, 2, 3$. The imaginary units i, j, k, satisfy $i^2 = j^2 = k^2 = -1$ and $ij = k,\ jk = i,\ ki = j$.

The quaternionic conjugation and the norm of any quaternion a are defined, respectively, by $\bar{a} := a_0 - a_1 i - a_2 j - a_3 k$ and $|a| := \sqrt{a_0^2 + a_1^2 + a_2^2 + a_3^2}$.

For any quaternion $a = a_0 + a_1 i + a_2 j + a_3 k$ the real number a_0 is called its scalar part, and $a_1 i + a_2 j + a_3 k$ is called its vectorial part.

The real subspace of $\mathbb{H}$ formed by quaternions whose scalar parts are zero is isometric and isomorphic to the usual three-dimensional Euclidean space $\mathbb{R}^3$, thus they are denoted by the same symbol.

The Moisil-Théodoresco operator $\mathcal{D}_{MT}$ and the right-Moisil-Théodoresco operator $\mathcal{D}_{MT,r}$ act on functions of class C^1 with quaternionic values and defined on domains in $\mathbb{R}^3$ as follows:

$$\mathcal{D}_{MT}[f] := i\frac{\partial f}{\partial x_1} + j\frac{\partial f}{\partial x_2} + k\frac{\partial f}{\partial x_3}, \qquad \mathcal{D}_{MT,r}[f] := \frac{\partial f}{\partial x_1}i + \frac{\partial f}{\partial x_2}j + \frac{\partial f}{\partial x_3}k.$$

They have the fundamental property: $-\mathcal{D}_{MT}^2 = -\mathcal{D}_{MT,r}^2 = \Delta_{\mathbb{R}^3}$, the Laplace operator.

Definition. 3.1.1. The right-$\mathbb{H}$-linear space of hyperholomorphic functions on Ω is defined by $\mathfrak{M}_{MT}(\Omega) := \ker \mathcal{D}_{MT}$, and the space of right-hyperholomorphic functions on Ω is defined by $\mathfrak{M}_{MT,r}(\Omega) := \ker \mathcal{D}_{MT,r}$.

Notation. 3.1.2.

- In order to simplify the notation, we will omit the prefix and the subindex MT in future notations, when there is no ambiguity; therefore, $\mathfrak{M}_{MT}(\Omega) = \mathfrak{M}(\Omega)$ and $\mathfrak{M}_{MT,r}(\Omega) = \mathfrak{M}_r(\Omega)$.
- Also sometimes we denote $e_0 = 1,\ e_1 = i,\ e_2 = j,\ e_3 = k$.

Note that $\mathfrak{M}_r(\Omega)$ is a left-$\mathbb{H}$-linear space.

The reader can find much more helpful information in, e.g., [15]; we give below a few facts only which will be used immediately.

Let $\Omega \subset \mathbb{R}^3$ be a domain with smooth boundary. Let $f \in \mathfrak{M}(\Omega)$ be a continuous function up to $\partial\Omega$, then the hyperholomorphic Cauchy integral formula for f is

$$\int_{\partial\Omega} \mathcal{K}(\zeta - x)\sigma_\zeta^{(2)} f(\zeta) = \begin{cases} f(x), & \text{if } x \in \Omega, \\ 0, & \text{if } x \notin \overline{\Omega}, \end{cases} \tag{17}$$

where the function

$$\mathcal{K}(\zeta - x) := -\frac{\zeta - x}{4\pi|\zeta - x|^2}$$

is called the hyperholomorphic Cauchy kernel, and the differential form

$$\sigma_x^{(2)} = \sum_{n=1}^{3}(-1)^{n+1} e_n d\hat{x}^n$$

is such that its restriction onto $\partial\Omega$ coincides with $\vec{n}\, ds$, where $\vec{n}$ is the unit normal vector to $\partial\Omega$. Each term $d\hat{x}^n$ is the wedge product $dx_1 \wedge dx_2 \wedge dx_3$ with the factor dx_n omitted for $1 \leq n \leq 3$.

Moreover, the operator T_r is defined on Lebesgue integrable functions on Ω by

$$T_r[f](x) := \int_\Omega f(\zeta)\mathcal{K}(\zeta - x)d\mu_\zeta. \tag{18}$$

where $d\mu$ is the differential form of the volume. Among its properties the identity

$$\mathcal{D}_r \circ T_r = I \tag{19}$$

on the spaces $\mathcal{L}_2(\Omega, \mathbb{H})$ and $C(\Omega, \mathbb{H})$, see [15], means that T_r is a right-inverse to $\mathcal{D}_r$ on these spaces.

3.2. Möbius transformations in $\mathbb{R}^3$ in quaternionic terms

It was discovered by K.Th. Vahlen in 1902 and recently rediscovered by L.V. Ahlfors, that, like in the complex analysis case, the Möbius transformations in $\mathbb{R}^4$ can be represented by quaternionic fractional-linear transformations:

$$F(x) = (ax + b)(cx + d)^{-1}, \quad \text{with} \quad a, b, c, d \in \mathbb{H}.$$

In our case we are interested in the Möbius transformations in $\mathbb{R}^3$. Their analytic description is obtained by restricting some of the above quaternionic fractional-linear transformations onto $\mathbb{R}^3$ which leads to the representations of the basic Möbius transformations:

Translation: $T_1(x) = x + q, \quad \forall x \in \mathbb{R}^3, \quad \text{with} \quad q \in \mathbb{R}^3.$

Rotation: $T_2(x) = cx\bar{c}, \quad \forall x \in \mathbb{R}^3, \quad \text{with} \quad c \in \mathbb{H}$ and $|c| = 1.$

Inversion: $T_3(x) = x^{-1} = -\dfrac{x}{|x|^2}, \quad \forall x \in \mathbb{R}^3 \setminus \{0\}, \quad 0^{-1} := \infty,$ (20)
$\infty^{-1} := 0.$

Dilation: $T_4(x) = \lambda x, \quad \forall x \in \mathbb{R}^3, \quad \text{with} \ \lambda$ a positive real number.

What is more, working with these four transformations we obtain that any Möbius transformation in $\mathbb{R}^3$ is expressed as follows:

$$T(x) = (ax + b)(cx + d)^{-1}, \quad x \in \mathbb{R}^3, \tag{21}$$

with $a, b, c, d \in \mathbb{H}$ such that

a) $bd^{-1} \in \mathbb{R}^3$, $d^{-1} = \bar{a}$ if $c = 0$,
b) $d(b - ac^{-1}d)^{-1}$, $ac^{-1} \in \mathbb{R}^3$, $(b - ac^{-1}d)^{-1} = \bar{c}$ if $c \neq 0$.

3.3. Möbius transformations and hyperholomorphy

According to the chain rule, the composition of two holomorphic functions is again a holomorphic function. In our case, the composition of two hyperholomorphic functions not always is defined but even if it is, it is not in general hyperholomorphic.

Example. 3.3.1. Any function given by $f_{i,j}(x) = f_{i,j}(x_1, x_2, x_3) := x_i e_j + x_j e_i$, $i, j \in \{1, 2, 3\}$ with $i < j$, is hyperholomorphic in the whole $\mathbb{R}^3$; however, the composition of any pair of these functions is not hyperholomorphic. For instance, the functions

$$f_{1,2} \circ f_{2,3}(x) = x_2 e_1, \quad f_{1,2} \circ f_{1,3}(x) = x_1 e_2, \quad f_{1,2} \circ f_{1,2}(x) = x_1 e_1 + x_2 e_2$$

are not hyperholomorphic in $\mathbb{R}^3$ because

$$\mathcal{D}[f_{1,2} \circ f_{2,3}](x) = -e_3, \quad \mathcal{D}[f_{1,2} \circ f_{1,3}](x) = e_3, \quad \mathcal{D}[f_{1,2} \circ f_{1,2}](x) = -2.$$

There exists a "weakened analogue" of the possibility to compose hyperholomorphic functions, namely, composing a hyperholomorphic function with a Möbius transformation in $\mathbb{R}^3$ we get an "almost" hyperholomorphic function, it lacks a special coefficient only that depends on the Möbius transformation in $\mathbb{R}^3$.

Theorem. 3.3.2. *Let $\Xi, \Omega \subset \mathbb{R}^3$ be domains such that there exists a Möbius transformation in $\mathbb{R}^3$, T, given by (21) with $\Omega = T(\Xi)$. Denote $y := T(x)$ and define*

$$\begin{aligned} \alpha_T(x) &:= \begin{cases} \bar{a}, & \text{if } c = 0; \\ -\bar{c}\dfrac{cx\bar{c} + d\bar{c}}{|cx\bar{c} + d\bar{c}|^3}, & \text{if } c \neq 0, \end{cases} \qquad \forall x \in \Xi, \\ \beta_T(y) &:= \begin{cases} |a|^2 \bar{a}, & \text{if } c = 0; \\ |c|^2 |y - ac^{-1}|^3 (y - ac^{-1})\bar{c}, & \text{if } c \neq 0, \end{cases} \qquad \forall y \in \Omega. \end{aligned} \tag{22}$$

Then

$$\mathcal{D}_x[\alpha_T f \circ T] = (\beta_T \circ T)\, \mathcal{D}_y[f] \circ T, \quad \forall f \in C^1(\Omega, \mathbb{H}). \tag{23}$$

Proof. It is enough to consider the Möbius transformations given by (20).

Let x^* be an arbitrary point of Ξ.

1. Translation: $y = T(x) = x + u$, where $u \in \mathbb{R}^3$. By direct calculation

$$\mathcal{D}_x[f \circ T](x^*) = \mathcal{D}_y[f](T(x^*)).$$

2. Dilation: $y = T(x) = \lambda x$, with $\lambda > 0$; we have that

$$\mathcal{D}_x[f \circ T](x^*) = \lambda \mathcal{D}_y[f](T(x^*)).$$

3. Rotation. Firstly, for quaternionic rotations of the kind $y = T(x) = ax\bar{a}$ where a is a unitary vector one obtains:

$$\mathcal{D}_x[af \circ T] = \sum_{i=1}^{3} e_i \frac{\partial}{\partial x_i} af \circ T = \sum_{i=1}^{3} e_i a \sum_{j=1}^{3} \frac{\partial f}{\partial y_j} \frac{\partial y_j}{\partial x_i}$$

$$= \sum_{i=1}^{3} \sum_{j=1}^{3} (-2) a_i a_j e_i a \frac{\partial f}{\partial y_j} \circ T + \sum_{i=1}^{3} e_i a \frac{\partial f}{\partial y_j} \circ T = \bar{a} \mathcal{D}_y[f] \circ T.$$

Furthermore, for any quaternion c there exist vectors $a, b \in \mathbb{R}^3$ such that $c = ab$, see [10], and using the above formula two times we obtain: if $y = T(x) = cx\bar{c}$, with c a unitary quaternion, then

$$\mathcal{D}_x[\bar{c}f \circ T](x^*) = \bar{c}\, \mathcal{D}_y[f](T(x^*)).$$

4. Inversion: $y = T(x) = x^{-1}$; we have that

$$\mathcal{D}_x[\frac{x}{|x|^3} f \circ T](x^*) = \sum_{i=0}^{3} e_i \frac{\partial}{\partial x_i} \frac{x}{|x|^3} f \circ T(x^*)$$

$$= -\sum_{i=1}^{3} e_i \frac{x}{|x|^5} \frac{\partial f}{\partial y_i} (T(x^*)) + 2 \sum_{j=1}^{3} \sum_{i=1}^{3} \frac{x_j x_i}{|x|^7} e_j x \frac{\partial f}{\partial y_i} (T(x^*))$$

$$= -|T(x^*)|^3 T(x^*) \mathcal{D}_y[f] \circ T(x^*).$$

These four formulas are particular cases of (23) and the general case reduces to a combination of them. □

Notation. 3.3.3.

- Given two domains $\Xi, \Omega \subset \mathbb{R}^3$, let $\alpha : \Xi \to \Omega$ be a one-to-one correspondence; if f belongs to a function space on Ω then α generates the operator of the change of variable denoted as $W_\alpha : f \mapsto f \circ \alpha$.
- For the same function space, if α and β are $\mathbb{H}$-valued functions then ${}^{\alpha}M$, M^{β} and ${}^{\alpha}M^{\beta}$ denote, respectively, the operators of the multiplication on the left, on the right, and on both sides:

$${}^{\beta}M : f \mapsto \beta f, \quad M^{\beta} : f \mapsto f\beta, \quad {}^{\alpha}M^{\beta} : f \mapsto \alpha f \beta.$$

- For a fixed $x \in \Omega$, denote by Φ_x the evaluation functional defined on a function space as $\Phi_x[f] := f(x)$.

Theorem 3.3.2, in terms of the last notation, means that

$$\mathcal{D}_x \circ {}^{\alpha_T}M \circ W_T = W_T \circ {}^{\beta_T}M \circ \mathcal{D}_y \quad \text{on the space} \quad C^1(\Omega, \mathbb{H}). \tag{24}$$

The above fact complements a series of results, see, e.g., [7], [12], [13], and references therein meaning that the operator $\mathcal{D}$ is conformally covariant.

Corollary. 3.3.4. *Let $f \in C^1(\Omega, \mathbb{H})$, then $f \in \mathfrak{M}(\Omega)$ if and only if $\alpha_T \cdot f \circ T \in \mathfrak{M}(\Xi)$; here α_T is defined by (22).*

Proposition. 3.3.5 (Isomorphism between right-linear spaces of hyperholomorphic functions emerging under Möbius transformations). *Under conditions of Theorem 3.3.2 the operator* ${}^{\alpha_T}M \cdot W_T : \mathfrak{M}(\Omega) \longrightarrow \mathfrak{M}(\Xi)$ *is an isomorphism of quaternionic right-linear spaces.*

3.4. Some properties of quaternionic Hilbert spaces

The concept of a quaternionic Hilbert space is well known and can be found in many sources. In this subsection we give the quaternionic generalizations of the two facts of the theory of complex Hilbert spaces which can be found in [11].

Denote by E a quaternionic Hilbert space, and recall that a subset F of a real linear space is called convex if for any $a, b \in F$ there holds $bt + a(1-t) \in F, \quad \forall t \in [0,1]$.

Proposition. 3.4.1. *Let $F \subset E$ be a closed and convex set. Then given $g \in E$ there exists a unique element $f \in F$ such that*

$$\|g - f\|_E = \operatorname{dist}(g, F) := \inf\{\|g - h\|_E \mid h \in F\}.$$

Proof. Let $\{h_n\} \subset F$ be a sequence such that $\lim_{n\to\infty} \|g - h_n\|_E = \operatorname{dist}(g, F)$. From the parallelogram identity we obtain:

$$\frac{1}{2}\|h_m - h_n\|_E^2 = \|g - h_n\|_E^2 + \|g - h_m\|_E^2 - \|g - \frac{1}{2}(h_n + h_m)\|_E^2,$$

which implies that there exists $f \in F$ such that $\lim_{n\to\infty} \|f - h_n\|_E = 0$ and also from the parallelogram identity one obtains that f is the unique element of F satisfying:

$$\|g - f\|_E = \operatorname{dist}(g, F).$$

□

Proposition. 3.4.2. *If F is a closed subspace of E, then for a pair g and f from the above proposition there holds: $g - f \in F^\perp$.*

Proof. Given $h \in F$ and $q \in \mathbb{H}$, we have that

$$0 \le \|g - (f + hq)\|_E^2 - \|g - f\|_E^2 = -\langle g - f, h\rangle_E\, q - \bar{q}\,\langle h, g - f\rangle_E + |q|^2\|h\|_E^2.$$

Let $q = r\,\langle h, g - f\rangle_E$, with r an arbitrary positive real number , then

$$0 \le -2r\|\,\langle h, g - f\rangle_E\,\|_E^2 + r^2\|\,\langle h, g - f\rangle_E\,\|_E^2\|h\|_E^2.$$

If we suppose that $\langle h, g - f\rangle_E \neq 0$ then $\frac{1}{2r} \le \|h\|_E^2$, which is not possible. Therefore $\langle h, g - f\rangle_E = 0$. □

Note. 3.4.3. We shall work mostly with particular quaternionic Hilbert spaces, namely, with the Hilbert space of the Lebesgue-measurable functions $f : \Omega \subset \mathbb{R}^3 \longrightarrow \mathbb{H}$ such that $\int_\Omega |f|^2 d\mu < +\infty$. It forms a quaternionic bi-linear space. We'll use the same notation $\mathcal{L}_2(\Omega, \mathbb{H})$ for both the left-linear and the right-linear cases. If $\mathcal{L}_2(\Omega, \mathbb{H})$ is seen as a right-linear space then it becomes a quaternionic right Hilbert space endowed with the inner product $\langle f, g\rangle_{\mathcal{L}_2(\Omega,\mathbb{H})} := \int_\Omega \bar{f} g\, d\mu, \quad f, g \in \mathcal{L}_2(\Omega, \mathbb{H})$, and with the norm denoted by $\|\cdot\|_{\mathcal{L}_2(\Omega,\mathbb{H})}$.

A more general example is a weighted $\mathcal{L}_2$-space with weight $\rho : \Omega \longrightarrow \mathbb{R}^+$, it is denoted by $\mathcal{L}_{2,\rho}(\Omega, \mathbb{H})$, meaning that $\int_\Omega |f|^2 \rho d\mu < +\infty$ for $f \in \mathcal{L}_{2,\rho}(\Omega, \mathbb{H})$. Now, the inner product is

$$\langle f, g\rangle_{\mathcal{L}_{2,\rho}(\Omega,\mathbb{H})} := \int_\Omega \bar{f} g \rho d\mu, \quad f, g \in \mathcal{L}_{2,\rho}(\Omega, \mathbb{H}),$$

and the corresponding norm is denoted by $\|\cdot\|_{\mathcal{L}_{2,\rho}(\Omega,\mathbb{H})}$.

3.5. The hyperholomorphic Bergman space

Let $\Omega \subset \mathbb{R}^3$ be a domain.

Definition. 3.5.1. By weighted hyperholomorphic Bergman space, with weight $\rho : \Omega \to \mathbb{R}^+$, associated to Ω we will understand the space $\mathcal{A}_\rho(\Omega) := \mathfrak{M}(\Omega) \cap \mathcal{L}_{2,\rho}(\Omega, \mathbb{H})$, equipped with the inner product and the norm inherited from $\mathcal{L}_{2,\rho}(\Omega, \mathbb{H})$, denoted, respectively, by $\langle \cdot, \cdot \rangle_{\mathcal{A}_\rho(\Omega)}$ and $\|\cdot\|_{\mathcal{A}_\rho(\Omega)}$.

The weighted right-hyperholomorphic Bergman space associated to Ω is defined by $\mathcal{A}_{\rho,r}(\Omega) := \mathfrak{M}_r(\Omega) \cap \mathcal{L}_{2,\rho}(\Omega, \mathbb{H})$, with the inner product and the norm inherited from $\mathcal{L}_{2,\rho}(\Omega, \mathbb{H})$, and denoted respectively by $\langle \cdot, \cdot \rangle_{\mathcal{A}_{\rho,r}(\Omega)}$ and $\|\cdot\|_{\mathcal{A}_{\rho,r}(\Omega)}$.

In case, $\rho \equiv 1$ the space $\mathcal{A}_\rho(\Omega)$ will be denoted as $\mathcal{A}(\Omega)$ and it will be called hyperholomorphic Bergman space. In the same way, the space $\mathcal{A}_{\rho,r}(\Omega) = \mathcal{A}_r(\Omega)$ will be called right-hyperholomorphic Bergman space.

Proposition. 3.5.2. *Let $\mathcal{C} \subset \Omega$ be a compact set, then*

$$\sup\{|f(x)| \mid x \in \mathcal{C}\} \leq k_{\mathcal{C}} \|f\|_{\mathcal{A}(\Omega)}, \quad \forall f \in \mathcal{A}(\Omega), \tag{25}$$

where $k_{\mathcal{C}} > 0$ depends on $\mathcal{C}$.

Proof. Let $r_c > 0$ be such that $\mathbb{B}(x, r_c) \subset \mathcal{C}$, and let $f \in \mathcal{A}(\Omega)$, then the hyperholomorphic Cauchy integral formula implies that

$$f(x) = \frac{1}{4\pi r_c^2} \int_{\mathbb{S}(x,r_c)} (\zeta - x) \sigma_\zeta^{(2)} f(\zeta).$$

Applying Stokes' Theorem we obtain:

$$|f(x)| \leq \frac{3}{4\pi r_c^2} \int_{\mathbb{B}(x,r_c)} |f(\zeta)| d\mu_\zeta.$$

Using the Hölder inequality for $f, 1 \in \mathcal{L}_2(\mathbb{B}(x, r_c), \mathbb{H})$, from the last expression we arrive at

$$|f(x)| \leq m_c \left(\int_{\mathbb{B}(x,r_c)} |f|^2 d\mu \right)^{\frac{1}{2}} \leq m_c \|f\|_{\mathcal{A}(\Omega)},$$

where $m_c = \dfrac{3}{4\pi r_c^2} \mu(\mathbb{B}(x, r_c))^{\frac{1}{2}}$. Finally, as $\mathcal{C}$ is a compact set, one obtains that there exists a constant k_c, such that

$$\sup\{|f(x)| \mid x \in \mathcal{C}\} \leq k_c \|f\|_{\mathcal{A}(\Omega)}. \qquad \square$$

Corollary. 3.5.3. *Given $x \in \Omega$, the evaluation functional ϕ_x is bounded on $\mathcal{A}(\Omega)$.*

Proposition. 3.5.4. *The space $(\mathcal{A}(\Omega), \|\cdot\|_{\mathcal{A}(\Omega)})$ is a right-$\mathbb{H}$-linear Banach space.*

Proof. Let $\{f_n\} \subset \mathcal{A}(\Omega)$ be a Cauchy sequence, therefore there exists $f \in \mathcal{L}_2(\Omega, \mathbb{H})$ such that $f_n \underset{n\to\infty}{\overset{\mathcal{L}_2}{\longrightarrow}} f$. By (25) we note that there exists a function f^* defined on Ω such that $\{f_n\} \to f^*$ uniformly on compact sets.

Given $\epsilon > 0$ sufficiently small the Cauchy integral formula for any f_n is

$$f_n(x) = \int_{\partial\mathbb{B}(x,\epsilon)} \mathcal{K}(\zeta - x)\sigma_\zeta^{(2)} f_n(\zeta) d\mu_\zeta, \quad \forall n \in \mathbb{N}.$$

As $f_n \to f^*$ uniformly on $\partial\mathbb{B}(x,\epsilon)$, then

$$f^*(x) = \int_{\partial\mathbb{B}(x,\epsilon)} \mathcal{K}(\zeta - x)\sigma_\zeta^{(2)} f^*(\zeta) d\mu\zeta,$$

therefore $f^* \in \mathfrak{M}(\Omega)$. Since

$$\left(\int_{\mathcal{C}} |f^* - f|^2 d\mu\right)^{\frac{1}{2}} \le \left(\int_{\mathcal{C}} |f^* - f_n|^2 d\mu\right)^{\frac{1}{2}} + \left(\int_{\mathcal{C}} |f - f_n|^2 d\mu\right)^{\frac{1}{2}} \to 0,$$

when $n \to \infty$ for any $\mathcal{C} \subset \Omega$ compact, we obtain that $f^* = f \in \mathcal{A}(\Omega)$. □

Of course, the above three facts are crucial since they allow us to introduce the hyperholomorphic Bergman kernel as follows.

By the Riesz representation theorem for quaternionic right-Hilbert spaces, see [1], given a fixed point $x \in \Omega$ there exists $B_x \in \mathcal{A}(\Omega)$ such that $\phi_x(f) := f(x) = \langle B_x, f\rangle_{\mathcal{A}(\Omega)}$, for all $f \in \mathcal{A}(\Omega)$. The function $\mathcal{B}_\Omega(x,\cdot) := \overline{B_x(\cdot)}$ will be called the hyperholomorphic Bergman kernel associated to Ω, and its definition implies that

$$f(x) = \int_\Omega \mathcal{B}_\Omega(x,\cdot) f d\mu, \quad \forall f \in \mathcal{A}(\Omega). \tag{26}$$

Theorem. 3.5.5 (General properties of the hyperholomorphic Bergman kernel).

a) *$\mathcal{B}_\Omega(\cdot,\cdot)$ is quaternionic hermitian: $\mathcal{B}_\Omega(x,\xi) = \overline{\mathcal{B}_\Omega(\xi,x)}, \quad \forall x, \xi \in \Omega$.*

b) *$\mathcal{B}_\Omega(\cdot,\cdot)$ is hyperholomorphic in the first variable and right-hyperholomorphic in the second variable.*

c) *For a fixed $\xi \in \Omega$, the function $\mathcal{B}_\Omega(\cdot,\xi)$ is the unique element of $\mathcal{A}(\Omega)$ which is hermitian and is reproducing; i.e., it satisfies* (26).

Proof. a) Directly, one has:

$$\mathcal{B}_\Omega(\xi,x) = \overline{B_\xi(x)} = \overline{\int_\Omega \overline{B_x(\varsigma)} B_\xi(\varsigma) d\mu(\varsigma)} = B_x(\xi) = \overline{\mathcal{B}_\Omega(x,\xi)}.$$

b) $B_x \in \mathcal{A}(\Omega)$ and $\mathcal{B}_\Omega(\cdot,x) = B_x(\cdot)$; hence $\mathcal{B}_\Omega(\cdot,x)$ is hyperholomorphic on Ω in the first variable, and the identity

$$\mathcal{D}_r[\mathcal{B}_\Omega(x,\cdot)] = -\overline{\mathcal{D}[\mathcal{B}_\Omega(\cdot,x)]} = 0,$$

implies that $\mathcal{B}_\Omega(x,\cdot)$ is a right-hyperholomorphic function.

c) Let $H(\cdot,\cdot)$ be a function on $\Omega \times \Omega$ such that given $\xi \in \Omega$, the function $H(\cdot,\xi) \in \mathcal{A}(\Omega)$ is reproducing and hermitian. Then

$$H(x,\xi) = \int_\Omega \mathcal{B}_\Omega(x,\zeta)H(\zeta,\xi)d\mu_\zeta = \overline{\int_\Omega \overline{H(\zeta,\xi)}\ \overline{\mathcal{B}_\Omega(x,\zeta)}d\mu_\zeta}$$
$$= \overline{\int_\Omega H(\xi,\zeta)\mathcal{B}_\Omega(\zeta,x)d\mu_\zeta} = \overline{\mathcal{B}_\Omega(\xi,x)} = \mathcal{B}_\Omega(x,\xi). \qquad \square$$

The following proposition is a deep analogue to a fact known in one complex variable theory, see [3], and in quaternionic analysis for the Fueter operator, see [5].

Proposition. 3.5.6 (The normalized MT-hyperholomorphic Bergman kernel as a solution of a variational problem). *Let $\xi \in \Omega$ be a fixed point, define $F_\xi := \{f \in \mathcal{A}(\Omega) \mid f(\xi) = 1\}$, then the function f_0 defined by*

$$f_0(x) := \frac{\mathcal{B}_\Omega(x,\xi)}{\mathcal{B}_\Omega(\xi,\xi)}, \qquad x \in \Omega,$$

belongs to F_ξ and it is the only solution to the variational problem of finding $\inf\{\|f\|_{\mathcal{A}(\Omega)} \mid f \in F_\xi\}$:

$$\|f_0\|_{\mathcal{A}(\Omega)} = \inf\{\|f\|_{\mathcal{A}(\Omega)} \mid f \in F_\xi\}.$$

Proof. Using (26) we arrive at the following:

$$\mathcal{B}_\Omega(\xi,\xi) = \int_\Omega \mathcal{B}_\Omega(\xi,\zeta)\mathcal{B}_\Omega(\zeta,\xi)d\mu_\zeta = \|\mathcal{B}_\Omega(\cdot,\xi)\|^2_{\mathcal{A}(\Omega)} < \infty,$$

implying, in particular, that $f_0 = \dfrac{\mathcal{B}_\Omega(\cdot,\xi)}{\mathcal{B}_\Omega(\xi,\xi)} \in F_\xi$. Furthermore, given any $f \in F_\xi$ there holds:

$$1 = f(\xi) = \int_\Omega \mathcal{B}_\Omega(\xi,\zeta)f(\zeta)d\mu_\zeta.$$

Applying the Cauchy-Buniakovsky inequality to the latter expression we obtain:

$$\frac{1}{\|\mathcal{B}_\Omega(\cdot,\xi)\|_{\mathcal{A}(\Omega)}} \le \|f\|_{\mathcal{A}(\Omega)}$$

and from the shown facts one concludes that

$$\|\frac{\mathcal{B}_\Omega(\cdot,\xi)}{\mathcal{B}_\Omega(\xi,\xi)}\|_{\mathcal{A}(\Omega)} \le \|f\|_{\mathcal{A}(\Omega)}.$$

This inequality is true for any $f \in F_\xi$, therefore f_0 is the solution of the variational problem; it is easy to see that f_0 is unique. $\square$

We complement the above proposition with some facts which seem to be unknown both in one-dimensional complex analysis and in quaternionic Fueter analysis settings.

Proposition. 3.5.7. *Let $g \in \mathcal{A}(\Omega)$ then there exists a unique function $f \in F_\xi$ such that*

$$\|g - f\|_{\mathcal{A}(\Omega)} = \mathrm{dis}(g, F_\xi) := \inf\{\|g - h\|_{\mathcal{A}(\Omega)} \mid h \in F_\xi\}.$$

Proof. Since F_ξ is a convex and closed set the proof is a direct application of Proposition 3.4.1, where $F = F_\xi$ and $E = \mathcal{A}(\Omega)$. □

3.5.8. In particular, taking $g = \mathcal{B}_\Omega(\cdot,\xi)$ and denoting by $f^\odot \in F_\xi$ the corresponding function, that is,

$$\|\mathcal{B}_\Omega(\cdot,\xi) - f^\odot\|_{\mathcal{A}(\Omega)} = \operatorname{dis}(\mathcal{B}_\Omega(\cdot,\xi), F_\xi),$$

we observe the following properties of $f^\odot$.

Corollary. 3.5.9.

$$\|\mathcal{B}_\Omega(\cdot,\xi) - f^\odot\|_{\mathcal{A}(\Omega)} \leq \|f\|_{\mathcal{A}(\Omega)} |\mathcal{B}_\Omega(\xi,\xi) - 1| \quad \forall f \in F_\xi.$$

Proof. As $\dfrac{\mathcal{B}_\Omega(\cdot,\xi)}{\mathcal{B}_\Omega(\xi,\xi)} \in F_\xi$, there holds:

$$\|\mathcal{B}_\Omega(\cdot,\xi) - f^\odot\|_{\mathcal{A}(\Omega)} \leq \|\mathcal{B}_\Omega(\cdot,\xi) - \frac{\mathcal{B}_\Omega(\cdot,\xi)}{\mathcal{B}_\Omega(\xi,\xi)}\|_{\mathcal{A}(\Omega)} \leq \|f\|_{\mathcal{A}(\Omega)} \|\mathcal{B}_\Omega(\xi,\xi) - 1\|_{\mathcal{A}(\Omega)},$$

for all $f \in F_\xi$. □

What is more, one has

Corollary. 3.5.10.

1. *If* $\mathcal{B}_\Omega(\xi,\xi) > 1$, *then* $\dfrac{\|\mathcal{B}_\Omega(\cdot,\xi)\|_{\mathcal{A}(\Omega)}}{\mathcal{B}_\Omega(\xi,\xi)} \leq \|f^\odot\|_{\mathcal{A}(\Omega)} \leq \|\mathcal{B}_\Omega(\cdot,\xi)\|_{\mathcal{A}(\Omega)} \mathcal{B}_\Omega(\xi,\xi)$.
2. *If* $\mathcal{B}_\Omega(\xi,\xi) \leq 1$, *then* $f^\odot = \dfrac{\mathcal{B}_\Omega(\cdot,\xi)}{\mathcal{B}_\Omega(\xi,\xi)}$, *i.e., the solution* $f^\odot$ *is the normalization of the Bergman kernel.*

Definition 3.5.11 (Hyperholomorphic Bergman projection). The operator $\mathfrak{B}_\Omega$ is defined on the space $\mathcal{L}_2(\Omega,\mathbb{H})$ by

$$\mathfrak{B}_\Omega[f](x) := \int_\Omega \mathcal{B}_\Omega(x,\xi) f(\xi) d\mu_\xi.$$

For an arbitrary $f \in \mathcal{L}_2(\Omega,\mathbb{H})$ there holds: $f = f_1 + f_2$ with the unique $f_1 \in \mathcal{A}(\Omega)$ and $f_2 \in \mathcal{A}(\Omega)^\perp$. Since by (26) we have that $\mathfrak{B}_\Omega[f_1] = f_1$, and as $\mathfrak{B}_\Omega[f_2](x) = \int_\Omega \overline{\mathcal{B}_\Omega(\xi,x)} f_2(\xi) d\mu_\xi = 0$, for a.e. $x \in \Omega$, we conclude that $\mathfrak{B}_\Omega[f] = f_1$ and $\mathfrak{B}_\Omega\mathfrak{B}_\Omega[f] = \mathfrak{B}_\Omega[f]$. Then $\mathfrak{B}_\Omega[\mathcal{L}_2(\Omega,\mathbb{H})] = \mathcal{A}(\Omega)$ and $\mathfrak{B}_\Omega^2 = \mathfrak{B}_\Omega$.

Moreover, the operator $\mathfrak{B}_\Omega$ is symmetric, i.e.,

$$\langle \mathfrak{B}_\Omega[f], g\rangle_{\mathcal{L}_2(\Omega,\mathbb{H})} = \langle f, \mathfrak{B}_\Omega[g]\rangle_{\mathcal{L}_2(\Omega,\mathbb{H})} \quad \forall f,g \in \mathcal{L}_2(\Omega,\mathbb{H}).$$

Then $\mathfrak{B}_\Omega$ satisfies the Closed Graph Theorem therefore $\mathfrak{B}_\Omega$ is a continuous operator. By previous reasonings $\mathfrak{B}_\Omega$ is called hyperholomorphic Bergman projection associated to the space $\mathcal{A}(\Omega)$.

Observation. 3.5.12. The space $\mathcal{A}_r(\Omega)$ has properties similar to those of $\mathcal{A}(\Omega)$.This can be obtained directly or using the ($\mathbb{R}$-linear) operator $Z_{\mathbb{H}} : \mathcal{A}(\Omega) \to \mathcal{A}_r(\Omega)$, defined by $Z_{\mathbb{H}}(f) := \bar{f}$, $\forall f \in \mathcal{A}(\Omega)$. For example, $\mathcal{D} = Z_{\mathbb{H}} \circ \overline{\mathcal{D}}_r \circ Z_{\mathbb{H}}$ leads to

$$\mathcal{B}_{\Omega,r}(x,\zeta) = \overline{\mathcal{B}_{\Omega}(x,\zeta)} = \mathcal{B}_{\Omega}(\zeta, x).$$

3.6. Möbius transformations in $\mathbb{R}^3$ and function spaces

Let $\Omega, \Xi \subset \mathbb{R}^3$ be domains such that there exists a Möbius transformation in $\mathbb{R}^3$, satisfying $T(\Xi) = \Omega$. We are interested in the relations between some function spaces on Ξ and on Ω.

Definition. 3.6.1. Let T be Möbius transformations in $\mathbb{R}^3$ given by (21), such that $T(\Xi) = \Omega$. Let $x \in \Xi$, define:

$$C_T(x) := \begin{cases} |a|^2\bar{a} & \text{if } c = 0, \\ -|c|^2\bar{c}\dfrac{cx\bar{c} + d\bar{c}}{|cx\bar{c} + d\bar{c}|^3}, & \text{if } c \neq 0, \end{cases}$$

$$\rho_T(x) := \begin{cases} 1, & \text{if } c = 0, \\ \dfrac{1}{|cx\bar{c} + d\bar{c}|^2}, & \text{if } c \neq 0. \end{cases}$$

Theorem. 3.6.2. *The operator* ${}^{C_T}M \circ W_T : \mathcal{L}_2(\Omega, \mathbb{H}) \longrightarrow \mathcal{L}_{2,\rho_T}(\Xi, \mathbb{H})$ *is an isometric isomorphism of quaternionic Hilbert spaces and its inverse is* $W_{T^{-1}} \circ {}^{C_T^{-1}}M$.

Proof. Observe that C_T does not vanish in Ξ and $({}^{C_T}M \circ W_T)^{-1} = W_{T^{-1}} \circ {}^{C_T^{-1}}M$, and that the function C_T^{-1} is defined at each point $x \in \Xi$ by $C_T^{-1}(x) := (C_T(x))^{-1}$, while the transformation T^{-1} is the inverse transformation of T.

For the basic Möbius transformations we have:

1. The dilatation $T(x) = \lambda x = y$ implies that given $f, g \in \mathcal{L}_2(\Omega, \mathbb{H})$ there holds:

$$\langle f, g\rangle_{\mathcal{L}_2(\Omega,\mathbb{H})} = \int_{\Omega} \overline{f(y)}g(y)d\mu_y = \int_{\Xi} \overline{\lambda^{\frac{3}{2}} f \circ T(x)}\lambda^{\frac{3}{2}} g \circ T(x) d\mu_x. \tag{27}$$

When $f = g$ then $\displaystyle\int_{\Omega} |f(y)|^2 d\mu_y = \int_{\Xi} |f(T(x))|^2\lambda^3 d\mu_x$, meaning that

$$^{C_T}M \circ W_T : \mathcal{L}_2(\Omega, \mathbb{H}) \to \mathcal{L}_{2,\rho}(\Xi, \mathbb{H}),$$

and it is isometric. Hence, (27) may be written now as

$$\langle f, g\rangle_{\mathcal{L}_2(\Omega,\mathbb{H})} = \left\langle {}^{\lambda^{\frac{3}{2}}} M \circ W_T[f], {}^{\lambda^{\frac{3}{2}}} M \circ W_T[g] \right\rangle_{\mathcal{L}_2(\Xi,\mathbb{H})},$$

and this is all.

In the other cases we give the analogues of the formula (27) only.

2. For the inversion $T(x) = x^{-1} = y$, there holds:
$$\langle f, g\rangle_{\mathcal{L}_2(\Omega,\mathbb{H})} = \int_\Xi \overline{\frac{\bar{x}}{|x|^3} f \circ T(x)} \frac{\bar{x}}{|x|^3} g \circ T(x) \rho(x) d\mu_x,$$
where $\rho(x) = \dfrac{1}{|x|^2}$.
3. For the translation $T(x) = x + e = y$, where $e \in \mathbb{R}^3$, one has:
$$\langle f, g\rangle_{\mathcal{L}_2(\Omega,\mathbb{H})} = \int_\Xi \overline{f \circ T(x)} g \circ T(x) d\mu_x.$$
4. For the rotation $T(x) = cx\bar{c} = y$, where $c \in \mathbb{H}$, $|c| = 1$, we obtain:
$$\langle f, g\rangle_{\mathcal{L}_2(\Omega)} = \int_\Xi \overline{\bar{c} f \circ T(x)} \bar{c} g \circ T(x) d\mu_x.$$

Of course, the general case of an arbitrary T is just a combination of the above four. □

Corollary. 3.6.3. *The operator:* ${}^{C_T}M \circ W_T : \mathcal{A}(\Omega) \longrightarrow \mathcal{A}_{\rho_T}(\Xi)$ *is an isometric isomorphism of quaternionic right linear Hilbert spaces.*

Proof. It is due to the fact that the coefficients C_T in Theorem 3.6.1 and α_T in the equation (22) differ by a positive constant factor only. □

Proposition 4 in [8] shows a fact similar to Corollary 3.6.3 for the hyperholomorphic Hardy space theory.

Corollary. 3.6.4. *The space* $(\mathcal{A}_{\rho_T}(\Xi), \|\cdot\|_{\mathcal{A}_{\rho_T}(\Xi)})$ *is a quaternionic right-Hilbert space and the evaluation functional is bounded on the space* $\mathcal{A}_{\rho_T}(\Xi)$.

Hence one can develop now for weighted spaces the rest of the theory we have constructed for the spaces without weight. We'll use this freely in what follows.

Corollary. 3.6.5. *The mapping* $\mathcal{J} : \mathfrak{L}(\mathcal{A}(\Omega)) \longrightarrow \mathfrak{L}(\mathcal{A}_{\rho_T}(\Xi))$ *given by*
$$P \longmapsto {}^{C_T}M \circ W_T \circ P \circ W_{T^{-1}} \circ {}^{C_T^{-1}}M, \quad \forall P \in \mathfrak{L}(\mathcal{A}(\Omega)),$$
is an isomorphism of real algebras.

Notation. 3.6.6. For any $\Upsilon \in (\mathcal{A}(\Omega))^*$, let $g \in \mathcal{A}(\Omega)$ be the unique element which corresponds to Υ by the Riesz theorem for quaternionic right-Hilbert space, thus Υ is denoted by Υ_g, the same for $(\mathcal{A}_{\rho_T}(\Xi))^*$.

Corollary. 3.6.7. *The mapping* $\mathfrak{F} : (\mathcal{A}(\Omega))^* \longrightarrow (\mathcal{A}_{\rho_T}(\Xi))^*$ *defined by*
$$\Upsilon_g \longmapsto \Upsilon_{{}^{C_T}M \circ W_T[g]}, \quad \forall \Upsilon_g \in (\mathcal{A}(\Omega))^*,$$
is an isometric isomorphism of quaternionic linear spaces. What is more, there holds:
$$\Upsilon = \mathfrak{F}[\Upsilon] \circ {}^{C_T}M \circ W_T, \quad \forall \Upsilon \in (\mathcal{A}(\Omega))^*,$$
$$\Upsilon = \mathfrak{F}^{-1}[\Upsilon] \circ W_{T^{-1}} \circ {}^{C_T^{-1}}M, \quad \forall \Upsilon \in (\mathcal{A}_{\rho_T}(\Xi))^*.$$

Additionally, as a consequence of Corollary 3.6.3 we obtain

Proposition. 3.6.8. *The hyperholomorphic Bergman kernel of* $\mathcal{A}_{\rho_T}(\Xi)$ *can be expressed as* $\mathcal{B}_{\Xi,\rho_T}(x,\xi) = C_T(x)\mathcal{B}_\Omega(T(x),T(\xi))\overline{C_T(\xi)}$.

Proof. Given $f \in \mathcal{A}_{\rho_T}(\Xi)$ then $W_{T^{-1}} \circ {}^{C_T^{-1}}M[f] \in \mathcal{A}(\Omega)$. Therefore

$$C_T^{-1}(T^{-1}(y))f \circ T^{-1}(y) = \int_\Omega \mathcal{B}_\Omega(y,\xi)C_T^{-1}(T^{-1}(\xi))f \circ T^{-1}(\xi)d\mu_\xi.$$

Applying the operator ${}^{C_T}M \circ W_T$ to the last equality we get

$$f(x) = \int_\Xi C_T(x)\mathcal{B}_\Omega(T(x),T(\xi))\overline{C_T(\xi)}f(\xi)\rho(\xi)d\mu_\xi.$$

It suffices now to allude to the uniqueness of the Bergman kernel. □

Corollary. 3.6.9. *The hyperholomorphic Bergman projection associated to the domain* Ξ *is*

$$\mathfrak{B}_{\Xi,\rho_T} = {}^{C_T}M \circ W_T \circ \mathfrak{B}_\Omega \circ W_{T^{-1}} \circ {}^{C_T^{-1}}M.$$

Observation. 3.6.10. In analogy to the papers [5], [8], [12], [13], and others, the property $\mathfrak{B}_{\Xi,\rho_T} = {}^{C_T}M \circ W_T \circ \mathfrak{B}_\Omega \circ W_{T^{-1}} \circ {}^{C_T^{-1}}M$ means the hyperholomorphic Bergman projection may be called conformally covariant. Similarly, the equality $\mathcal{B}_{\Xi,\rho_T}(z,\xi) = C_T(z)\mathcal{B}_\Omega(T(z),T(\xi))\overline{C_T(\xi)}$ means that the hyperholomorphic Bergman kernel can be called conformally invariant.

3.7. Decomposition of the space $\mathcal{L}_2(\Omega,\mathbb{H})$

Consider $\mathcal{L}_2(\Omega,\mathbb{H})$ as a right-linear quaternionic Hilbert space, then according to the general theory of such spaces, $\mathcal{A}(\Omega)$ has a right-linear orthogonal complement:

$$\mathcal{L}_2(\Omega,\mathbb{H}) = \mathcal{A}(\Omega) \oplus \mathcal{A}(\Omega)^\perp.$$

Of course, in such a form the statement is trivial but the next proposition gives a description of this orthogonal complement in terms of the Bergman kernel.

Proposition. 3.7.1. *Let* Ω *be a bounded domain with the smooth boundary. Given* $f \in \mathcal{L}_2(\Omega,\mathbb{H})$, *it is in* $\mathcal{A}(\Omega)^\perp$ *if and only if*

$$\int_\Omega \overline{f(x)}\mathcal{B}_\Omega(x,\zeta)d\mu_x = 0, \quad \forall \zeta \in \Omega.$$

Proof. ($\Longrightarrow$) It is due to $\mathcal{B}_\Omega(\cdot,\zeta) \in \mathcal{A}(\Omega)$.

($\Longleftarrow$) For all $g \in \mathcal{A}(\Omega)$ the inner product $\langle f,g\rangle_{\mathcal{L}_2(\Omega,\mathbb{H})}$ is equal to

$$\int_\Omega \overline{f(x)} \int_\Omega \mathcal{B}_\Omega(x,\zeta)g(\zeta)d\mu_\zeta d\mu_x = \int_\Omega \left(\int_\Omega \overline{f(x)}\,\mathcal{B}_\Omega(x,\zeta)d\mu_x\right) g(\zeta)d\mu_\zeta = 0. \quad \square$$

What is more, using the above fact one can prove that $\mathcal{A}(\Omega)^{\perp} = \mathcal{D}_M \dot{\mathbf{W}}_2^1(\Omega, \mathbb{H})$, where $\dot{\mathbf{W}}_2^1(\Omega, \mathbb{H})$ represents the set of $\mathbb{H}$-valued functions whose components are in the Sobolev space $\dot{\mathbf{W}}_2^1(\Omega, \mathbb{R})$, thus obtaining the decomposition $\mathcal{L}_2(\Omega, \mathbb{H}) = \mathcal{A}(\Omega) \oplus \mathcal{D}_{MT} \dot{\mathbf{W}}_2^1(\Omega, \mathbb{H})$. We do not enter into the details since the decomposition itself, with a different proof, the reader can find in the book [6] of K. Gürlebeck and W. Sprössig.

4. The SI- and hyperholomorphic Bergman spaces

4.1. Preliminaries

In order to emphasize the relation between the quaternions whose scalar part is zero with the vectors, both will be denoted in the same way using arrows over the letters.

For any pair of quaternions $a = a_0 + \vec{a},\ b = b_0 + \vec{b} \in \mathbb{H}$, we have the following relations between some quaternionic and vectorial operations:

- $(a_0 + \vec{a}) + (b_0 + \vec{b}) = (a_0 + b_0) + (\vec{a} + \vec{b})$,
- $(a_0 + \vec{a})(b_0 + \vec{b}) = a_0 b_0 + b_0 \vec{a} + a_0 \vec{b} - \left\langle \vec{a}, \vec{b} \right\rangle_{\mathbb{R}^3} + \left[\vec{a}, \vec{b} \right]_{\mathbb{R}^3}$,
- $\overline{a_0 + \vec{a}} = a_0 - \vec{a}$,
- $|a|^2 = |a_0|^2 + \|\vec{a}\|^2_{\mathbb{R}^3}$.

Besides, let $\vec{x},\ \vec{y}$ be vectors, their quaternionic product has a direct relation with the scalar and the cross products:

$$\vec{x}\vec{y} = -\langle \vec{x}, \vec{y} \rangle_{\mathbb{R}^3} + [\vec{x}, \vec{y}]_{\mathbb{R}^3},$$

which implies immediately:

- $\langle \vec{x}, \vec{y} \rangle_{\mathbb{R}^3} = -\frac{1}{2}(\vec{x}\vec{y} + \vec{y}\vec{x})$,
- $[\vec{x}, \vec{y}]_{\mathbb{R}^3} = \frac{1}{2}(\vec{x}\vec{y} - \vec{y}\vec{x})$.

Moreover, the norm of a vector is the negative of its square in the sense of quaternionic multiplication:

$$\|\vec{x}\|_{\mathbb{R}^3} = \bar{\vec{x}}\vec{x} = -\vec{x}\vec{x} = -\vec{x}^2.$$

All these properties will be widely used in the proofs of this section.

Let $\Omega \subset \mathbb{R}^3$ be a domain. The application of the Moisil-Théodoresco operator $\mathcal{D}$ to any function $f = f_0 + \vec{f} \in C^1(\Omega, \mathbb{H})$ gives us:

$$\mathcal{D}[f] = \mathcal{D}[f_0 + \vec{f}] = \text{grad } f_0 - \text{div } \vec{f} + \text{rot } \vec{f}.$$

Moreover,

$$\mathcal{D}[f] = 0 \quad \Longleftrightarrow \quad \begin{cases} \text{grad } f_0 = -\text{rot } \vec{f}, \\ \text{div } \vec{f} = 0, \end{cases}$$

$$\mathcal{D}_r[f] = 0 \quad \Longleftrightarrow \quad \begin{cases} \text{grad } f_0 = \text{rot } \vec{f}, \\ \text{div } \vec{f} = 0, \end{cases}$$

which implies

$$\mathbb{R} \oplus \vec{\mathfrak{M}}(\Omega) = \mathfrak{M}(\Omega) \cap \mathfrak{M}_r(\Omega), \tag{28}$$

and one notes that any SI-vector field generates a hyperholomorphic function and a right-hyperholomorphic function.

Observation. 4.1.1. From the equality (28) we obtain that

- If Ω is a bounded domain then

$$\mathbb{R} \oplus \vec{\mathcal{A}}(\Omega) = \mathcal{A}(\Omega) \cap \mathcal{A}_r(\Omega).$$

- If Ω is unbounded domain then

$$\vec{\mathcal{A}}(\Omega) = \mathcal{A}(\Omega) \cap \mathcal{A}_r(\Omega).$$

Observation. 4.1.2. Note that the $\mathbb{R}$-linear space $\vec{\mathcal{A}}(\Omega)$ is a closed subset of $\mathcal{A}(\Omega)$. Therefore Proposition 3.4.1 implies that given $\vec{x} \in \Omega$, there exists a unique function $\vec{g}_{\vec{x}} \in \vec{\mathcal{A}}(\Omega)$ such that

$$\operatorname{dist}(\mathcal{B}_\Omega(\cdot, \vec{x}), \vec{\mathcal{A}}(\Omega)) = \|\mathcal{B}_\Omega(\cdot, \vec{x}) - \vec{g}_{\vec{x}}\|_{\mathcal{A}(\Omega)}.$$

Proposition. 4.1.3. *The function $\vec{g}_{\vec{x}}$ given in the previous observation is identically zero.*

Proof. Note that

$$\|\mathcal{B}_\Omega(\cdot, \vec{x}) - \vec{g}_{\vec{x}}\|^2_{\mathcal{A}(\Omega)} \leq \|\mathcal{B}_\Omega(\cdot, \vec{x}) - \vec{f}\|^2_{\mathcal{A}(\Omega)} \quad \forall \vec{f} \in \vec{\mathcal{A}}(\Omega).$$

In particular if we choose $\vec{f} = \vec{0}$, then there holds:

$$\|\mathcal{B}_\Omega(\cdot, \vec{x}) - \vec{g}_{\vec{x}}\|^2_{\mathcal{A}(\Omega)} \leq \|\mathcal{B}_\Omega(\cdot, \vec{x})\|^2_{\mathcal{A}(\Omega)}.$$

Denoting $\mathcal{B}_\Omega = \mathcal{B}_{\Omega,0} + \vec{\mathcal{B}}_\Omega$, we obtain that

$$\|\vec{\mathcal{B}}_\Omega(\cdot, \vec{x}) - \vec{g}_{\vec{x}}\|^2_{\mathcal{L}(\Omega,\mathbb{R}^3)} \leq \|\vec{\mathcal{B}}_\Omega(\cdot, \vec{x})\|^2_{\mathcal{L}_2(\Omega,\mathbb{R}^3)},$$

and applying the property

$$\int_\Omega \left\langle \vec{\mathcal{B}}_\Omega(\vec{\zeta}, \vec{x}), \vec{g}_{\vec{x}}(\vec{\zeta}) \right\rangle_{\mathbb{R}^3} d\mu_{\vec{\zeta}} = 0,$$

see Theorem 2.2.1, we arrive at

$$\|\vec{g}_{\vec{x}}\|_{\mathcal{L}_2(\Omega,\mathbb{R}^3)} \leq 0.$$ □

4.1.3.1 In conclusion, the element of $\vec{\mathcal{A}}(\Omega)$ closest to $\mathcal{B}_\Omega(\cdot, \vec{x})$ is the function $\vec{0}$, i.e.,

$$\operatorname{dist}(\mathcal{B}_\Omega(\cdot, \vec{x}), \vec{\mathcal{A}}(\Omega)) = \|\mathcal{B}_\Omega(\cdot, \vec{x})\|_{\mathcal{A}(\Omega)}.$$

4.2. Proofs of the statements of Section 2

Proof of Theorem 2.2.1. The fields $\mathcal{B}_{\Omega,0}$ and $\vec{\mathcal{B}}_\Omega$ are, respectively, the scalar part and the vectorial part of MT-hyperholomorphic Bergman kernel $\mathcal{B}_\Omega$ given in (26). Their uniqueness as well as their properties are consequences of Theorem 3.5.5.

On the other hand, it is easy to see that $\hat{\mathcal{L}}_2(\Omega,\mathbb{R}^3)$ is a real-linear space and using the inequality

$$\left|\int_\Omega \left\langle \vec{\mathcal{B}}_\Omega(\vec{x},\vec{v}), \vec{f}(\vec{v})\right\rangle_{\mathbb{R}^3} d\mu_{\vec{v}} - \int_\Omega \left\langle \vec{\mathcal{B}}_\Omega(\vec{x},\vec{v}), \vec{g}(\vec{v})\right\rangle_{\mathbb{R}^3} d\mu_{\vec{v}}\right|$$

$$\leq \left(\int_\Omega |\vec{\mathcal{B}}_\Omega(\vec{x},\vec{v})|^2 d\mu_{\vec{v}}\right)^{\frac{1}{2}} \|\vec{f}-\vec{g}\|_{\mathcal{L}_2(\Omega,\mathbb{R}^3)} \quad \forall \vec{f},\vec{g} \in \mathcal{L}_2(\Omega,\mathbb{R}^3),$$

we obtain that $\hat{\mathcal{L}}_2(\Omega,\mathbb{R}^3)$ is a complete space.

Besides, due to $\vec{\mathfrak{M}}(\Omega)\cap\mathcal{L}_2(\Omega,\mathbb{H}) \subset \mathcal{A}(\Omega)$ we see that any element $\vec{g} \in \vec{\mathfrak{M}}(\Omega)\cap \mathcal{L}_2(\Omega,\mathbb{H})$ satisfies (26), which implies that

$$\int_\Omega \left\langle \vec{\mathcal{B}}_\Omega(\vec{x},\vec{v}), \vec{g}(\vec{v})\right\rangle_{\mathbb{R}^3} d\mu_{\vec{v}} = 0, \quad \text{for } \textit{a.e. } \vec{x} \in \Omega.$$

Moreover $\vec{\mathfrak{M}}(\Omega)\cap\mathcal{L}_2(\Omega,\mathbb{H}) \neq \hat{\mathcal{L}}_2(\Omega,\mathbb{R}^3)$, since the elements of $\mathcal{A}(\Omega)^\perp$ with vectorial values also belong to $\hat{\mathcal{L}}_2(\Omega,\mathbb{R}^3)$. □

Proof of Proposition 2.2.4. Let $\{\vec{f}_n\} \subset \vec{\mathcal{A}}(\Omega)$ be a Cauchy sequence, then there exists $f \in \mathcal{A}(\Omega)$ such that $\vec{f}_n \to f = f_0 + \vec{f}$ in the norm $\|\cdot\|_{\mathcal{A}(\Omega)}$. Proposition 3.5.2 implies that $f_n \to f$ uniformly on compact subsets of Ω, then for any compact set $C \subset \Omega$ there holds:

$$\int_C |f_0|^2 d\mu + \int_C |\vec{f}-\vec{f}_n|^2 d\mu = \int_C |f-\vec{f}_n|^2 d\mu \to 0, \quad \text{when } n \to \infty.$$

Therefore $f_0 = 0$ and $f = \vec{f} \in \vec{\mathcal{A}}(\Omega)$.

Note that equation (1) is a direct consequence of the equation (26).

Finally, we will prove the equation (5) which is equivalent to (2). Thus for any $\vec{f} \in \hat{\mathcal{L}}_2(\Omega,\mathbb{R}^3)$ there holds

$$\int_\Omega \vec{f}(\vec{v}) \odot \left(\mathcal{B}_{\Omega,0}(\vec{x},\vec{v}), -\vec{\mathcal{B}}_\Omega(\vec{x},\vec{v})\right) d\mu_{\vec{v}} = \frac{1}{2}(\mathfrak{B}_\Omega + \mathfrak{B}_{\Omega,r})[\vec{f}].$$

Then

($\Longleftarrow$) It is due to the fact that $\vec{\mathcal{A}}(\Omega) \subseteq \mathcal{A}(\Omega) \cap \mathcal{A}_r(\Omega)$.

($\Longrightarrow$) If $\vec{f} = \frac{1}{2}(\mathfrak{B}_\Omega + \mathfrak{B}_{\Omega,r})[\vec{f}]$, then

$$\mathfrak{B}_\Omega[\vec{f}](\vec{x}) - \vec{f}(\vec{x}) = -\int_\Omega \left\langle \vec{\mathcal{B}}(\vec{x},\vec{v}), \vec{f}(\vec{v})\right\rangle_{\mathbb{R}^3} d\mu_{\vec{v}}$$

and

$$\mathfrak{B}_{\Omega,r}[\vec{f}] - \vec{f}(\vec{x}) = \int_\Omega \left\langle \vec{\mathcal{B}}(\vec{x},\vec{v}), \vec{f}(\vec{v})\right\rangle_{\mathbb{R}^3} d\mu_{\vec{v}}.$$

Therefore

$$\int_\Omega \left\langle \vec{\mathcal{B}}(\vec{x},\vec{v}), \vec{f}(\vec{v}) \right\rangle_{\mathbb{R}^3} d\mu_{\vec{v}} \in \mathcal{A}_\Omega(\Omega)^\perp \cap \mathcal{A}_{\Omega,r}(\Omega)^\perp$$

and

$$\mathfrak{B}_\Omega[\vec{f}](\vec{x}) - \mathfrak{B}_{\Omega,r}[\vec{f}](\vec{x}) = -2 \int_\Omega \left\langle \vec{\mathcal{B}}(\vec{x},\vec{v}), \vec{f}(\vec{v}) \right\rangle_{\mathbb{R}^3} d\mu_{\vec{v}}.$$

Thus $\mathfrak{B}_\Omega[\vec{f}] - \mathfrak{B}_r[\vec{f}] \in \mathcal{A}_\Omega(\Omega)^\perp \cap \mathcal{A}_{\Omega,r}(\Omega)^\perp$ implies:

$$\left\langle \mathfrak{B}_\Omega[\vec{f}] - \mathfrak{B}_{\Omega,r}[\vec{f}], \mathfrak{B}_\Omega[\vec{f}] \right\rangle_{\mathcal{A}_\Omega(\Omega)} = 0,$$
$$\left\langle \mathfrak{B}_\Omega[\vec{f}] - \mathfrak{B}_{\Omega,r}[\vec{f}], \mathfrak{B}_{\Omega,r}[\vec{f}] \right\rangle_{\mathcal{A}_{\Omega,r}(\Omega)} = 0,$$

and using these equations we have that $\mathfrak{B}_\Omega[\vec{f}](\vec{x}) = \vec{f}(\vec{x}) = \mathfrak{B}_{\Omega,r}[\vec{f}]$, i.e., $\vec{f} \in \vec{\mathcal{A}}(\Omega)$. □

Proof of Proposition 2.2.5. Let $\{\vec{f}_n\}_n \subset \vec{F}_{\vec{u},\hat{\imath}}$ be a sequence such that

$$\lim_{n\to\infty} \|\vec{f}_n\|_{\vec{\mathcal{A}}(\Omega)} = \sigma.$$

As $\frac{1}{2}(\vec{f}_k + \vec{f}_m) \in \vec{F}_{\vec{u},\hat{\imath}}$ for all k, m and using the parallelogram inequality one obtains the following:

$$\|\vec{f}_k - \vec{f}_m\|^2_{\vec{\mathcal{A}}(\Omega)} \leq 2\|\vec{f}_k\|^2_{\vec{\mathcal{A}}(\Omega)} + 2\|\vec{f}_m\|^2_{\vec{\mathcal{A}}(\Omega)} - 4\sigma^2,$$

which implies that $\{\vec{f}_n\}_n$ is a Cauchy sequence, hence there exists $\vec{f}_* \in \vec{\mathcal{A}}(\Omega)$ such that $\{\vec{f}_n\}_n \to \vec{f}_*$. As $\|\vec{f}_n\|_{\vec{\mathcal{A}}(\Omega)} \to \|\vec{f}_*\|_{\vec{\mathcal{A}}(\Omega)}$ then $\|\vec{f}_*\|_{\vec{\mathcal{A}}(\Omega)} = \sigma$.

Proposition 3.5.2 implies that $f_n \to f$ uniformly on compact subsets of Ω and using the MT-hyperholomorphic Cauchy formula we arrive at the fact that $\vec{f}_*(\vec{u}) = \hat{\imath}$, then $\vec{f}_* \in \vec{F}_{\vec{u},\hat{\imath}}$.

The uniqueness of $\vec{f}_*$ is obtained applying the parallelogram inequality as above, and the inequality

$$\|\vec{f}_*\|^2_{\vec{\mathcal{A}}(\Omega)} > \frac{\displaystyle\int_\Omega |\mathcal{B}_{\Omega,0}(\vec{v},\vec{u})|^2 d\mu_{\vec{v}} + \|\vec{\mathcal{B}}_\Omega(\cdot,\vec{u})\|_{\mathcal{L}_2(\Omega,\mathbb{R}^3)}}{\left(\mathcal{B}_{\Omega,0}(\vec{u},\vec{u})\right)^2},$$

which is a direct consequence of Proposition 3.5.6. □

Proof of Proposition 2.3.1. It is a direct consequence of the properties of the MT-hyperholomorphic Bergman projection $\mathfrak{B}_\Omega$, see Definition 3.5.11. □

Proof of Proposition 2.4.1. Given any element $\vec{f} \in \hat{\mathcal{L}}_2(\Omega,\mathbb{R}^3)$, then what the theory presented in Subsection 3.7 says is that there exist unique functions $g \in \mathcal{A}(\Omega)$

and $f \in \mathcal{D}(\dot{W}_2^1(\Omega,\mathbb{H}))$, such that $\vec{f} = g + h$; but the condition

$$\int_\Omega \left\langle \vec{\mathcal{B}}_\Omega(\vec{x},\vec{v}), \vec{f}(\vec{v}) \right\rangle_{\mathbb{R}^3} d\mu_{\vec{v}} = 0, \quad \text{for } a.e. \ \vec{x} \in \Omega,$$

implies that $g = \vec{g} \in \vec{\mathcal{A}}(\Omega)$ and $h = \nabla \ell_0 + \nabla \times \vec{\ell}$, with

$$(\ell_0, \vec{\ell}) \in \dot{W}_2^1(\Omega,\mathbb{R}) \times \dot{W}_2^1(\Omega,\mathbb{R}^3).$$ □

Proof of Proposition 2.6.1. Firstly, note that formulas (8) and (9) are the vectorial versions of the quaternionic representation of the Möbius transformations in $\mathbb{R}^3$ given in (21).

Equations (10) and (12) are obtained from Theorem 3.3.2 as follows: the case $c = 0$ implies that

$$\mathcal{D}_{\vec{x}}[\bar{a}(f \circ T)a] = |a|^2 \bar{a}(\mathcal{D}_{\vec{y}}[f] \circ T)a, \quad \forall f \in C^1(\Omega,\mathbb{H}),$$

and applying Notation 3.3.3 we have that

$$\mathcal{D}_{\vec{x}}[{}^{\bar{a}}M^a \circ W_T[f]] = |a|^2 \, {}^{\bar{a}}M^a \circ W_T[\mathcal{D}_{\vec{y}}[f]], \quad \forall f \in C^1(\Omega,\mathbb{H}).$$

Now, using the identity

$$P_{r,\vec{a}}[\vec{f} \circ T] = {}^{\bar{a}}M^a \circ W_T[\vec{f}], \quad \forall \vec{f} \in C^1(\Omega,\mathbb{R}^3), \tag{29}$$

in the previous equation and separating the scalar and vectorial parts we arrive at the expressions (10) and (12).

Working similarly with the case $c \neq 0$ and with the same theorem we obtain the equations (11) and (13). □

Proof of Corollary 2.6.2. It is due to the fact that equations (10), (11), (12) and (13) relate the zeros of the operators div and rot. □

Proof of Corollary 2.6.5. It is a consequence of the steps 1) and 2) of Corollary 2.6.2. □

Proof of Proposition 2.6.6. Using the identity (29) one notes that the proof is a consequence of Theorem 3.6.2. □

Proof of Corollary 2.6.7. It is a direct consequence of Corollary 3.6.9 and (29). □

Proof of Corollary 2.6.8. It is a consequence of Corollaries 3.6.5, 3.6.9, and of the identity (29). □

References

[1] R. Delanghe, F. Brackx, *Hypercomplex function theory and Hilbert modules with reproducing kernel*, Proc. London Math. Soc. 1978, (3) (37), 545–576.

[2] B.A. Dubrovin, S.P. Novikov, A.T. Fomenko, *Modern Methods and Applications*, Springer, 1984, Vol. 1.

[3] P.L. Duren and A. Schuster, *Bergman Spaces*, Mathematical Surveys and Monographs, AMS, 2004, Vol. 100.

[4] J.O. González-Cervantes, M.E. Luna-Elizarrarás, M. Shapiro, *On some categories and functors in the theory of quaternionic Bergman spaces*, Adv. Appl. Clifford Alg. No. 19 (2009), 325–338.

[5] J.O. González-Cervantes, M. Shapiro, *Hyperholomorphic Bergman Spaces and Bergman Operators Associated with Domains in* $\mathbb{C}^2$. Compl. anal. oper. theory, 2008, Vol. 2, # 2, 361–382.

[6] K. Gürlebeck, W. Sprössig, *Quaternionic analysis and elliptic boundary value problems*, Birkhäuser Verlag, 1990.

[7] K. Gürlebeck, W. Sprössig, *Quaternionic and Clifford calculus for physicists and engineers*, John Wiley and Sons, 1997.

[8] X. Ji, T. Qian and J. Ryan *Fourier theory under Möbius transformations*, in the book Clifford Algebras and their Applications in Mathematical Physics: Clifford Analysis. Birkhäuser, 2000, Vol. 2, 57–80.

[9] A. Meziani, *On real analytic planar vector fields near the characteristic set*, Contemp. Math., 2000, Vol. 251, 429–438.

[10] I. Porteous, *Topological Geometry*, Cambridge University, 1981, 2nd ed.

[11] J.R. Retherford, *Hilbert Space: Compact Operators and the Trace Theorem*. Cambridge Press, 1998.

[12] J. Ryan, *The conformal covariance of Huygens' principle-type integral formulae in Clifford analysis*, in the book Clifford Algebras and Spinor Structures, Kluwer Academic Publishers, 1995, 301–310.

[13] J. Ryan, *Some applications of conformal covariance in Clifford analysis*, in the book Clifford Algebras in Analysis and Related Topics, CRC Press, 1996, 129–155.

[14] M.V. Shapiro, *On the conjugate harmonic functions of M. Riesz – E. Stein – G. Weiss*. Topics in Complex Analysis, Differential Geometry and Mathematical Physics, World Scientific, 1997, 8–32.

[15] M.V. Shapiro, N.L. Vasilevski, *Quaternionic* ψ*-hyperholomorphic functions, singular integral operators and boundary value problems. I.* ψ*-hyperholomorphic function theory.* Compl. Var. Theory Appls. 1995, Vol. 27, 17–46.

[16] M.V. Shapiro, N.L. Vasilevski, *Quaternionic* ψ*-hyperholomorphic functions, singular operators and boundary value problems. II. Algebras of singular integral operators and Riemann type boundary value problems.* Compl. Var. Theory Appls., 1995, Vol. 27, 67–96.

[17] M.V. Shapiro, N.L. Vasilevski, *On the Bergman Kernel Function in Hyperholomorphic Analysis*, Acta Appl. Math., 1997, 1–27.

[18] E.M. Stein, G. Weiss, *Introduction to Fourier Analysis on Euclidean Spaces*, Princeton Univ. Press, Princeton, N.J., 1971.

[19] A. Sudbery, *Quaternionic analysis.* Math. Proc. Camb. Phil. Soc. 1979, 85, 199–225.

[20] S. Thangavelu, *Harmonic analysis on the Heisenberg group*, Springer, 1998.

[21] N.L. Vasilevski, *Toeplitz operators on the Bergman spaces: inside-the-domain effects*, Contemp. Math., 2001, Vol. 289, 79–145.

José Oscar González-Cervantes, María Elena Luna-Elizarrarás
and Michael Shapiro
Instituto Politécnico Nacional
E.S.F.M.
Av. Instituto Politécnico Nacional, S/N
U.P. Adolfo López Mateos
Edif. 9
C. P. 07738
Mexico, D. F., Mexico
e-mail: jogc200678@yahoo.com.mx
eluna@esfm.ipn.mx
shapiro@esfm.ipn.mx

Operator Theory:
Advances and Applications, Vol. 210, 107–122

Weighted Estimates of Generalized Potentials in Variable Exponent Lebesgue Spaces on Homogeneous Spaces

Mubariz G. Hajibayov and Stefan G. Samko

To Nikolai Vasilevski on the occasion of his 60th birthday

Abstract. For generalized potential operators with the kernel $\frac{a[\varrho(x,y)]}{[\varrho(x,y)]^N}$ on bounded measure metric space (X,μ,ϱ) with doubling measure μ satisfying the upper growth condition $\mu B(x,r) \leq Cr^N, N \in (0,\infty)$, we prove weighted estimates in the case of radial type power weight $w = [\varrho(x,x_0)]^\nu$. Under some natural assumptions on $a(r)$ in terms of almost monotonicity we prove that such potential operators are bounded from the weighted variable exponent Lebesgue space $L^{p(\cdot)}(X,w,\mu)$ into a certain weighted Musielak-Orlicz space $L^\Phi(X,w^{\frac{1}{p(x_0)}},\mu)$ with the N-function $\Phi(x,r)$ defined by the exponent $p(x)$ and the function $a(r)$.

Mathematics Subject Classification (2000). Primary 43A85, 46E30; Secondary 47B38.

Keywords. Weighted estimates, generalized potential, variable exponent, variable Lebesgue space, metric measure space, space of homogeneous type, Musielak-Orlicz space, Matuszewska-Orlicz indices.

1. Introduction

The Lebesgue spaces $L^{p(\cdot)}$ with variable exponent were intensively investigated during the last years, we refer to the papers [35], [23] for the basic properties of these spaces. The growing interest to such spaces is caused by applications to various problems, for instance, in image restoration, fluid dynamics, elasticity theory and differential equations with non-standard growth conditions (see, e.g., [2], [28],

---

The first author partially supported by the INTAS grant Ref. Nr 06-1000015-5777. The second author supported by INTAS grant Ref. Nr 06-1000017-8792.

[36]). The spaces $L^{p(\cdot)}$ with variable exponent are special cases of Orlicz-Musielak spaces, see [24] for these spaces. We refer to [3], where the maximal operator was studied in the context of Orlicz-Musielak spaces. A significant progress has already been made been made in the study of classical integral operators in the context of the $L^{p(\cdot)}$ spaces, see for instance the surveying papers [4], [18] and[34].

The spaces $L^{p(\cdot)}$ on measure quasimetric spaces and maximal and potential operators in such spaces were studied in [1], [6], [15], [16], [14], [17], [21].

We study the generalized Riesz potential operators

$$I_a f(x) := \int\limits_X \mathcal{K}(x,y) f(y) d\mu(y), \quad \mathcal{K}(x,y) = \frac{a(\varrho(x,y))}{[\varrho(x,y)]^N} \tag{1.1}$$

over a bounded measure space X with quasimetric ϱ, where N is the upper Ahlfors dimension of X. In [13], under some assumptions on the function $a(\varrho)$ there was proved a Sobolev-type theorem on the boundedness of the operator I_a from $L^{p(\cdot)}(X)$ into a certain Orlicz-Musielak space. In this paper we extend this result to the weighted case. We deal with the case of power weights

$$w(x) = [\varrho(x, x_0)]^\nu, \quad x_0 \in X.$$

Note that the interest to the case of power weights is caused not only by the fact that such weights are first of all important in various applications, but also because in the case of variable exponents it is a problem to derive the result for concrete weights from the existing forms of general conditions on weights. (Recall that even in the case of constant exponents the belongness of these or other special weights to the Muckenhoupt type classes was first not checked directly, but obtained from the necessity of the Muckenhoupt condition.)

An extension to the weighted case proved to be a non-easy task within the frameworks of variable exponents even for power weights, the difficulties being caused both by the variability of the exponent and non-homogeneity of the kernel. This extension is based on the technique of weighted norm estimation of kernels of truncated potentials given and applied in [29], [32], [31], [33], which is developed in this paper for non-homogeneous kernels.

The generalized Riesz potential operators I_a attracted attention last years, we refer in particular to [12], [25], where such potentials were studied in Orlicz spaces in the case $X = \mathbb{R}^n$ and Euclidean metric, and to [26], where homogeneous spaces with constant dimension were admitted. We refer also to [27] for the study of the similar generalized potentials in the Euclidean setting in rearrangement invariant spaces. For "standard" potentials (that is, potentials with the kernel of the form $\frac{1}{d(x,y)^{N-\alpha}}$ or $\frac{d(x,y)^\alpha}{B(x,d(x,y))}$) on metric measure spaces, we refer to [5], [7], [8], [9], [10], [11], [19], [20] and references therein.

The main results are formulated in Section 2 and proved in Section 5. The main technical tool is provided by Lemma 4.1 in Section 3.

2. Formulation of the main result

In the sequel (X, ϱ, μ) always stands for a bounded quasimetric space with quasidistance $\varrho(x,y) = \varrho(y,x)$:

$$\varrho(x,y) \leq k[\varrho(x,z) + \varrho(z,y)], \quad k \geq 1 \tag{2.1}$$

and Borel regular measure μ. We denote $d = \operatorname{diam} X$. The measure μ is supposed to satisfy the growth condition

$$\mu\left(B\left(x,r\right)\right) < Kr^N. \tag{2.2}$$

Definition 2.1. A function $\Phi : X \times [0,\infty) \to [0,+\infty)$ is said to be an N-function, if

1. for every $x \in X$ the function $\Phi\left(x,t\right)$ is convex, nondecreasing and continuous in $t \in [0,\infty)$,
2. $\Phi\left(x,0\right) = 0, \quad \Phi\left(x,t\right) > 0$ for every $t > 0$,
3. $\Phi\left(x,t\right)$ is a μ-measurable function of x for every $t \geq 0$.

Definition 2.2. Let Φ be an N-function and w a weight. The weighted Orlicz-Musielak space $L^{\Phi}(X,w)$ is defined as the set of all real-valued μ-measurable functions f on X such that

$$\int_X \Phi\left(x, \frac{w(x)f(x)}{\lambda}\right) d\mu(x) < \infty$$

for some $\lambda > 0$. We equip it with the norm

$$\|f\|_{\Phi,w} = \inf\left\{\lambda > 0 : \int_X \Phi\left(x, \frac{w(x)f(x)}{\lambda}\right) d\mu(x) \leq 1\right\}.$$

In particular, $\Phi(x,t) = t^{p(x)}$, where $1 \leq p(x) < \infty$, is an N-function and the corresponding space is the variable exponent Lebesgue space $L^{p(\cdot)}(X,w)$.

Everywhere in the sequel, when dealing with the space $L^{p(\cdot)}(X,w)$, we suppose that

$$1 < p_- \leq p(x) \leq p_+ < +\infty, \tag{2.3}$$

$$|p(x) - p(y)| \leq \frac{A}{ln\frac{1}{\varrho(x,y)}}, \quad \varrho(x,y) < \frac{1}{2} \tag{2.4}$$

and denote

$$w^{\nu} = [\varrho(x,x_0)]^{\nu}, \quad x_0 \in X.$$

The function $a : [0,d] \to [0,\infty)$ is assumed to satisfy the assumptions

1) $a(r)$ is continuous, almost increasing, positive for $r > 0$ and $a(0) = 0$,

2) $\int\limits_0^d \frac{a(r)}{r} dr < \infty$.

We denote

$$A(r) = \int_0^r \frac{a(t)}{t}\, dt.$$

In the following theorem we make use of the notion of the lower dimension of X defined by

$$\underline{\mathfrak{dim}}(X) = \sup_{t>1} \frac{\ln\left(\liminf\limits_{r\to 0}\inf\limits_{x\in X}\frac{\mu B(x,rt)}{\mu B(x,r)}\right)}{\ln t}.$$

as introduced in [30]. It is clear that $\underline{\mathfrak{dim}}(X) = N$ in the cases where X has constant dimension N, that is, $c_1 r^N \leq \mu B(x,r) \leq c_2 r^N$. In general, if X has the property that

$$0 < \underline{\mathfrak{dim}}(X) < \infty,$$

then X satisfies the growth condition with every

$$0 < N < \underline{\mathfrak{dim}}(X). \tag{2.5}$$

This follows from the inequality

$$\mu B(x,r) \leq C r^{\underline{\mathfrak{dim}}(X)-\varepsilon}, \tag{2.6}$$

where $\varepsilon > 0$ is arbitrarily small and $C = C(\varepsilon) > 0$ does not depend on x, which is easily derived from the results in [30], Subsection 2.1.

Theorem 2.3. *Let (X, ϱ, μ) be quasimetric space with doubling measure and positive finite lower dimension $\underline{\mathfrak{dim}}(X)$, and let p fulfill assumptions (2.3)–(2.4) and*

$$0 \leq \nu < \frac{\underline{\mathfrak{dim}}(X)}{p'(x_0)}.$$

Suppose that there exists a $\beta \in \left(0, \frac{\underline{\mathfrak{dim}}(X)}{p_+}\right)$ such that

$$\frac{a(r)}{r^\beta} \quad \text{is almost decreasing.} \tag{2.7}$$

Then the operator I_a is bounded from the space $L^{p(\cdot)}(X, w^\nu)$ into the weighted Orlicz-Musielak space $L^\Phi(X, w^{\nu_1})$, where $\nu_1 = \frac{\nu}{p(x_0)}$ and the N-function Φ is defined by its inverse (for every fixed $x \in X$)

$$\Phi^{-1}(x,r) = \int_0^r A\left(t^{-\frac{1}{N}}\right) t^{-\frac{1}{p'(x)}}\, dt. \tag{2.8}$$

The proof of Theorem 2.3 will be based on Lemma 4.1 and the following statement proved in [21], [22].

Theorem 2.4. *Let X be a bounded doubling measure quasimetric space and $p(x)$ satisfy assumptions (2.3)–(2.4). The maximal operator*

$$Mf(x) = \sup_{r>0} \frac{1}{B(x,r)} \int_{B(x,r)} |f(y)|\, d\mu(y)$$

is bounded in $L^{p(\cdot)}(X, w^\nu)$, if $-\frac{\underline{\mathfrak{dim}}(X)}{p(x_0)} < \nu < \frac{\underline{\mathfrak{dim}}(X)}{p'(x_0)}$.

We will also use the following lemma proved in [13] (see Lemma 4.9 in [13]).

Lemma 2.5. *Let $p(x)$ satisfy condition (2.3) and $a(r)$ be a non-negative almost increasing continuous on $[0,d], 0<d<\infty$ function such that the function $\frac{a(t)}{t^{\frac{N}{p_+}-\varepsilon}}$ is almost decreasing for some $\varepsilon>0$. Then there exist constants $C_1>0, C_2>0$ not depending on x and r such that*

$$C_1\frac{A(r)}{r^{\frac{N}{p(x)}}} \leq \Phi^{-1}\left(x, \frac{1}{r^N}\right) \leq C_2\frac{A(r)}{r^{\frac{N}{p(x)}}}. \tag{2.9}$$

3. Auxiliary estimates

To prove our weighted generalized Sobolev-type theorem for the potential I_a via Hedberg approach, we need to estimate the integral

$$\mathcal{J}(x,r) := \int\limits_{X\backslash B(x,r)} \left(\frac{a\left(\varrho\left(x,y\right)\right)}{\varrho\left(x,y\right)^N}\right)^{p(x)} \varrho(y,x_0)^b d\mu\left(y\right), \quad x_0\in X.$$

Lemma 3.1. *Let X satisfy the growth condition (2.2), let the function $a(r)$ be non-negative and almost decreasing on $[0,d]$ and $\gamma(x)$ be an arbitrary bounded function on Ω. Then the estimate*

$$\int\limits_{X\backslash B(x,r)} \left(\frac{a\left[\varrho(x,y)\right]}{\varrho(x,y)^{\gamma(x)}}\right)^{p(x)} d\mu(y) \leq C\int\limits_r^d t^{N-1}\left[\frac{a(t)}{t^{\gamma(x)}}\right]^{p(x)} dt, \quad 0<r<\frac{d}{2}, \tag{3.1}$$

holds, where $C>0$ does not depend on x and r.

Lemma 3.1 was proved in [13] in the case $\gamma(x)=N$, the proof based on the binary decomposition is the same for an arbitrary bounded $\gamma(x)$ in view of the monotonicity of the power function $t^{\gamma(x)}$.

Lemma 3.2. *Let X satisfy the growth condition (2.2), Suppose that the function $a(r):(0,d)\to(0,+\infty)$ is almost increasing and the function $\frac{a(r)}{r^N}$ is almost decreasing. Then for $0<r<\frac{d}{2}$ the estimate*

$$\mathcal{J}(x,r) \leq C\mathcal{G}(x,r) := C\begin{cases} \int\limits_r^d t^{N-1}\left[\frac{a(t)}{t^N}\right]^{p(x)} t^b\, dt, & \text{if } \varrho(x_0,x)\leq r \\ \varrho(x_0,x)^b \int\limits_r^d t^{N-1}\left[\frac{a(t)}{t^N}\right]^{p(x)} dt, & \text{if } \varrho(x_0,x)>r \end{cases} \tag{3.2}$$

holds, where $p:X\to(1,+\infty)$, $1\leq p(x)<p_+<+\infty$, $b>-N$ and $C>0$ does not depend on x and r.

Proof. Consider separately the cases $\varrho(x_0,x)\le\frac{r}{2k}$, $\frac{r}{2k}\le\varrho(x_0,x)\le 2kr$, $\varrho(x_0,x)\ge 2kr$, where k is the constant from the triangle inequality (2.1).

The case $\varrho(x_0,x)\le\frac{r}{2k}$.

We have $\dfrac{\varrho(x_0,y)}{\varrho(x,y)}\le\dfrac{k\left(\varrho(x,y)+\varrho(x_0,x)\right)}{\varrho(x,y)}\le k\left(1+\dfrac{\varrho(x_0,x)}{r}\right)\le 2k$ and $\dfrac{\varrho(x_0,y)}{\varrho(x,y)}\ge\dfrac{1}{k}-\dfrac{\varrho(x_0,x)}{r}\ge\dfrac{1}{2k}$. Hence $\dfrac{1}{2k}\le\dfrac{\varrho(x_0,y)}{\varrho(x,y)}\le 2k$. Consequently,

$$\mathcal{J}(x,r)\le C\int\limits_{X\setminus B(x,r)}\left(\frac{a\left(\varrho\left(x,y\right)\right)}{\varrho\left(x,y\right)^{N-\frac{b}{p(x)}}}\right)^{p(x)}d\mu\left(y\right).$$

Then by Lemma 3.1

$$\mathcal{J}(x,r)\le C\int\limits_r^d t^{N-1}\left[\frac{a(t)}{t^N}\right]^{p(x)}t^b\,dt.$$

Therefore

$$\mathcal{J}(x,r)\le C\mathcal{G}(x,r),\qquad \varrho(x_0,x)\le\frac{r}{2k}\tag{3.3}$$

The case $\frac{r}{2k}\le\varrho(x_0,x)\le 2kr$.

We split the integration in $\mathcal{J}(x,r)$ as follows

$$\mathcal{J}(x,r):=\int\limits_{r<\varrho(x,y)<2k\varrho(x_0,x)}\left(\frac{a\left(\varrho\left(x,y\right)\right)}{\varrho\left(x,y\right)^N}\right)^{p(x)}\varrho(x_0,y)^b d\mu\left(y\right)$$

$$+\int\limits_{\varrho(x,y)>2k\varrho(x_0,x)}\left(\frac{a\left(\varrho\left(x,y\right)\right)}{\varrho\left(x,y\right)^N}\right)^{p(x)}\varrho(x_0,y)^b d\mu\left(y\right):=J_1+J_2.$$

Since $\frac{a(r)}{r^N}$ is almost decreasing, we obtain

$$J_1\le C\left(\frac{a\left(r\right)}{r^N}\right)^{p(x)}\int\limits_{r<\varrho(x,y)<2k\varrho(x_0,x)}\varrho(x_0,y)^b d\mu\left(y\right)$$

When $\varrho(x,y)>r$ and $\varrho(x_0,x)<2kr$, then $\varrho(x_0,y)\le k\left(\varrho(x,y)+\varrho(x_0,x)\right)\le k\left(\varrho(x,y)+2kr\right)\le 3k^2\varrho(x,y)$. Consequently,

$$\le C\left(\frac{a\left(r\right)}{r^N}\right)^{p(x)}\int\limits_{\substack{\varrho(x,y)<2k\varrho(x_0,x)\\ \varrho(x_0,y)\le 3k^2\varrho(x,y)}}\varrho(x_0,y)^b d\mu\left(y\right)$$

$$\le C\left(\frac{a\left(r\right)}{r^N}\right)^{p(x)}\int\limits_{\varrho(x_0,y)\le 6k^3\varrho(x_0,x)}\varrho(x_0,y)^b d\mu\left(y\right)$$

We make use of the known estimate

$$\int\limits_{\varrho(x,y)\le R} \varrho(x,y)^b d\mu(y) \le CR^{b+N}, \quad b > -N \tag{3.4}$$

valid for quasimetric spaces with the growth condition (2.2), see for instance [8], Lemma 1 (actually $C = \frac{K2^N}{2^{N+b}-1}$ in (3.4), where K is the constant from (2.2)), which yields

$$J_1 \le C\left(\frac{a\,(r)}{r^N}\right)^{p(x)} \varrho(x_0,x)^{b+N}.$$

It is easily seen that then

$$J_1 \le Ca\,(r)^{p(x)}\,\varrho(x_0,x)^{b+N}\int\limits_r^d t^{-Np(x)-1}dt \le C\int\limits_r^d t^{N-1}\left[\frac{a(t)}{t^N}\right]^{p(x)} t^b\,dt,$$

so that

$$J_1 \le C\mathcal{G}(x,r).$$

The estimate for $J_2 = \mathcal{J}(x,2k\varrho(x_0,x))$ is contained in (3.3) with $r = 2k\varrho(x_0,x)$. Hence

$$\mathcal{J}(x,r) \le C\mathcal{G}(x,r), \qquad \frac{r}{2k} \le \varrho(x_0,x) \le 2kr \tag{3.5}$$

The case $\varrho(x_0,x) \ge 2kr$.

We have

$$\mathcal{J}(x,r) = \int\limits_{r<\varrho(x,y)<\frac{\varrho(x_0,x)}{2k}} \left(\frac{a\,(\varrho\,(x,y))}{\varrho\,(x,y)^N}\right)^{p(x)} \varrho(x_0,y)^b d\mu\,(y)$$
$$+ \int\limits_{\varrho(x,y)>\frac{\varrho(x_0,x)}{2k}} \left(\frac{a\,(\varrho\,(x,y))}{\varrho\,(x,y)^N}\right)^{p(x)} \varrho(x_0,y)^b d\mu\,(y) = J_3 + J_4.$$

For the term J_3 we have $\varrho(x_0,y) \ge \frac{1}{k}\varrho(x_0,x) - \varrho(x,y) \ge \frac{1}{k}\varrho(x_0,x) - \frac{1}{2k}\varrho(x_0,x) = \frac{1}{2k}\varrho(x_0,x)$ and $\varrho(x_0,y) \le k(\varrho(x_0,x) + \varrho(x,y)) < 2k\varrho(x_0,x)$. Then

$$J_3 \le C\varrho(x_0,x)^b \int\limits_{r<\varrho(x,y)<\frac{1}{2k}\varrho(x_0,x)} \left(\frac{a\,(\varrho\,(x,y))}{\varrho\,(x,y)^N}\right)^{p(x)} d\mu\,(y).$$

By Lemma 3.1 we then obtain

$$J_3 \le C\varrho(x_0,x)^b \int\limits_r^d t^{N-1}\left[\frac{a(t)}{t^N}\right]^{p(x)} dt = C\mathcal{G}(x,r).$$

The term J_4, coincides with $\mathcal{J}\left(x, \frac{\varrho(x_0,x)}{2k}\right)$ and its estimate is contained in the preceding case $\frac{r}{2k} \leq \varrho(x_0, x) \leq 2kr$. Therefore,

$$\mathcal{J}(x,r) \leq C\mathcal{G}(x,r), \qquad \varrho(x_0,x) \geq 2kr. \tag{3.6}$$

Gathering estimates (3.3), (3.5), (3.6), we arrive at (3.2). □

4. Main lemma

We need to estimate the norm

$$\eta_{p,\gamma}(x,r) = \left\| \frac{a\left(\varrho(x,y)\right)}{\varrho(x,y)^N} \right\|_{L^{p(\cdot)}\left(X\setminus B(x,r), w^{\frac{\gamma-N}{p(x_0)}}\right)}, \tag{4.1}$$

where $w^{\frac{\gamma-N}{p(x_0)}}(y) = \varrho(x_0,y)^{\frac{\gamma-N}{p(x_0)}}$ and $\gamma > 0$.

Lemma 4.1. *Let (X, ϱ, μ) be a bounded quasimetric space with Borel regular measure μ: satisfying the growth condition* (2.2), $d =$ diam X *and let p satisfy assumptions* (2.3)–(2.4). *Suppose that the function $a(r) : (0,d) \to (1,+\infty)$ is almost increasing, there exists $0 < \beta < \min\left(\frac{N}{(p_-)'}, N - \frac{\gamma}{p_-}\right)$ such that*

$$\frac{a(r)}{r^\beta} \quad \textit{is almost decreasing.} \tag{4.2}$$

Then

$$\eta_{p,\gamma}(x,r) \leq C\frac{a(r)}{r^{\frac{N}{p'(x)}}}\left[\max\left(r, \varrho(x_0,x)\right)\right]^{\frac{\gamma-N}{p(x)}} \quad \textit{for} \quad 0 < r < \frac{d}{2}. \tag{4.3}$$

Proof. By definition of the norm

$$\int\limits_{\substack{y\in X\\ \varrho(x,y)>r}} \left(\frac{a(\varrho(x,y))}{\varrho(x,y)^N\eta_{p,\gamma}}\right)^{p(y)} \varrho(x_0,y)^{\gamma-N}d\mu(y) = 1. \tag{4.4}$$

1*st step.* Values $\eta_{p,\gamma} \geq 1$ are only of interest. This follows from the fact that the right-hand side of (4.3) is bounded from below.

$$\frac{a(r)}{r^{\frac{N}{p'(x)}}}\left[\max\left(r, \varrho(x_0,x)\right)\right]^{\frac{\gamma-N}{p(x)}} \geq \frac{a(r)}{r^{\frac{N}{p'(x)}}}\min\left(r^{\frac{\gamma-N}{p(x)}}, d^{\frac{\gamma-N}{p(x)}}\right)$$

$$= \min\left(\frac{a(r)}{r^{N-\frac{\gamma}{p(x)}}}, d^{\frac{\gamma-N}{p(x)}}\frac{a(r)}{r^{\frac{N}{p'(x)}}}\right) \geq C\frac{a(r)}{r^\beta} \geq C > 0,$$

the last inequality following from the fact that $\frac{a(r)}{r^\beta}$ is almost decreasing on $[0,d]$.
2*nd step.* Small values of r, say $0 < r < \frac{1}{2}$, are only of interest. To show that this assumption is possible, we have to check that the right-hand side of (4.3) is bounded from below and $\eta_{p,\gamma}(x,r)$ is bounded from above when $r \geq \frac{1}{2}$.

Let $r \geq \frac{1}{2}$. From the fact that $\eta_{p,\gamma} \geq 1$ it follows that

$$\int\limits_{\substack{y\in X:\\ \varrho(x,y)>1}} \left(\frac{a(\varrho(x,y))}{\varrho(x,y)^N}\right)^{p(y)} \frac{1}{\eta_{p,\gamma}} \varrho(x_0,y)^{\gamma-N} d\mu(y) \geq 1.$$

Hence

$$\eta_{p,\gamma} \leq \int\limits_{\substack{y\in X:\\ \varrho(x,y)>1}} \left(\frac{a(\varrho(x,y))}{\varrho(x,y)^N}\right)^{p(y)} \varrho(x_0,y)^{\gamma-N} d\mu(y) < C.$$

3rd step. Rough estimate. First we derive a weaker estimate

$$\eta_{p,\gamma}(x,r) \leq Cr^{-N}a(r) \tag{4.5}$$

which will be used later to obtain the final estimate (4.3). From (4.4) we have

$$1 \leq \int\limits_{\substack{y\in X\\ \varrho(x,y)>r}} \left[\left(\frac{a(\varrho(x,y))}{\varrho(x,y)^N\eta_{p,\gamma}}\right)^{p_-} + \left(\frac{a(\varrho(x,y))}{\varrho(x,y)^N\eta_{p,\gamma}}\right)^{p_+}\right] \varrho(x_0,y)^{\gamma-N} d\mu(y)$$

Since $\varrho(x,y) > r$, we obtain

$$\begin{aligned} 1 &\leq C\left[\left(\frac{a(r)}{r^N\eta_{p,\gamma}}\right)^{p_-} + \left(\frac{a(r)}{r^N\eta_{p,\gamma}}\right)^{p_+}\right] \int\limits_X \varrho(x_0,y)^{\gamma-N} d\mu(y) \\ &\leq C\left[\left(\frac{a(r)}{r^N\eta_{p,\gamma}}\right)^{p_-} + \left(\frac{a(r)}{r^N\eta_{p,\gamma}}\right)^{p_+}\right], \end{aligned}$$

where the convergence of the integral $\int\limits_X \varrho(x_0,y)^{\gamma-N} d\mu(y)$ with $\gamma > 0$ (see (3.4)) was taken into account.

If $\frac{a(r)}{r^N\eta_{p,\gamma}} \geq 1$, there is nothing to prove. When $\frac{a(r)}{r^N\eta_{p,\gamma}} < 1$, we obtain $1 \leq 2C\left(\frac{a(r)}{r^N\eta_{p,\gamma}}\right)^{p_-}$, which proves the estimate.

4th step. We split integration in (4.4) as follows

$$1 = \sum_{i=1}^{3} \int\limits_{X_i(x)} \left(\frac{a(\varrho(x,y))}{\varrho(x,y)^N\eta_{p,\gamma}}\right)^{p(y)} \varrho(x_0,y)^{\gamma-N} d\mu(y) := I_1 + I_2 + I_3,$$

where

$$X_1(x) = \left\{y \in X : r < \varrho(x,y) < \frac{1}{2},\ K(x,y) > \eta_{p,\gamma}\right\},$$

$$X_2(x) = \left\{y \in X : r < \varrho(x,y) < \frac{1}{2},\ K(x,y) < \eta_{p,\gamma}\right\},$$

$$X_3(x) = \left\{y \in X : \varrho(x,y) > \frac{1}{2}\right\}.$$

5th step. Estimation of I_1. We have

$$I_1 = \int\limits_{X_1(x)} \left(\frac{a(\varrho(x,y))}{\varrho(x,y)^N \eta_{p,\gamma}}\right)^{p(x)} \varrho(x_0,y)^{\gamma-N} u_r(x,y)d\mu(y),$$

where

$$u_r(x,y) = \left(\frac{a(\varrho(x,y))}{\varrho(x,y)^N \eta_{p,\gamma}}\right)^{p(y)-p(x)}.$$

We show that the function $u_r(x,y)$ is bounded from below and above uniformly in x,y and r. To this end, we make use of (2.4) and following estimations in [33], p. 432, and obtain

$$|\ln u_r(x,y)| \le C\frac{\ln \frac{a(\varrho(x,y))}{\varrho(x,y)^N \eta_{p,\gamma}}}{\ln \frac{1}{\varrho(x,y)}} = C\frac{\ln \frac{a(\varrho(x,y))}{\varrho(x,y)^N} - \ln \eta_{p,\gamma}}{\ln \frac{1}{\varrho(x,y)}},$$

where we took into account that $\frac{a(\varrho)}{\varrho^N \eta_{p,\gamma}} \ge 1$. Therefore,

$$|\ln u_r(x,y)| \le C\frac{\ln \frac{a(\varrho(x,y))}{\varrho(x,y)^N}}{\ln \frac{1}{\varrho(x,y)}} \le C\frac{|\ln a(\varrho(x,y)| + N\ln \frac{1}{\varrho(x,y)}}{\ln \frac{1}{\varrho(x,y)}} \le C.$$

Then

$$I_1 \le \frac{C}{\eta_{p,\gamma}^{p(x)}} \int\limits_{X_1(x)} \left(\frac{a(\varrho(x,y))}{\varrho(x,y)^N}\right)^{p(x)} \varrho(x_0,y)^{\gamma-N} d\mu(y).$$

By Lemma 3.2 we get

$$I_1 \le C\mathcal{F}(x,r,p(x)), \tag{4.6}$$

where

$$\mathcal{F}(x,r,q) = \begin{cases} \int\limits_r^d t^{\gamma-1}\left[\frac{a(t)}{\eta_{p,\gamma}t^N}\right]^q dt, & \text{if } \varrho(x_0,x) \le r \\ \varrho(x_0,x)^{\gamma-N}\int\limits_r^d t^{N-1}\left[\frac{a(t)}{\eta_{p,\gamma}t^N}\right]^q dt, & \text{if } \varrho(x_0,x) > r \end{cases}$$

6th step. Estimation of I_2. For I_2 we obtain

$$\begin{aligned} I_2 &\le \int\limits_{r<\varrho<1} \left(\frac{a(\varrho(x,y))}{\varrho(x,y)^N \eta_{p,\gamma}}\right)^{p_-} \varrho(x_0,y)^{\gamma-N} d\mu(y) \\ &\le \frac{1}{\eta_{p,\gamma}^{p_-}} \int\limits_{r<\varrho<1} \left(\frac{a(\varrho(x,y))}{\varrho(x,y)^N}\right)^{p_-} \varrho(x_0,y)^{\gamma-N} d\mu(y) \end{aligned}$$

and the application of Lemma 3.2 gives

$$I_2 \leq C\mathcal{F}(x,r,p_-). \tag{4.7}$$

7th step. Estimation of I_3. For I_3 we have

$$\begin{aligned} I_3 &\leq \frac{C}{\eta_{p,\gamma}^{p_-}} \int\limits_{\varrho>\frac{1}{2}} \left(\frac{\frac{a(\varrho(x,y))}{\sup\limits_{t\in(0,d)} a(t)}}{(2\varrho(x,y))^N} \right)^{p(y)} \varrho(x_0,y)^{\gamma-N} d\mu(y) \\ &\leq \frac{C}{\eta_{p,\gamma}^{p_-}} \int\limits_{\varrho>\frac{1}{2}} \left(\frac{\frac{a(\varrho(x,y))}{\sup\limits_{t\in(0,d)} a(t)}}{(2\varrho(x,y))^N} \right)^{p_-} \frac{d\mu(y)}{\varrho(x_0,y)^{N-\gamma}} \\ &\leq \frac{C}{\eta_{p,\gamma}^{p_-}} \int\limits_{\varrho>\frac{1}{2}} \left(\frac{a(\varrho(x,y))}{\varrho(x,y)^N} \right)^{p_-} \frac{d\mu(y)}{\varrho(x_0,y)^{N-\gamma}}, \end{aligned}$$

where the last integral is convergent and uniformly bounded with respect to x by Lemma 3.2. Hence

$$I_3 \leq \frac{C}{\eta_{p,\gamma}^{p_-}}. \tag{4.8}$$

8th step. By (4.6), (4.7), (4.8) we have

$$1 \leq C \left[\mathcal{F}(x,r,p(x)) + \mathcal{F}(x,r,p_-) + \frac{1}{\eta_{p,\gamma}^{p_-}} \right].$$

We may consider $\eta_{p,\gamma}(x,r)$ only for those x,r for which $\eta_{p,\gamma}(x,r)$ is sufficiently large: $\eta_{p,\gamma}(x,r) \geq (2C)^{\frac{1}{p_-}}$, where C is the constant from the last inequality. For such x,r we have $\frac{C}{\eta_{p,\gamma}} \leq \frac{1}{2}$ and we then obtain

$$\frac{1}{2} \leq C\left[\mathcal{F}(x,r,p(x)) + \mathcal{F}(x,r,p_-)\right]. \tag{4.9}$$

Taking into account that $\dfrac{Ca(t)}{t^N \eta_{p,\gamma}} \geq 1$ by (4.5), we have

$$\mathcal{F}(x,r,p_-) \leq C\mathcal{F}(x,r,p(x))$$

Then (4.9) yields the inequality $1 \leq C\mathcal{F}(x,r,p(x))$, that is,

$$\eta_{p,\gamma}^{p(x)} \leq C \begin{cases} \int\limits_r^d t^{\gamma-1} \left[\frac{a(t)}{t^N}\right]^{p(x)} dt, & \text{if } \varrho(x_0,x) \leq r \\ \varrho(x_0,x)^{\gamma-N} \int\limits_r^d t^{N-1} \left[\frac{a(t)}{t^N}\right]^{p(x)} dt, & \text{if } \varrho(x_0,x) > r \end{cases} \tag{4.10}$$

9th step. Final estimate of $\eta_{p,\gamma}$. Write (4.10) in the next form

$$\eta_{p,\gamma}^{p(x)} \le C \begin{cases} \int\limits_r^d t^{\beta p(x)-Np(x)+\gamma-1}\left[\frac{a(t)}{t^\beta}\right]^{p(x)} dt, & \text{if} \quad \varrho(x_0,x)\le r \\ \varrho(x_0,x)^{\gamma-N}\int\limits_r^d t^{\beta p(x)-Np(x)+N-1}\left[\frac{a(t)}{t^\beta}\right]^{p(x)} dt, & \text{if} \quad \varrho(x_0,x) > r \end{cases}.$$

By (4.2) we have

$$\eta_{p,\gamma}^{p(x)} \le C\left[\frac{a(r)}{r^\beta}\right]^{p(x)} \begin{cases} \int\limits_r^d t^{\beta p(x)-Np(x)+\gamma-1}\, dt, & \text{if} \quad \varrho(x_0,x)\le r \\ \varrho(x_0,x)^{\gamma-N}\int\limits_r^d t^{\beta p(x)-Np(x)+N-1}\, dt, & \text{if} \quad \varrho(x_0,x) > r \end{cases}.$$

Since $0<\beta<\min\left(\frac{N}{(p_-)'}, N-\frac{\gamma}{p_-}\right)$ we have $\beta p(x)-Np(x)+m<0$, where m can take two values: N or γ. Then

$$\int\limits_r^d t^{\beta p(x)-Np(x)+m-1}dt \le C r^{\beta p(x)-Np(x)+m}.$$

Therefore

$$\begin{aligned} \eta_{p,\gamma}^{p(x)} &\le C\left[\frac{a(r)}{r^N}\right]^{p(x)} \begin{cases} r^\gamma, & \text{if} \quad \varrho(x_0,x)\le r \\ \varrho(x_0,x)^{\gamma-N} r^N, & \text{if} \quad \varrho(x_0,x)>r \end{cases} \\ &= C\left[\frac{a(r)}{r^{\frac{N}{p'(x)}}}\right]^{p(x)} [\max(\varrho(x_0,x),r)]^{\gamma-N}, \end{aligned}$$

which proves (4.3). □

5. Proof of the main result

Proof. As usual, we may suppose that $f(x)\ge 0$ and $\|f\|_{L^{p(\cdot)}(X,w^\nu)}\le 1$ and show that

$$\int\limits_X \Phi\left[x, w(x)I^a f(x)\right] d\mu(x) \le C < \infty. \tag{5.1}$$

In accordance with Hedberg's trick, we split $I_a f(x)$ as follows

$$\begin{aligned} I_a f(x) &= \int\limits_{B(x,r)} \frac{a[\varrho(x,y)]}{\varrho(x,y)^N} f(y)d\mu(y) + \int\limits_{X\setminus B(x,r)} \frac{a\left(\varrho\left(x,y\right)\right)}{\varrho\left(x,y\right)^N} f(y)d\mu\left(y\right) \\ &= \mathcal{A}_r(x)+\mathcal{B}_r(x). \end{aligned}$$

The estimation of the first term via the maximal function well known in the case $a(r) = r^\alpha$, now holds in the form

$$\mathcal{A}_r(x) \le CA(r)Mf(x), \quad A(r) = \int_0^r \frac{a(t)}{t}\, dt, \tag{5.2}$$

see [13], Subsection 4.4.

For $\mathcal{B}_r(x)$, by the Hölder inequality

$$\left| \int_X f(x)g(x)d\mu(x) \right| \le k\|f\|_{L^{p(\cdot)}(X,\varrho)} \|g\|_{L^{p'(\cdot)}(X,\varrho^{-1})}$$

for variable exponents, we obtain

$$\mathcal{B}_r(x) \le C\, \|f\|_{L^{p(\cdot)}(X\backslash B(x,r),w^\nu)} \left\| \frac{a\left(\varrho\left(x,y\right)\right)}{\varrho\left(x,y\right)^N} \right\|_{L^{p'(\cdot)}(X\backslash B(x,r),w^{-\nu})},$$

where we denote $w = \varrho(\cdot, x_0)$ for brevity. Under notation (4.1) we obtain

$$\mathcal{B}_r(x) \le C\eta_{p',\gamma}(x,r) \quad \text{with} \quad \gamma = N - \nu p' x_0),$$

with N from the growth condition (one may take $N < \underline{\mathfrak{dim}}(X)$ arbitrarily close to $\underline{\mathfrak{dim}}(X)$ according to (2.5)–(2.6)).

We apply Lemma 4.1 and obtain

$$\begin{aligned} \eta_{p',\gamma}(x,r) &\le C\frac{a(r)}{r^{\frac{N}{p(x)}}}[\max(r, \varrho(x_0,x))]^{-\nu} \\ &\le C\frac{a(r)}{r^{\frac{N}{p(x)}}}\varrho(x_0,x)^{-\nu} \sim C\frac{a(r)}{r^{\frac{N}{p(x)}}}\varrho(x_0,x)^{-\nu} \end{aligned}$$

Therefore

$$\mathcal{B}_r(x) \le C\frac{a(r)}{r^{\frac{N}{p(x)}}}\varrho(x_0,x)^{-\nu}$$

and

$$\begin{aligned} I_a f(x) &\le C\left[A(r)Mf(x) + \frac{a(r)}{r^{\frac{N}{p(x)}}}\varrho(x_0,x)^{-\nu}\right] \\ &\le CA(r)\left[Mf(x) + \frac{1}{r^{\frac{N}{p(x)}}}\varrho(x_0,x)^{-\nu}\right], \end{aligned}$$

where we used the fact that $a(r \le CA(r))$ which follows from (4.2). Consequently, by Lemma 2.5 we get

$$I_a f(x) \le C\Phi^{-1}\left(x, \frac{1}{r^N}\right)\left[r^{\frac{N}{p(x)}} Mf(x) + \varrho(x,x_0)^{-\nu}\right]$$

Now we choose $r = \dfrac{1}{\varrho(x_0,x)^{\frac{\nu p(x)}{N}} Mf(x)^{\frac{p(x)}{N}}}$ and get

$$\frac{1}{C}[\varrho(x,x_0)]^{\nu} I_a f(x) \leq \Phi^{-1}\left(x, [\varrho(x,x_0)]^{\nu p(x)} Mf(x)^{p(x)}\right).$$

Hence

$$\int\limits_X \Phi\left(x, \frac{1}{C}[\varrho(x,x_0)]^{\nu} I_a f(x)\right) d\mu(x) \leq \int\limits_X [\varrho(x,x_0)]^{\nu p(x)} Mf(x)^{p(x)} d\mu(x).$$

Then the application of Theorem 2.4 completes the proof of (5.1), if we take into account property (2.5). □

References

[1] A. Almeida and S. Samko, *Pointwise inequalities in variable Sobolev spaces and applications.* Zeit. Anal. und ihre Anwend. **26(2)** (2007),179–193.

[2] Y. Chen, S. Levine, and Rao M, *Variable exponent, linear growth functionals in image restoration.* SIAM J. Appl. Math. **66** (2006), 1383–1406.

[3] L. Diening, *Maximal function on Musielak-Orlicz spaces and generalized Lebesgue spaces.* Bull. Sci. Math. **129(8)** (2005), 657–700.

[4] L. Diening, P. Hästö, and A. Nekvinda, *Open problems in variable exponent Lebesgue and Sobolev spaces.* In "Function Spaces, Differential Operators and Nonlinear Analysis", Proceedings of the Conference held in Milovy, Bohemian-Moravian Uplands, May 28–June 2, 2004. Math. Inst. Acad. Sci. Czech Republick, Praha.

[5] D.E. Edmunds, V. Kokilashvili, and A. Meskhi, *Bounded and compact integral operators*, volume 543 of *Mathematics and its Applications.* Kluwer Academic Publishers, Dordrecht, 2002.

[6] D.E. Edmunds, V. Kokilashvili, and A. Meskhi, *A trace inequality for generalized potentials in Lebesgue spaces with variable exponent.* J. Funct. Spaces Appl. **2(1)** (2004).

[7] J. García-Cuerva and A.E. Gatto, *Boundedness properties of fractional integral operators associated to non-doubling measures.* Studia Math. **162** (2004), 245–261.

[8] A.E. Gatto, *On fractional calculus associated to doubling and non-doubling measures.* In J. Marshall (ed.) et al. Ash, editor, Harmonic analysis. Calderon-Zygmund and beyond. A conference in honor of Stephen Vagi's retirement, Chicago, IL, USA, December 6–8, **411** (2002), 15–37. Providence, RI: American Mathematical Society, 2006. Contemporary Mathematics.

[9] A.E. Gatto, C. Segovia, and S. Vagi, *On fractional differentiation on spaces of homogeneous type.* Revista Mat. Iberoamer. **12(1)** (1996), 1–35.

[10] A.E. Gatto and S. Vagi, *Fractional integrals on spaces of homogeneous type.* In Cora Sadosky (ed.), Analysis and Partial Differential Equations **122**, 171–216. Marcel Dekker, New York and Basel, 1990. Lecture Notes in Pure and Appl. Math.

[11] A.E. Gatto and S. Vagi, *On functions arising as potentials on spaces of homogeneous type.* Proc. Amer. Math. Soc. **125(4)** (1997), 1149–1152.

[12] H. Gunawan, *Generalized fractional integral operators and their modified versions.* In Asian Math. Conf. AMC, Singapore, July 2005. 10 pp; accessible at http://ww1.math.nus.edu.sg/AMC/papers/Gunawan-Hendra.pdf.

[13] M.G. Hajibayov and S.G. Samko, *Generalized potentials in variable exponent Lebesgue spaces on homogeneous spaces.* Math. Nachr. submitted.

[14] P. Harjulehto, P. Hästö, and V. Latvala, *Sobolev embeddings in metric measure spaces with variable dimension.* Math. Z. **254(3)** (2006), 591–609.

[15] P. Harjulehto, P. Hästö, and M. Pere, *Variable exponent Lebesgue spaces on metric spaces: the Hardy-Littlewood maximal operator.* Real Anal. Exchange **30(1)** (2004/05), 87–103.

[16] P. Harjulehto, P. Hästö, and M. Pere, *Variable exponent Sobolev spaces on metric measure spaces.* Funct. Approx. Comment. Math. **36(1)** (2006), 79–94.

[17] M. Khabazi, *The maximal operator in spaces of homogenous type.* Proc. A. Razmadze Math. Inst. **138** (2005), 17–25.

[18] V. Kokilashvili, *On a progress in the theory of integral operators in weighted Banach function spaces.* In "Function Spaces, Differential Operators and Nonlinear Analysis", Proceedings of the Conference held in Milovy, Bohemian-Moravian Uplands, May 28–June 2, 2004. Math. Inst. Acad. Sci. Czech Republick, Praha.

[19] V. Kokilashvili and A. Meskhi, *Fractional integrals on measure spaces.* Fract. Calc. Appl. Anal. **4(1)** (2001), 1–24.

[20] V. Kokilashvili and A. Meskhi, *On some weighted inequalities for fractional integrals on nonhomogeneous spaces.* Zeitschr. Anal. Anwend. **24(4)** (2005), 871–885.

[21] V. Kokilashvili and S. Samko, *The maximal operator in weighted variable spaces on metric measure spaces.* Proc. A. Razmadze Math. Inst. **144** (2007), 137–144.

[22] V. Kokilashvili and S. Samko, *The maximal operator in weighted variable exponent spaces on metric spaces.* Georgian Math. J. 2008. to appear.

[23] O. Kováčik and J. Rákosník, *On spaces $L^{p(x)}$ and $W^{k,p(x)}$.* Czechoslovak Math. J. **41(116)** (1991), 592–618.

[24] J. Musielak, *Orlicz spaces and modular spaces.* Lecture Notes in Mathematics **1034** Springer-Verlag, Berlin, 1983.

[25] E. Nakai, *On generalized fractional integrals.* Taiwanese J. Math. **5(3)** (2001), 587–602.

[26] E. Nakai and H. Sumitomo, *On generalized Riesz potentials and spaces of some smooth functions.* Sci. Math. Jpn. **54(3)** (2001), 463–472.

[27] E.I. Pustylnik, *Generalized potential type operators on rearrangement invariant spaces.* Israel Math. Conf. Proc. **13** (1999), 161–171.

[28] M. Růžička, *Electroreological fluids: Modeling and mathematical theory.* Springer, Lecture Notes in Math. **1748** (2000).

[29] N. Samko, S. Samko, and Vakulov B, *Weighted Sobolev theorem in Lebesgue spaces with variable exponent.* J. Math. Anal. and Applic. **335** (2007), 560–583.

[30] N.G. Samko, *Parameter depending almost monotonic functions and their applications to dimensions in metric measure spaces.* J. Funct. Spaces and Appl. to appear.

[31] S.G. Samko, *Convolution and potential type operators in $L^{p(x)}$.* Integr. Transf. and Special Funct. **7(3-4)** (1998), 261–284.

[32] S.G. Samko, *Convolution type operators in $L^{p(x)}$*. Integr. Transf. and Special Funct. **7(1-2)** (1998), 123–144.

[33] S.G. Samko, *Hardy-Littlewood-Stein-Weiss inequality in the Lebesgue spaces with variable exponent.* Frac. Calc. and Appl. Anal. **6(4)** (2003), 421–440.

[34] S.G. Samko, *On a progress in the theory of Lebesgue spaces with variable exponent: maximal and singular operators.* Integr. Transf. and Spec. Funct. **16(5-6)** (2005), 461–482.

[35] I.I. Sharapudinov, *The topology of the space* $L^{p(t)}([0, 1])$. Mat. Zametki **26(4)** (1979), 613–632. Engl. transl. in Math. Notes. **26(3-4)** (1979), 796–806.

[36] V.V. Zhikov, *Meyer-type estimates for solving the non-linear Stokes system.* Differ. Equat. **33(1)** (1997), 108–115.

Mubariz G. Hajibayov
Institute of Mathematics and Mechanics
of NAS of Azerbaijan
9, F.Agaev str.
AZ1141 Baku, Azerbaijan
e-mail: hajibayovm@yahoo.com

Stefan G. Samko
Department of Mathematics
University of Algarve
Campus de Gambelas
8005-139 Faro, Portugal
e-mail: ssamko@ualg.pt

Operator Theory:
Advances and Applications, Vol. 210, 123–145

Wiener-Hopf Operators with Oscillating Symbols on Weighted Lebesgue Spaces

Yu.I. Karlovich and J. Loreto Hernández

To Professor N.L. Vasilevski on the occasion of his 60th birthday

Abstract. We establish Fredholm criteria for Wiener-Hopf operators $W(a)$ with oscillating symbols a, continuous on $\mathbb{R}$ and admitting mixed (slowly oscillating and semi-almost periodic) discontinuities at $\pm\infty$, on weighted Lebesgue spaces $L^p_N(\mathbb{R}_+, w)$ where $1 < p < \infty$, $N \in \mathbb{N}$, and w belongs to a subclass of Muckenhoupt weights. For $N > 1$ these criteria are conditional.

Mathematics Subject Classification (2000). Primary 47B35; Secondary 42A45, 47A53, 47G10.

Keywords. Wiener-Hopf operator, limit operator, weighted Lebesgue space, Muckenhoupt weight, slowly oscillating and semi-almost periodic matrix functions, local principle, symbol, Fredholmness.

1. Introduction

Let $\mathcal{B}(X)$ be the Banach algebra of all bounded linear operators on a Banach space X, and $\mathcal{K}(X)$ the closed two-sided ideal of all compact operators in $\mathcal{B}(X)$. An operator $A \in \mathcal{B}(X)$ is called *Fredholm* if $\operatorname{Im} A$ is closed in X and the numbers $n(A) := \dim \operatorname{Ker} A$ and $d(A) := \dim(X/\operatorname{Im} A)$ are finite (see, e.g., [7]). In that case

$$\operatorname{Ind} A := n(A) - d(A).$$

Given $1 \le p \le \infty$, let $L^p(\mathbb{R})$ be the usual Lebesgue space with norm denoted by $\|\cdot\|_p$. A measurable function $w : \mathbb{R} \to [0, \infty]$ is called a *weight* if $w^{-1}(\{0, \infty\})$ has Lebesgue measure zero. For $1 \le p < \infty$ and a weight w, we denote by $L^p(\mathbb{R}, w)$

Work was supported by the SEP-CONACYT Projects No. 25564 and 47220 (México). The second author was also sponsored by the CONACYT scholarship No. 163480.

the weighted Lebesgue space with the norm

$$\|f\|_{p,w} := \left(\int_{\mathbb{R}} |f(x)|^p w^p(x) dx \right)^{1/p}.$$

Given $N \in \mathbb{N}$, let $L^p_N(\mathbb{R}, w)$ be the Banach space of vector functions $f = (f_k)_{k=1}^N$ with entries $f_k \in L^p(\mathbb{R}, w)$ and the norm $\|f\|_{L^p_N(\mathbb{R},w)} = \left(\sum_{k=1}^N \|f_k\|^p_{p,w}\right)^{1/p}$. If $\mathcal{A}$ is a subalgebra of $L^\infty(\mathbb{R})$, then $\mathcal{A}_{N\times N}$ or $[\mathcal{A}]_{N\times N}$ denote the matrix functions $a : \mathbb{R} \to \mathbb{C}^{N\times N}$ whose entries belong to $\mathcal{A}$.

In what follows we assume that $1 < p < \infty$ and w is a *Muckenhoupt weight* (that is, $w \in A_p(\mathbb{R})$), which means that the Cauchy singular integral operator $S_{\mathbb{R}}$ given by

$$(S_{\mathbb{R}} f)(x) = \lim_{\varepsilon\to 0} \frac{1}{\pi i} \int_{\mathbb{R}\setminus(x-\varepsilon,x+\varepsilon)} \frac{f(t)}{t-x}\, dt, \quad x \in \mathbb{R}, \tag{1.1}$$

is bounded on the space $L^p(\mathbb{R}, w)$ or, equivalently (see [18] and also [13]),

$$\sup_I \left(\frac{1}{|I|} \int_I w^p(x)\, dx \right)^{1/p} \left(\frac{1}{|I|} \int_I w^{-q}(x)\, dx \right)^{1/q} < \infty,$$

where $1/p + 1/q = 1$, I ranges over all bounded intervals $I \subset \mathbb{R}$, and $|I|$ is the length of I.

Let $\mathcal{F} : L^2(\mathbb{R}) \to L^2(\mathbb{R})$ denote the *Fourier transform*,

$$(\mathcal{F}f)(x) := \int_{\mathbb{R}} f(t) e^{itx} dt, \quad x \in \mathbb{R}.$$

A function $a \in L^\infty(\mathbb{R})$ is called a *Fourier multiplier* on $L^p(\mathbb{R}, w)$ if the convolution operator $W^0(a) := \mathcal{F}^{-1} a \mathcal{F}$ maps $L^2(\mathbb{R}) \cap L^p(\mathbb{R}, w)$ into itself and extends to a bounded linear operator on $L^p(\mathbb{R}, w)$ (notice that $L^2(\mathbb{R}) \cap L^p(\mathbb{R}, w)$ is dense in $L^p(\mathbb{R}, w)$ if $w \in A_p(\mathbb{R})$). Let $[M_{p,w}]_{N\times N}$ stand for the Banach algebra of all Fourier multipliers a on $L^p_N(\mathbb{R}, w)$ equipped with the norm

$$\|a\|_{[M_{p,w}]_{N\times N}} := \left\|W^0(a)\right\|_{\mathcal{B}(L^p_N(\mathbb{R},w))}.$$

Let χ_+ be the characteristic function of $\mathbb{R}_+ = (0, \infty)$. By $L^p(\mathbb{R}_+, w)$ we understand the space $L^p(\mathbb{R}_+, w|\mathbb{R}_+)$. For $a \in M_{p,w}$, the Wiener-Hopf operator $W(a)$ is defined on the space $L^p(\mathbb{R}_+, w)$ by

$$W(a)f = \chi_+ W^0(a) \chi_+ f, \quad \text{for} \quad f \in L^p(\mathbb{R}_+, w).$$

Let $\dot{\mathbb{R}} = \mathbb{R} \cup \{\infty\}$, $\overline{\mathbb{R}} = [-\infty, +\infty]$, and let PC be the C^*-algebra of all functions on $\mathbb{R}$ having finite one-sided limits at every point $t \in \dot{\mathbb{R}}$. By Stechkin's inequality (see, e.g., [6, Theorem 17.1]), if $a \in PC$ has finite total variation $V_1(a)$, then $a \in M_{p,w}$ and

$$\|a\|_{M_{p,w}} \le \|S_{\mathbb{R}}\|_{\mathcal{B}(L^p(\mathbb{R},w))} \big(\|a\|_\infty + V_1(a)\big), \tag{1.2}$$

where $S_{\mathbb{R}}$ is given by (1.1). We denote by $C_{p,w}(\dot{\mathbb{R}})$ (resp. $C_{p,w}(\overline{\mathbb{R}})$) the closure in $M_{p,w}$ of the set of all functions $a \in C(\dot{\mathbb{R}})$ (resp. $a \in C(\overline{\mathbb{R}})$) with finite total variation. Obviously, $C_{p,w}(\dot{\mathbb{R}}) \subset C(\dot{\mathbb{R}})$, $C_{p,w}(\overline{\mathbb{R}}) \subset C(\overline{\mathbb{R}})$.

To study Wiener-Hopf operators with semi-almost periodic (SAP) symbols we need the set $A_p^0(\mathbb{R})$ consisting of all weights $w \in A_p(\mathbb{R})$ for which the functions $e_\lambda : x \mapsto e^{i\lambda x}$ belong to $M_{p,w}$ for all $\lambda \in \mathbb{R}$. Let $w \in A_p^0(\mathbb{R})$. Then the set AP^0 of all almost periodic polynomials $\sum_{\lambda\in\Lambda_0} c_\lambda e_\lambda$, where $c_\lambda \in \mathbb{C}$ and Λ_0 is a finite subset of $\mathbb{R}$, is contained in $M_{p,w}$. We define $AP_{p,w}$ as the closure of AP^0 in $M_{p,w}$. Clearly, $AP_{p,w}$ is a Banach subalgebra of $M_{p,w}$. Let $SAP_{p,w}$ denote the smallest closed subalgebra of $M_{p,w}$ that contains $C_{p,w}(\overline{\mathbb{R}})$ and $AP_{p,w}$. It is clear that

$$AP_{p,w} \subset AP := AP_{2,1} \subset L^\infty(\mathbb{R}), \quad SAP_{p,w} \subset SAP := SAP_{2,1} \subset L^\infty(\mathbb{R}).$$

Let $C_b(\mathbb{R})$ be the C^*-algebra of all bounded continuous functions $a : \mathbb{R} \to \mathbb{C}$. Following [25] we denote by SO the C^*-algebra of *slowly oscillating at* ∞ functions,

$$SO := \Big\{ f \in C_b(\mathbb{R}) : \lim_{x\to+\infty} \sup_{t,s\in[-2x,-x]\cup[x,2x]} |f(t) - f(s)| = 0 \Big\}. \tag{1.3}$$

Consider the commutative Banach algebra

$$SO^3 := \Big\{ a \in SO \cap C^3(\mathbb{R}) : \lim_{|x|\to\infty} (D^\gamma a)(x) = 0,\ \gamma = 1,2,3 \Big\} \tag{1.4}$$

equipped with the norm $\|a\|_{SO^3} := \max_{\gamma=0,1,2,3} \|D^\gamma a\|_{L^\infty(\mathbb{R})}$ where $(Da)(x) = xa'(x)$ for $x \in \mathbb{R}$. By [21, Corollary 2.10], $SO^3 \subset M_{p,w}$. For $1 < p < \infty$ and $w \in A_p(\mathbb{R})$, let $SO_{p,w}$ denote the closure of SO^3 in $M_{p,w}$. Clearly, $SO_{p,w}$ is a commutative Banach subalgebra of $M_{p,w}$. Since $M_{p,w} \subset M_2 = L^\infty(\mathbb{R})$, we conclude that $SO_{p,w} \subset SO$.

Let $[\mathcal{A}, \mathcal{B}]$ denote the smallest Banach algebra that contains Banach algebras $\mathcal{A}$ and $\mathcal{B}$. Then $[SO_{p,w}, SAP_{p,w}]$ is the Banach subalgebra of $M_{p,w}$ generated by all functions in $SO_{p,w}$ and $SAP_{p,w}$. If $w = 1$, we write SO_p, SAP_p, and so on.

Fredholm theories for Wiener-Hopf operators $W(a)$ with symbols $a \in PC_p$ on Lebesgue spaces $L^p(\mathbb{R}_+)$ and for the Banach subalgebras of $\mathcal{B}(L^p(\mathbb{R}))$ generated by the multiplication operators aI ($a \in PC$) and by the convolution operators $W^0(b)$ ($b \in PC_p$) were constructed by R.V. Duduchava (see [10] and [11]). Fredholmness and index formulas for such operators on weighted Lebesgue spaces and algebras generated by these operators were studied in [29] and [27] in the case of power weights, and in [8] and [9] for general Muckenhoupt weights.

Fredholmness for Banach algebras generated by all operators aI and $W^0(b)$ with $a \in [SO, PC]_{N\times N}$ and $b \in [SO_p, PC_p]_{N\times N}$ on unweighted Lebesgue spaces $L_N^p(\mathbb{R})$ was studied in [1], [2]. Wiener-Hopf operators with slowly oscillating matrix symbols on weighted Lebesgue spaces were investigated in [21].

Wiener-Hopf operators with semi-almost periodic symbols on the spaces $L^p(\mathbb{R}_+)$ ($1 < p < \infty$) were studied by R.V. Duduchava and A.I. Saginashvili [12] (for preceding results on integro-difference operators see [15], [16]). The Fredholm theory for Wiener-Hopf operators with semi-almost periodic matrix symbols on the spaces $L_N^p(\mathbb{R}_+)$ ($1 < p < \infty$, $N > 1$) based on the concept of almost periodic (AP) factorization was constructed by I.M. Spitkovsky and the first author (see [6] and the references therein). Wiener-Hopf operators with semi-almost periodic matrix symbols on weighted Lebesgue spaces were studied in [20].

A Fredholm theory for Toeplitz operators with oscillating matrix symbols $a \in [SO, SAP]_{N\times N}$ on Hardy spaces H^p_N was constructed in [3].

In the present paper we establish Fredholm criteria for Wiener-Hopf operators $W(a)$ with oscillating symbols $a \in [SO_{p,w}, SAP_{p,w}]$ on weighted Lebesgue spaces $L^p(\mathbb{R}_+, w)$ with Muckenhoupt weights $w \in A^0_p(\mathbb{R})$ and, under additional conditions, with matrix symbols $a \in [SO_{p,w}, SAP_{p,w}]_{N\times N}$ on the spaces $L^p_N(\mathbb{R}_+, w)$ with $N > 1$. To study mentioned Wiener-Hopf operators we apply the Allan-Douglas local principle (see, e.g., [7]), the notion of almost periodic factorization [6], and techniques of limit operators (see, e.g., [5], [3], [26]).

The paper is organized as follows. In Section 2 we collect results on algebras of slowly oscillating and semi-almost periodic functions, their maximal ideal spaces and associated limit operators, and also present invertibility criteria for Wiener-Hopf operators with almost periodic symbols.

Section 3 deals with geometric means of matrix functions $a \in [AP_p]_{N\times N}$ being symbols of Wiener-Hopf operators invertible on the space $L^p_N(\mathbb{R}_+)$.

In Section 4, applying techniques of limit operators, we obtain necessary Fredholm conditions for Wiener-Hopf operators $W(a)$ with oscillating matrix symbols $a \in [SO_{p,w}, SAP_{p,w}]_{N\times N}$ on the space $L^p_N(\mathbb{R}_+, w)$ in terms of invertibility of some Wiener-Hopf operators with symbols in the Banach algebra $[AP_{p,w}]_{N\times N}$.

In Section 5 we prove that all the regularizers of Wiener-Hopf operators with symbols in $SAP_{p,w}$ belong to the Banach algebra $\operatorname{alg} W(SAP_{p,w})$ generated by all operators $W(a)$ with symbols $a \in SAP_{p,w}$.

In Section 6 we study the invertibility in quotient algebras associated with the points of the maximal ideal space $\mathcal{M}(SO)$ of SO. Then, applying the Allan-Douglas local principle and the results of previous sections, we obtain a Fredholm criterion for Wiener-Hopf operators $W(a)$ with symbols $a \in [SO_{p,w}, SAP_{p,w}]$ on weighted Lebesgue spaces $L^p(\mathbb{R}_+, w)$ with $1 < p < \infty$ and $w \in A^0_p(\mathbb{R})$.

In Section 7 we establish a conditional Fredholm criterion for Wiener-Hopf operators $W(a)$ with oscillating matrix symbols $a \in [SO_{p,w}, SAP_{p,w}]_{N\times N}$ on the space $L^p_N(\mathbb{R}_+, w)$, and a Fredholm criterion for such operators on the space $L^p_N(\mathbb{R}_+)$ provided that their symbols belong to a Wiener subclass of $[SO_p, SAP_p]_{N\times N}$.

2. Auxiliary results

2.1. Slowly oscillating Fourier multipliers on $L^p(\mathbb{R}, w)$

Consider the commutative C^*-algebra SO of slowly oscillating functions defined by (1.3). Clearly, SO is a subalgebra of $L^\infty(\mathbb{R})$ which contains all functions in $C(\dot{\mathbb{R}})$. Identifying the points $t \in \dot{\mathbb{R}}$ with the evaluation functionals δ_t on $\dot{\mathbb{R}}$, $\delta_t(f) = f(t)$, we see that the maximal ideal space $\mathcal{M}(SO)$ of SO is of the form

$$\mathcal{M}(SO) = \mathbb{R} \cup \mathcal{M}_\infty(SO), \quad \text{where } \mathcal{M}_\infty(SO) := \Big\{\xi \in \mathcal{M}(SO) : \xi|_{C(\dot{\mathbb{R}})} = \delta_\infty\Big\}$$

is the fiber of $\mathcal{M}(SO)$ over ∞. By [3, Proposition 5], $\mathcal{M}_\infty(SO) = (\operatorname{clos}_{SO^*} \mathbb{R}) \setminus \mathbb{R}$ where $\operatorname{clos}_{SO^*} \mathbb{R}$ is the weak-star closure of $\mathbb{R}$ in SO^*, the dual space of SO. Thus, any functional $\xi \in \mathcal{M}_\infty(SO)$ is the limit of a net $t_\alpha \in \mathbb{R}$ that does not converge to functionals $t \in \mathbb{R}$, that is, $f(\xi) := \xi(f) = \lim_\alpha f(t_\alpha)$ for every $f \in SO$.

Proposition 2.1. [3, Proposition 6] *Let $\{a_k\}_{k=1}^\infty$ be a countable subset of SO. If $\xi \in \mathcal{M}_\infty(SO)$, then there exists a sequence $g = \{g_n\} \subset \mathbb{R}$ such that $g_n \to \infty$ and*

$$\xi(a_k) = \lim_{n\to\infty} a_k(g_n), \quad k \in \mathbb{N}. \tag{2.1}$$

Conversely, if $g_n \in \mathbb{R}$, $g_n \to \infty$, and the limits $\lim_{n\to\infty} a_k(g_n)$ exist for all k, then there is a $\xi \in \mathcal{M}_\infty(SO)$ such that (2.1) holds.

Given $1 < p < \infty$ and $w \in A_p(\mathbb{R})$, we consider the Banach subalgebra $SO_{p,w}$ of $M_{p,w}$ being the closure in $M_{p,w}$ of the Banach algebra SO^3 defined by (1.4).

Lemma 2.2. [21, Lemma 3.4] *If $1 < p < \infty$ and $w \in A_p(\mathbb{R})$, then the maximal ideal spaces of $SO_{p,w}$ and SO coincide as sets, that is, $\mathcal{M}(SO_{p,w}) = \mathcal{M}(SO)$.*

Lemma 2.2 and the Gelfand theory immediately give the following assertion.

Corollary 2.3. *If $1 < p < \infty$ and $w \in A_p(\mathbb{R})$, then the Banach algebra $SO_{p,w}$ is inverse closed in the C^*-algebras SO and $L^\infty(\mathbb{R})$.*

2.2. Limit operators and the fiber $\mathcal{M}_\infty([SO, SAP])$

Let $1 < p < \infty$. Given $\lambda \in \mathbb{R}$, we introduce the shift operator U_λ on $L^p(\mathbb{R})$ by

$$(U_\lambda f)(x) = f(x - \lambda), \quad \text{for } x \in \mathbb{R}.$$

Clearly, the operators U_λ are invertible, $U_\lambda^{-1} = U_{-\lambda}$, $(U_\lambda)^* = U_\lambda^{-1}$, and $\|U_\lambda\| = 1$.

Let $A \in \mathcal{B}(L^p(\mathbb{R}))$. The operator $A_h \in \mathcal{B}(L^p(\mathbb{R}))$ is called the *limit operator* of A with respect to operators U_λ and a sequence $h = \{h_m\}$ of real numbers tending to infinity (respectively, to $+\infty$ or to $-\infty$) if (cf. [3]) there exist the strong limits

$$A_h = \operatorname*{s-lim}_{m\to\infty} \left(U_{-h_m} A U_{h_m}\right) \quad \text{and} \quad B_h = \operatorname*{s-lim}_{m\to\infty} \left(U_{-h_m} A U_{h_m}\right)^*$$

in $\mathcal{B}(L^p(\mathbb{R}))$ and $\mathcal{B}(L^q(\mathbb{R}))$, respectively, where $1/p + 1/q = 1$. Then $B_h = (A_h)^*$.

Let BUC be the C^*-algebra of all bounded uniformly continuous functions $a : \mathbb{R} \to \mathbb{C}$. By [3, Proposition 3], for every function $a \in BUC$ and every sequence h tending to ∞ (respectively, to $+\infty$, $-\infty$) there exists a subsequence g of h such that the limit operator $(aI)_g \in \mathcal{B}(L^p(\mathbb{R}))$ exists and $(aI)_g = a_g I$ where $a_g \in BUC$. Since $[SO, SAP] \subset BUC$, the limit operators for the operator aI with $a \in [SO, SAP]$ always exist for a subsequence $g \subset h$. By [3, Proposition 4], $a_g \in AP$ if $a \in AP$ and a_g is a constant if $a \in SO$. Moreover, by Proposition 2.1, for every $a \in SO$ there exists a $\xi \in \mathcal{M}_\infty(SO)$ such that $a_g = \xi(a)$.

Let us define the limit operators for aI when $a \in [SO, SAP]$ (see [3]). By [28], any function $a \in SAP$ can be represented in the form

$$a = a_+ u_+ + a_- u_- + a_0$$

where the functions $a_\pm \in AP$, $a_0 \in C(\dot{\mathbb{R}})$, $a_0(\infty) = 0$, $u_\pm(x) = (1 \pm \tanh x)/2$, and the mappings $\nu_\pm : a \mapsto a_\pm$ are C^*-algebra homomorphisms of SAP onto AP.

According to [25, Section 3], the C^*-algebras SO and SAP are asymptotically independent, which means the following.

Proposition 2.4. *The fiber $\mathcal{M}_\infty([SO,\, SAP])$ is naturally homeomorphic to the set $\mathcal{M}_\infty(SO)\times\mathcal{M}_\infty(SAP)$, that is, for every character $\mu\in\mathcal{M}_\infty([SO,\, SAP])$ there are characters $\xi\in\mathcal{M}_\infty(SO)$ and $\nu\in\mathcal{M}_\infty(SAP)$ such that $\mu|_{SO}=\xi$ and $\mu|_{SAP}=\nu$.*

Identifying $\mu\in\mathcal{M}_\infty([SO,\, SAP])$ with pairs $(\xi,\nu)\in\mathcal{M}_\infty(SO)\times\mathcal{M}_\infty(SAP)$ due to Proposition 2.4, for every $\xi\in\mathcal{M}_\infty(SO)$ we obtain a natural homomorphism

$$\beta_\xi : [SO,\, SAP]\to SAP|_{\mathcal{M}_\infty(SAP)},\quad (\beta_\xi\,\varphi)(\nu)=(\xi,\nu)\varphi \quad \text{for } \nu\in\mathcal{M}_\infty(SAP).$$

Hence, for every $\varphi\in[SO,\, SAP]$ there exists a non-unique function $\varphi_\xi\in SAP$ with uniquely determined almost periodic representatives $\varphi_{\xi,\pm}$ at $\pm\infty$ such that

$$\beta_\xi\,\varphi=\varphi_\xi\big|_{\mathcal{M}_\infty(SAP)}. \tag{2.2}$$

Since the fiber $\mathcal{M}_\infty(AP)$ is homeomorphic to $\mathcal{M}(AP)$, identifying $\mathcal{M}_\infty(SAP)$ and $\mathcal{M}_\infty(AP)\times\mathcal{M}_\infty(AP)$, we conclude that the maps

$$\gamma_\pm : \varphi_\xi\,|_{\mathcal{M}_\infty(SAP)}\mapsto\varphi_{\xi,\pm}|_{\mathcal{M}_\infty(AP)}\mapsto\varphi_{\xi,\pm}$$

are Banach algebra homomorphisms of $SAP|_{\mathcal{M}_\infty(SAP)}$ onto AP. Thus the maps

$$\nu_{\xi,\pm}=\gamma_\pm\circ\beta_\xi : [SO,\, SAP]\to AP,\quad \nu_{\xi,\pm}\,\varphi=\varphi_{\xi,\pm} \tag{2.3}$$

are well defined Banach algebra homomorphisms for every $\xi\in\mathcal{M}_\infty(SO)$.

The C^*-algebra $[SO,\, SAP]$ consists of all functions of the form

$$c=\lim_{k\to\infty}\sum\nolimits_{i=1}^{m_k} b_{i,k}\,a_{i,k} \tag{2.4}$$

where $b_{i,k}\in SO$, $a_{i,k}\in SAP$, and limit is taken in the norm $\|\cdot\|_{L^\infty(\mathbb{R})}$. Therefore, for every $\xi\in\mathcal{M}_\infty(SO)$, the maps $\nu_{\xi,\pm} : [SO,\, SAP]\to AP$ act by the rule

$$\nu_{\xi,\pm}c=\lim_{k\to\infty}\sum\nolimits_{i=1}^{m_k}\xi(b_{i,k})\nu_\pm(a_{i,k}). \tag{2.5}$$

On the other hand, by [3, Section 4], for every $c\in[SO,\, SAP]$ and every $\xi\in\mathcal{M}_\infty(SO)$ there exist sequences $g_\pm=\{g_n^\pm\}\to\pm\infty$ such that for $x\in\mathbb{R}$,

$$\lim_{n\to\infty} b_{i,k}(x+g_n^\pm)=\xi(b_{i,k}),\quad \lim_{n\to\infty} a_{i,k}(x+g_n^\pm)=(\nu_\pm a_{i,k})(x),$$

where for all i,k the convergence is uniform on $\mathbb{R}$ for $a_{i,k}$ and is uniform on compact subsets of $\mathbb{R}$ for $b_{i,k}$. Consequently, for the function (2.4) we obtain

$$(\nu_{\xi,\pm}c)(x)=\lim_{n\to\infty} c(x+g_n^\pm),\quad \text{for } x\in\mathbb{R}. \tag{2.6}$$

In what follows for every $a\in AP$ we use the notation (see [6]):

$$M(a):=\lim_{x\to\infty}\frac{1}{x}\int_0^x a(t)dt,\quad \Omega(a):=\{\lambda\in\mathbb{R} : M(ae_{-\lambda})\neq 0\};$$

$$\kappa(a):=\lim_{x\to\infty}\left(x^{-1}\arg a(x)\right) \text{ if } a^{\pm1}\in AP,\quad \mathbf{d}(a):=e^{M(\ln a)} \text{ if } \ln a\in AP.$$

2.3. Wiener-Hopf operators with almost periodic symbols

Let APW be the Banach algebra of all functions in AP of the form $a = \sum_\lambda a_\lambda e_\lambda$ with $a_\lambda \in \mathbb{C}$, $\lambda \in \mathbb{R}$, and the norm $\|a\|_W := \sum_\lambda |a_\lambda| < \infty$.

Let $APW^\pm$ be the closure in APW of the set AP^0 of all almost periodic polynomials $\sum_\lambda a_\lambda e_\lambda$ with $\pm\lambda \geq 0$. Thus, $APW^\pm$ are Banach subalgebras of APW.

Given $1 < p < \infty$ and $w \in A_p^0(\mathbb{R})$, let $APW_{p,w}$ be the Banach subalgebra of $M_{p,w}$ composed by the series $a = \sum_\lambda a_\lambda e_\lambda$ with coefficients $a_\lambda \in \mathbb{C}$ and the norm

$$\|a\|_W := \sum\nolimits_\lambda |a_\lambda| \, \|e_\lambda\|_{M_{p,w}},$$

where $\|e_\lambda\|_{M_{p,w}} = \|v_\lambda\|_{L^\infty(\mathbb{R})}$ for $\lambda \in \mathbb{R}$ and the functions $v_\lambda(x) = \frac{w(x+\lambda)}{w(x)}$ are in $L^\infty(\mathbb{R})$ for weights $w \in A_p^0(\mathbb{R})$ (see [20, Proposition 2.3]).

Let $AP_{p,w}^\pm$ be the $M_{p,w}$ closure of the set of all almost periodic polynomials $\sum_\lambda a_\lambda e_\lambda$ with $\pm\lambda \geq 0$. Along with the Banach subalgebras $AP_{p,w}^\pm$ of $M_{p,w}$ we consider the Banach subalgebras $APW_{p,w}^\pm := APW_{p,w} \cap AP_{p,w}^\pm$ of $APW_{p,w}$. Clearly,

$$APW_{p,w} \subset AP_{p,w} \subset AP, \quad APW_{p,w}^\pm \subset AP_{p,w}^\pm \subset AP^\pm.$$

Let $G\mathcal{A}$ denote the group of all invertible elements of a unital algebra $\mathcal{A}$.

Theorem 2.5. [20, Theorem 4.2] *If $p \in (1,\infty)$, $w \in A_p^0(\mathbb{R})$ and $a \in APW_{p,w}$, then the following three assertions are equivalent:*

(i) *the operator $W(a)$ is Fredholm on the space $L^p(\mathbb{R}_+, w)$;*
(ii) *the operator $W(a)$ is invertible on the space $L^p(\mathbb{R}_+, w)$;*
(iii) *the function a has a canonical right $APW_{p,w}$ factorization, that is,*

$$a = a_- a_+ \quad \text{with} \quad a_\pm \in GAPW_{p,w}^\pm. \tag{2.7}$$

For general functions $a \in AP_{p,w}$ we obtain the following.

Theorem 2.6. *If $p \in (1,\infty)$, $w \in A_p^0(\mathbb{R})$ and $a \in AP_{p,w}$, then the Wiener-Hopf operator $W(a)$ is invertible on the space $L^p(\mathbb{R}_+, w)$ if and only if the function a is invertible in $L^\infty(\mathbb{R})$ (equivalently, in $AP_{p,w}$) and its mean motion $\kappa(a) = 0$. Moreover, in that case the geometric mean value $\mathbf{d}(a)$ of a is not zero.*

Proof. For arbitrary weights $w \in A_p^0(\mathbb{R})$, the first assertion follows from [20, Theorem 4.1] (for the case $w = 1$, also see [6, Theorem 2.28 and p. 376]). Thus, if $a \in AP_{p,w}$ and the operator $W(a)$ is invertible on the space $L^p(\mathbb{R}_+, w)$, then the functions a and $1/a$ are in $AP_{p,w}$ and $\kappa(a) = \kappa(1/a) = 0$. Hence $\ln a \in AP_{p,w}$ due to [14, § 13], and therefore the geometric mean values $\mathbf{d}(a) = e^{M(\ln a)}$ and $\mathbf{d}(1/a) = e^{-M(\ln a)}$ are mutually inverse complex numbers. □

Consider now $W(a)$ with $a \in [APW_{p,w}]_{N\times N}$. Let $p \in (1,\infty)$, $w \in A_p^0(\mathbb{R})$ and

$$\lim_{|t|\to\infty} \operatorname*{ess\,sup}_{x,y\in[t,\, t+1]} \big|\ln w(x) - \ln w(y)\big| = 0. \tag{2.8}$$

According to [20, Example 2.4], if the parameters $\delta, \nu, \eta \in \mathbb{R}$ satisfy the relations

$$-1/p < \delta - |\nu|\sqrt{\eta^2+1} \leq \delta + |\nu|\sqrt{\eta^2+1} < 1/q,$$

then the following weight with different indices of powerlikeness at ∞ (see [4, Section 3.6]) gives an example of weights in $A_p^0(\mathbb{R})$ possessing the property (2.8):

$$w(x) = \begin{cases} e^{(\delta+\nu\sin(\eta\log(\log|x|)))\log|x|} & \text{if } |x| \geq e, \\ e^{\delta} & \text{if } |x| < e. \end{cases}$$

If $w \in A_p^0(\mathbb{R})$, then, due to [24], we can replace w by an equivalent weight $\omega \in A_p^0(\mathbb{R})$ continuous on $\mathbb{R}$ (see [20, Subsection 2.2]). If $w \in A_p^0(\mathbb{R})$ satisfies (2.8), then, replacing w by ω, we may assume that $v_\lambda \in C(\dot{\mathbb{R}})$ and $v_\lambda(\infty) = 1$ for all $\lambda \in \mathbb{R}$. In that case $APW_{p,w} \subset APW$ and $APW_{p,w}^{\pm} \subset APW^{\pm}$.

Combining [20, Theorem 6.1] and [6, Corollary 19.11], we get the following.

Theorem 2.7. *Let $1 < p < \infty$, $w \in A_p^0(\mathbb{R})$, $N \in \mathbb{N}$, and let (2.8) hold. If $a \in [APW_{p,w}]_{N\times N}$, then the following three assertions are equivalent:*

(i) *the operator $W(a)$ is invertible on the space $L_N^p(\mathbb{R}_+, w)$;*
(ii) *the operator $W(a)$ is invertible on the space $L_N^p(\mathbb{R}_+)$;*
(iii) *the matrix function a admits a canonical right APW factorization*

$$a = a_- a_+ \quad \textit{with} \quad a_\pm \in GAPW_{N\times N}^{\pm}.$$

3. Geometric mean for matrix functions in $[AP_p]_{N\times N}$

For $1 < p < \infty$ and $w \in A_p(\mathbb{R})$, define $H_{p,w,\pm}^{\infty} := H_{\pm}^{\infty} \cap M_{p,w}$. It is easily seen that $H_{p,w,\pm}^{\infty}$ are closed subalgebras of $M_{p,w}$. Let $D_{p,w,\pm} := \mathrm{alg}\{C_{p,w}(\dot{\mathbb{R}}), H_{p,w,\pm}^{\infty}\}$ be the minimal Banach subalgebras of $M_{p,w}$ that contain $C_{p,w}(\dot{\mathbb{R}})$ and $H_{p,w,\pm}^{\infty}$.

Let $\chi_\pm$ denote the characteristic functions of $\mathbb{R}_\pm$.

Lemma 3.1. *Let $1 < p < \infty$ and $w \in A_p(\mathbb{R})$.*

(i) *If $a_\pm \in H_{p,w,\pm}^{\infty}$, then $\chi_- W^0(a_+)\chi_+ I$ and $\chi_+ W^0(a_-)\chi_- I$ are the zero operators on the space $L^p(\mathbb{R}, w)$.*
(ii) *If $a_\pm \in D_{p,w,\pm}$, then the operators $\chi_- W^0(a_+)\chi_+ I$ and $\chi_+ W^0(a_-)\chi_- I$ are compact on the space $L^p(\mathbb{R}, w)$.*

Proof. If $a_+ \in H_{p,w,+}^{\infty}$, then from [6, Section 2.5] it follows that $\chi_- W^0(a_+)\chi_+ I = 0$ on the space $L^2(\mathbb{R})$. Extending $\chi_- W^0(a_+)\chi_+ I$ by continuity from $L^2(\mathbb{R}) \cap L^p(\mathbb{R}, w)$ to $L^p(\mathbb{R}, w)$, we infer that $\chi_- W^0(a_+)\chi_+ I = 0$ on the space $L^p(\mathbb{R}, w)$ too. Analogously, if $a_- \in H_{p,w,-}^{\infty}$, then $\chi_+ W^0(a_-)\chi_- I = 0$ on the space $L^p(\mathbb{R}, w)$.

Let $a \in C_{p,w}(\dot{\mathbb{R}})$. Then there is a sequence $\{a_n\} \subset C(\dot{\mathbb{R}})$ with $V_1(a_n) < \infty$ that converges to a in $M_{p,w}$. By inequality (1.2), the operators $\chi_- W^0(a_n)\chi_+ I$ and $\chi_+ W^0(a_n)\chi_- I$ are bounded on all the spaces $L^r(\mathbb{R}, \omega)$ with $r \in (1, \infty)$ and $\omega \in A_p(\mathbb{R})$. Since these operators are compact on the space $L^2(\mathbb{R})$ (see, e.g., [6, Theorem 2.18]), by analogy with [23, Theorem 3.10] and [22, Theorem 3.2], we conclude that the operators $\chi_- W^0(a_n)\chi_+ I$ and $\chi_+ W^0(a_n)\chi_- I$ are compact on every space $L^r(\mathbb{R}, \omega)$ and, in particular, on $L^p(\mathbb{R}, w)$. Hence, their limits $\chi_- W^0(a)\chi_+ I$ and $\chi_+ W^0(a)\chi_- I$ in $\mathcal{B}(L^p(\mathbb{R}, w))$ are also compact on the space $L^p(\mathbb{R}, w)$.

Consequently, the operators $\chi_- W^0(a_+)\chi_+ I$ and $\chi_+ W^0(a_-)\chi_- I$ are compact on the space $L^p(\mathbb{R}, w)$ for all $a_\pm \in D_{p,w,\pm} = \mathrm{alg}\{C_{p,w}(\dot{\mathbb{R}}), H^\infty_{p,w,\pm}\}$. □

Under the conditions of Theorem 2.7, the matrix $\mathbf{d}(a) := M(a_-)M(a_+)$ is defined uniquely. We now extend $\mathbf{d}(a)$ by analogy with [6, Theorem 18.12].

Theorem 3.2. *Let $1 < p < \infty$, $N \in \mathbb{N}$ and*

$$\mathcal{G} := \Big\{ a \in [AP_p]_{N\times N} : W(a) \ \ \textit{is invertible on} \ \ L^p_N(\mathbb{R}_+) \Big\}.$$

If $a \in \mathcal{G}$ and $\{a_n\} \subset AP^0_{N\times N}$ is any sequence of almost periodic matrix polynomials which converges to a in $[AP_p]_{N\times N}$, then the matrix polynomials a_n admit canonical right APW factorizations for all sufficiently large n and the limit $\mathbf{d}(a) := \lim_{n\to\infty} \mathbf{d}(a_n)$ exists and is a matrix in $G\mathbf{C}^{N\times N}$. The map

$$\mathcal{G} \to G\mathbf{C}^{N\times N}, \quad a \mapsto \mathbf{d}(a) \tag{3.1}$$

is continuous.

Proof. Fix a sequence a_n of almost periodic matrix polynomials which converges to a in the Banach algebra $[AP_p]_{N\times N}$. Since $W(a)$ is invertible on the space $L^p_N(\mathbb{R}_+)$, the operators $W(a_n)$ are also invertible on this space for all sufficiently large n and $\big\|[W(a_n)]^{-1} - [W(a)]^{-1}\big\|_{\mathcal{B}(L^p_N(\mathbb{R}_+))} \to 0$ as $n \to \infty$. Without loss of generality assume that $W(a_n)$ are invertible for all n. By [6, Corollary 19.11], the matrix functions a_n admit canonical right APW factorizations $a_n = a_n^- a_n^+$ with $a_n^\pm \in GAPW^\pm_{N\times N}$. Consequently, applying Lemma 3.1(i), we get

$$\begin{aligned}\chi_+[W(a_n)]^{-1}\chi_+ &= \chi_+ W^0([a_n^+]^{-1})\chi_+ W^0([a_n^-]^{-1})\chi_+ I\\ &= W^0([a_n^+]^{-1})\chi_+ W^0([a_n^-]^{-1}).\end{aligned} \tag{3.2}$$

Let $\chi(x) := \tanh x$. Then, by [6, Lemma 19.13], we obtain

$$\chi[a_n^\pm]^{-1} = \chi M^{-1}(a_n^\pm) + f_n^\pm, \tag{3.3}$$

where $f_n^\pm \in [C_p(\dot{\mathbb{R}}) + H^\infty_{p,\pm}]_{N\times N}$. As $C_p(\dot{\mathbb{R}}) + H^\infty_{p,\pm} = D_{p,1,\pm}$, we deduce from Lemma 3.1(ii) that the operators $\chi_- W^0(f_n^+)\chi_+ I$ and $\chi_+ W^0(f_n^-)\chi_- I$ are compact on the space $L^p_N(\mathbb{R})$. Hence from (3.2) and (3.3) it follows that

$$\begin{aligned}&\chi_- W^0(\chi)\chi_+[W(a_n)]^{-1}\chi_+ W^0(\chi)\chi_- I\\ &= \chi_- W^0(\chi[a_n^+]^{-1})\chi_+ W^0(\chi[a_n^-]^{-1})\chi_- I\\ &\simeq \chi_- W^0(\chi M^{-1}(a_n^+))\chi_+ W^0(\chi M^{-1}(a_n^-))\chi_- I\\ &= M^{-1}(a_n^+)M^{-1}(a_n^-)\chi_- W^0(\chi)\chi_+ W^0(\chi)\chi_- I\\ &= \mathbf{d}^{-1}(a_n)\chi_- W^0(\chi)\chi_+ W^0(\chi)\chi_- I,\end{aligned}$$

where $A \simeq B$ means that $A - B$ is a compact operator. Therefore,

$$\begin{aligned}&[\mathbf{d}^{-1}(a_n) - \mathbf{d}^{-1}(a_m)]\chi_- W^0(\chi)\chi_+ W^0(\chi)\chi_- I\\ &\simeq \chi_- W^0(\chi)\chi_+\big([W(a_n)]^{-1} - [W(a_m)]^{-1}\big)\chi_+ W^0(\chi)\chi_- I,\end{aligned}$$

and passing to the (j,k)-entry on each side of this formula we obtain

$$\begin{aligned}&\big|\big(\mathbf{d}^{-1}(a_n)-\mathbf{d}^{-1}(a_m)\big)_{jk}\big|\,\big\|\chi_-W^0(\chi)\chi_+W^0(\chi)\chi_-I\big\|_{\mathrm{ess}}\\&\le\big\|W^0(\chi)\big\|^2\big\|[W(a_n)]^{-1}-[W(a_m)]^{-1}\big\|\end{aligned}\tag{3.4}$$

where $\|A\|_{\mathrm{ess}}:=\inf\{\|A+K\|:K\in\mathcal{K}(L^p(\mathbb{R}))\}$. Following [6, Section 17.3], let us calculate $\operatorname{Sym}H$ for $H=\chi_-W^0(\chi)\chi_+W^0(\chi)\chi_-I$ at the points $(0,\infty,\mu)$ for $\mu\in\mathcal{L}_p:=\big\{2^{-1}+2^{-1}\coth[\pi(x+i/p)]:x\in\overline{\mathbb{R}}\big\}$. Setting $\rho=\sqrt{\mu(1-\mu)}$, we get

$$\begin{aligned}(\operatorname{Sym}H)(0,\infty,\mu)&=\begin{bmatrix}0&0\\0&1\end{bmatrix}\begin{bmatrix}1-2\mu&2\rho\\2\rho&2\mu-1\end{bmatrix}\begin{bmatrix}1&0\\0&0\end{bmatrix}\begin{bmatrix}1-2\mu&2\rho\\2\rho&2\mu-1\end{bmatrix}\begin{bmatrix}0&0\\0&1\end{bmatrix}\\&=\operatorname{diag}\{0,4\rho^2\}.\end{aligned}$$

As $\operatorname{Sym}H$ is not identically zero, we deduce from [6, Theorem 17.12(b)] that the operator $\chi_-W^0(\chi)\chi_+W^0(\chi)\chi_-I$ is not compact. Hence from (3.4) it follows that

$$\big|\big(\mathbf{d}^{-1}(a_n)-\mathbf{d}^{-1}(a_m)\big)_{jk}\big|\le c_0^{-1}\big\|W^0(\chi)\big\|^2\big\|[W(a_n)]^{-1}-[W(a_m)]^{-1}\big\|\tag{3.5}$$

where $c_0:=\big\|\chi_-W^0(\chi)\chi_+W^0(\chi)\chi_-I\big\|_{\mathrm{ess}}>0$. Thus, the limit $b:=\lim\limits_{n\to\infty}\mathbf{d}^{-1}(a_n)$ exists. Hence the limit $\mathbf{d}(a)=\lim\limits_{n\to\infty}\mathbf{d}(a_n)$ will exist if $b\in G\mathbf{C}^{N\times N}$. To prove that $b\in G\mathbf{C}^{N\times N}$, it suffices to verify that there is an $M\in(0,\infty)$ such that

$$\limsup_{n\to\infty}\big|(\mathbf{d}(a_n))_{jk}\big|\le M\quad\text{for all }\ j,k=1,2,\dots,N.\tag{3.6}$$

Indeed, (3.6) implies that $|\det\mathbf{d}(a_n)|\le N!M^N$, and therefore

$$|\det b|=\lim_{n\to\infty}|\det\mathbf{d}^{-1}(a_n)|\ge 1/(N!M^N)>0.$$

We now prove (3.6). Without loss of generality assume that $a_n^{\pm}$ are almost periodic matrix polynomials. Fix a number $n\in\mathbb{N}$. Then there is an $\alpha=\alpha_n>0$ such that the Bohr-Fourier spectra $\Omega(a_n^{\pm})$ of $a_n^{\pm}$ possess the property:

$$\Omega(a_n^-)\cap(-\alpha,0)=\emptyset,\quad \Omega(a_n^+)\cap(0,\alpha)=\emptyset.\tag{3.7}$$

Let $\chi_0:=\chi_{[0,\alpha]}$ and $\chi_1:=\chi_{[\alpha,+\infty)}$. Since the operator $W(a_n)$ is invertible on the space $L_N^p(\mathbb{R}_+)$, it is evident that the operator $\chi_1W(a_n)\chi_1I$ is invertible on the space $L_N^p([\alpha,+\infty))$. Consider the operator

$$B_n=\chi_0W(a_n)\chi_0I-\chi_0W(a_n)\chi_1\big(\chi_1W(a_n)\chi_1I\big)^{-1}\chi_1W(a_n)\chi_0I.\tag{3.8}$$

Because $\lim\limits_{n\to\infty}\|W(a_n)-W(a)\|=0$ and $\lim\limits_{n\to\infty}\big\|(W(a_n))^{-1}-(W(a))^{-1}\big\|=0$, there exists an $M\in(0,\infty)$ such that for all $n\in\mathbb{N}$,

$$\|B_n\|\le\|W(a_n)\|+\|W(a_n)\|^2\|(W(a_n))^{-1}\|\le M<\infty.\tag{3.9}$$

Since $\chi_1W(a_n)\chi_1I=\chi_1W^0(a_n^-)\chi_1W^0(a_n^+)\chi_1I$, we conclude similarly to (3.2) that

$$(\chi_1W(a_n)\chi_1I)^{-1}=W^0((a_n^+)^{-1})\chi_1W^0((a_n^-)^{-1}).\tag{3.10}$$

Taking into account the equality $\chi_0 W(a_n)\chi_0 I = \chi_0 W^0(a_n^-)\chi_+ W^0(a_n^+)\chi_0 I$, we infer from (3.8) and (3.10) that

$$\begin{aligned} B_n &= \chi_0 W^0(a_n^-)\chi_+ W(a_n^+)\chi_0 I - \chi_0 W^0(a_n^-)\chi_1 W(a_n^+)\chi_0 I \\ &= \chi_0 W^0(a_n^-)\chi_0 W(a_n^+)\chi_0 I. \end{aligned} \tag{3.11}$$

On the other hand, from (3.7) it follows that $a_n^\pm = M(a_n^\pm) + g_n^\pm$, where the matrix polynomials $g_n^\pm \in [APW^\pm]_{N\times N}$ are of the form $g_n^\pm = \sum_{\pm\lambda\geq\alpha} c_\lambda e_\lambda$. Hence

$$\chi_0 W^0(g_n^\pm)\chi_0 I = \chi_{[0,\alpha]} \sum\nolimits_{\pm\lambda\geq\alpha} c_\lambda \chi_{[\lambda,\lambda+\alpha]} U_\lambda = 0, \tag{3.12}$$

where $W^0(e_\lambda) = U_\lambda$ and $(U_\lambda f)(x) = f(x-\lambda)$ for $x \in \mathbb{R}$. Applying (3.12), we infer from (3.11) that

$$\begin{aligned} B_n &= \chi_0 W^0(M(a_n^-) + g_n^-)\chi_0 W(M(a_n^+) + g_n^+)\chi_0 I \\ &= \chi_0 M(a_n^-)\chi_0 M(a_n^+)\chi_0 I = \mathbf{d}(a_n)\chi_0 I. \end{aligned} \tag{3.13}$$

Finally, from (3.13) and (3.9) it follows that

$$\limsup_{n\to\infty} \left|(\mathbf{d}(a_n))_{jk}\right| \leq \|B_n\| \leq M < \infty,$$

and hence $\mathbf{d}(a) = \lim_{n\to\infty} \mathbf{d}(a_n)$.

We now prove the continuity of the map (3.1). Fix $a \in \mathcal{G}$ and $\epsilon > 0$. Then there is a $\delta > 0$ such that $b \in \mathcal{G}$ and $\left\|(W(a))^{-1} - (W(b))^{-1}\right\| < \varepsilon$ whenever $b \in [AP_p]_{N\times N}$ and $\|b-a\|_{[M_p]_{N\times N}} < \delta$. Fix this b and choose $a_n, b_n \in AP^0_{N\times N}$ such that $a_n \to a$ and $b_n \to b$ in $[M_p]_{N\times N}$. As in the proof of (3.5), we obtain

$$\left\|\mathbf{d}^{-1}(a_n) - \mathbf{d}^{-1}(b_n)\right\| \leq C\left\|(W(a_n))^{-1} - (W(b_n))^{-1}\right\|$$

with some constant $C < \infty$. Passing to the limit as $n \to \infty$, we get

$$\left\|\mathbf{d}^{-1}(a) - \mathbf{d}^{-1}(b)\right\| \leq C\left\|(W(a))^{-1} - (W(b))^{-1}\right\| < C\varepsilon.$$

Hence the map $a \mapsto \mathbf{d}^{-1}(a)$ is continuous, which implies that the map $a \mapsto \mathbf{d}(a)$ is also continuous. □

4. Necessary Fredholm conditions

Below we need the following weighted analogue of [7, p. 487].

Lemma 4.1. [21, Lemma 5.2] *If $p \in (1,\infty)$ and $w \in A_p(\mathbb{R})$, then the Banach subalgebra $\operatorname{alg} W(C_{p,w}(\dot{\mathbb{R}}))$ of $\mathcal{B}(L^p(\mathbb{R}_+, w))$ generated by all Wiener-Hopf operators $W(c)$ with symbols $c \in C_{p,w}(\dot{\mathbb{R}})$ contains the closed two-sided ideal $\mathcal{K}(L^p(\mathbb{R}_+, w))$ of all compact operators in $\mathcal{B}(L^p(\mathbb{R}_+, w))$.*

Let $1 < p < \infty$, $w \in A_p^0(\mathbb{R})$, $N \in \mathbb{N}$, and let

$$\begin{aligned} \mathfrak{A} &:= \operatorname{alg}(\operatorname{sgn} x, W^0([SO_{p,w}, SAP_{p,w}]_{N\times N})), \\ \mathfrak{B} &:= \operatorname{alg}(\operatorname{sgn} x, W^0([AP_{p,w}]_{N\times N})) \end{aligned} \tag{4.1}$$

be the Banach subalgebras of $\mathcal{B}(L^p_N(\mathbb{R},w))$ generated by the operators $(\operatorname{sgn} x)I$ and $W^0(a)$ where, respectively, $a \in [SO_{p,w}, SAP_{p,w}]_{N\times N}$ and $a \in [AP_{p,w}]_{N\times N}$.

Lemma 4.1 and [20, Lemma 3.8] (also see [6, Lemma 10.1]) give the following.

Lemma 4.2. *If $1 < p < \infty$, $w \in A^0_p(\mathbb{R})$ and $N \in \mathbb{N}$, then the ideal $\mathcal{K}(L^p_N(\mathbb{R},w))$ is contained in the Banach algebra $\mathfrak{A}$ given by (4.1), and for every $K \in \mathcal{K}(L^p_N(\mathbb{R},w))$ and every real-valued sequence h tending to ∞ (respectively, to $+\infty$, $-\infty$), the limit operator $K_h := \operatorname*{s-lim}_{h_n\to\pm\infty} (e_{-h_n} A\, e_{h_n} I)$ exists and is the zero operator.*

For every $\xi \in \mathcal{M}_\infty(SO)$, we introduce the mappings $\mu_{\xi,\pm} : A \mapsto A_{\xi,\pm}$ defined on the generators of the Banach algebra $\mathfrak{A}$ by

$$\mu_{\xi,\pm}((\operatorname{sgn} x)I) := (\operatorname{sgn} x)I, \tag{4.2}$$

$$\mu_{\xi,\pm}(W^0(a)) := W^0(a_{\xi,\pm}) \quad \text{for } a \in [SO_{p,w}, SAP_{p,w}]_{N\times N}, \tag{4.3}$$

where $a_{\xi,\pm} := \nu_{\xi,\pm} a \in [AP_{p,w}]_{N\times N}$ and the map $\nu_{\xi,\pm}$ is defined by (2.3) or (2.5).

Theorem 4.3. *For every $\xi \in \mathcal{M}_\infty(SO)$, the mapping $\mu_{\xi,\pm} : A \mapsto A_{\xi,\pm}$, defined on the generators of the Banach algebra $\mathfrak{A}$ by (4.2) and (4.3), extends to a Banach algebra homomorphism of the Banach algebra $\mathfrak{A}$ onto the Banach algebra $\mathfrak{B}$. Moreover, for every $A \in \mathfrak{A}$ and all $\xi \in \mathcal{M}_\infty(SO)$,*

$$\|A_{\xi,\pm}\| \le \|A\|_{\mathrm{ess}} := \inf\big\{\|A+K\| : K \in \mathcal{K}(L^p_N(\mathbb{R},w))\big\}. \tag{4.4}$$

Proof. First we consider the operators of the form

$$A = \sum\nolimits_i \prod\nolimits_j A_{i,j} \tag{4.5}$$

where $A_{i,j} \in \big\{(\operatorname{sgn} x)I,\ W^0(a) : a \in [SO_{p,w}, SAP_{p,w}]_{N\times N}\big\}$, the indices i and j run through finite sets, and the products in (4.5) are ordered. For such A, we put

$$\mu_{\xi,\pm}(A) = \sum\nolimits_i \prod\nolimits_j \mu_{\xi,\pm}(A_{i,j}).$$

On the other hand, for every $\xi \in \mathcal{M}_\infty(SO)$ there exists a sequence $h_n \to \pm\infty$ as $n \to \pm\infty$, such that (see (2.6))

$$\mu_{\xi,\pm}(A) = \operatorname*{s-lim}_{n\to\pm\infty} (e_{-h_n} A e_{h_n} I). \tag{4.6}$$

Indeed, all limit operators for the operator $(\operatorname{sgn} x)I \in \mathcal{B}(L^p_N(\mathbb{R},w))$ coincide with $(\operatorname{sgn} x)I$ and, according to Subsection 2.2, the limit operators for the convolution operator $W^0(a)$ with matrix symbol $a \in [SO_{p,w}, SAP_{p,w}]_{N\times N}$ coincide with $W^0(a_{\xi,\pm})$ for $\xi \in \mathcal{M}_\infty(SO)$, where $a_{\xi,\pm} = \nu_{\xi,\pm} a \in [AP_{p,w}]_{N\times N}$. Applying then Proposition 2.1 and (2.6), we arrive at (4.6).

Lemma 4.2 implies now that

$$\mu_{\xi,\pm}(A) = \operatorname*{s-lim}_{n\to\pm\infty} \big(e_{-h_n}(A+K)e_{h_n} I\big) \quad \text{for all } K \in \mathcal{K}(L^p_N(\mathbb{R},w)).$$

Hence, for all $A \in \mathfrak{A}$ of the form (4.5), from [5, Proposition 6.1] it follows that

$$\|\mu_{\xi,\pm}(A)\| \le \inf_{K\in\mathcal{K}(L^p_N(\mathbb{R},w))} \|A+K\| = \|A\|_{\mathrm{ess}}. \tag{4.7}$$

As the set of all operators of the form (4.5) is dense in the Banach algebra $\mathfrak{A}$, we infer from (4.7) that the homomorphisms $\mu_{\xi,\pm}$ extend by continuity to the whole algebra $\mathfrak{A}$. Setting $A_{\xi,\pm} := \mu_{\xi,\pm}(A)$ for every $A \in \mathfrak{A}$, we obtain (4.4). $\square$

Applying limit operators of the form $A_h = \underset{h_n\to\pm\infty}{\text{s-lim}}\,(e_{-h_n} A e_{h_n} I)$ and slightly modifying the proof of [6, Corollary 18.11] (cf. also the proof of Lemma 6.2 below), we obtain the following.

Theorem 4.4. *If an operator $A \in \mathfrak{A}$ is Fredholm on the space $L^p_N(\mathbb{R}, w)$, then for every $\xi \in \mathcal{M}_\infty(SO)$ the operators $A_{\xi,\pm} = \mu_{\xi,\pm}(A) \in \mathfrak{B}$, where the homomorphisms $\mu_{\xi,\pm}$ are given by Lemma 4.3, are invertible on the space on $L^p_N(\mathbb{R}, w)$, and*

$$\sup_{\xi\in\mathcal{M}_\infty(SO)} \max \left\|(A_{\xi,\pm})^{-1}\right\| < \infty.$$

Lemma 4.3 and Theorem 4.4 imply the following corollary.

Corollary 4.5. *If $p \in (1,\infty)$, $w \in A^0_p(\mathbb{R})$, $N \in \mathbb{N}$, and the Wiener-Hopf operator $W(a)$ with a symbol $a \in [SO_{p,w}, SAP_{p,w}]_{N\times N}$ is Fredholm on the space $L^p_N(\mathbb{R}_+, w)$, then for every $\xi \in \mathcal{M}_\infty(SO)$ the operators $W(a_{\xi,\pm})$ with symbols $a_{\xi,\pm} = \nu_{\xi,\pm} a \in [AP_{p,w}]_{N\times N}$ are invertible on the space $L^p_N(\mathbb{R}_+, w)$ and the norms of their inverses are uniformly bounded.*

5. Regularizers of Wiener-Hopf operators with symbols in $SAP_{p,w}$

Lemma 5.1. *If $1 < p < \infty$, $w \in A^0_p(\mathbb{R})$ and the Wiener-Hopf operator $W(a)$ with a symbol $a \in SAP_{p,w}$ is Fredholm on the space $L^p(\mathbb{R}_+, w)$, then every its regularizer $(W(a))^{(-1)}$ belongs to the Banach algebra* $\operatorname{alg} W(SAP_{p,w})$.

Proof. By [20, Theorem 3.1], every function $a \in SAP_{p,w}$ can be represented in the form

$$a = a_- u_- + a_+ u_+ + a_0, \tag{5.1}$$

where $a_\pm \in AP_{p,w}$, the functions $u_\pm : x \mapsto (1 \pm \tanh x)/2$ are in $C_{p,w}(\overline{\mathbb{R}})$, and $a_0 \in C_{p,w}(\dot{\mathbb{R}})$, $a_0(\infty) = 0$. Hence, in view of the stability of Fredholmness under small perturbations and by the density of the almost periodic polynomials in $AP_{p,w}$, it is sufficient to prove the lemma for the Fredholm Wiener-Hopf operators $W(a)$ with symbols (5.1) such that $a_\pm$ are almost periodic polynomials.

Since the operator $W(a)$ is Fredholm on the space $L^p(\mathbb{R}_+, w)$, we infer from [20, Theorem 3.7 and Corollary 3.11] that $a \in GSAP_{p,w}$ and the operators $W(a_\pm)$ are invertible on the space $L^p(\mathbb{R}_+, w)$. Because $a_\pm \in APW_{p,w}$, from Theorem 2.5 it follows that the functions $a_\pm$ admit canonical right $APW_{p,w}$ factorizations (2.7), which can be written in the form $a_\pm = a_\pm^- \mathbf{d}(a_\pm) a_\pm^+$, where

$$a_\pm^- \in GAPW^-_{p,w}, \quad a_\pm^+ \in GAPW^+_{p,w}, \quad M(a_\pm^-) = M(a_\pm^+) = 1 \tag{5.2}$$

and, according to Theorem 2.6, $\mathbf{d}(a_\pm) \neq 0$. Then, by [20, Theorem 3.15], there exist functions $b \in GC_{p,w}(\overline{\mathbb{R}})$ and $g^\pm \in GSAP_{p,w} \cap GD_{p,w,\pm}$ such that $b(\pm\infty) = \mathbf{d}(a_\pm)$ and $a = g^- b g^+$. Consequently, we deduce from [20, Theorem 2.9] that

$$W(a) = W(g^-)\,W(b)\,W(g^+) + K, \tag{5.3}$$

where $W(g^\pm)$ are Fredholm operators because $g^\pm \in GD_{p,w,\pm}$, and K is a compact operator on the space $L^p(\mathbb{R}_+, w)$. Therefore, the operator $W(b)$ is Fredholm along with $W(a)$.

According to [6, Theorem 17.9], every regularizer $(W(b))^{(-1)}$ of the Fredholm operator $W(b)$ belongs to the Banach algebra $\operatorname{alg} W(C_{p,w}(\overline{\mathbb{R}})) \subset \operatorname{alg} W(SAP_{p,w})$. On the other hand, since $g^\pm \in GSAP_{p,w} \cap GD_{p,w,\pm}$, all the regularizers $(W(g^\pm))^{(-1)}$ of the operators $W(g^\pm)$ are in the Banach algebra $\operatorname{alg} W(SAP_{p,w})$. Hence, applying (5.3), we infer that the regularizers

$$(W(a))^{(-1)} = (W(g^+))^{(-1)}(W(b))^{(-1)}(W(g^-))^{(-1)}$$

of the operator $W(a)$ belong to the Banach algebra $\operatorname{alg} W(SAP_{p,w})$. □

Approximating Wiener-Hopf operators $W(a) \in \mathcal{B}(L^p_N(\mathbb{R}_+))$ with matrix symbols $a \in [SAP_p]_{N\times N}$ by Wiener-Hopf operators $W(a_n)$, where $a_n \in [SAP_p]_{N\times N}$ and their almost periodic representatives $(a_n)_\pm \in AP^0_{N\times N}$, and applying Theorem 2.7 in the case $w = 1$, we obtain by analogy with Lemma 5.1 its unweighted matrix analogue.

Lemma 5.2. *If $1 < p < \infty$, $N \in \mathbb{N}$, and the Wiener-Hopf operator $W(a)$ with a matrix symbol $a \in [SAP_p]_{N\times N}$ is Fredholm on the space $L^p_N(\mathbb{R}_+)$, then every its regularizer $(W(a))^{(-1)}$ belongs to the Banach algebra* $\operatorname{alg} W([SAP_p]_{N\times N})$.

6. Wiener-Hopf operators with symbols in $[SO_{p,w}, SAP_{p,w}]$

Given $p \in (1,\infty)$ and $w \in A^0_p(\mathbb{R})$, we consider the following Banach subalgebras

$$\mathcal{D} := \operatorname{alg} W([SO_{p,w},\, SAP_{p,w}]), \quad \mathcal{Z} := \operatorname{alg} W(SO_{p,w})$$

of $\mathcal{B}(L^p(\mathbb{R}_+, w))$. Both these algebras contain the ideal $\mathcal{K} := \mathcal{K}(L^p(\mathbb{R}_+, w))$.

By [21, Lemma 5.3], the commutators of operators aI ($a \in PC$) and convolution operators $W^0(b)$ ($b \in SO_{p,w}$) are compact on the space $L^p(\mathbb{R}, w)$. Hence

$$W(a)W(b) \simeq W(ab) \simeq W(b)W(a) \quad \text{for all } a \in M_{p,w} \text{ and all } b \in SO_{p,w}, \tag{6.1}$$

where $A \simeq B$ means that the operator $A - B$ is compact on the space $L^p(\mathbb{R}_+, w)$.

Let $\Lambda := \Lambda(\mathcal{Z})$ denote the Banach subalgebra of $\mathcal{B} = \mathcal{B}(L^p(\mathbb{R}_+, w))$ that consists of all operators of local type (with respect to $\mathcal{Z}$), that is,

$$\Lambda := \Big\{A \in \mathcal{B}(L^p(\mathbb{R}_+, w)) : W(c)A - AW(c) \in \mathcal{K} \text{ for all } c \in SO_{p,w}\Big\}.$$

The quotient Banach algebra $\Lambda^\pi = \Lambda/\mathcal{K}$ is inverse closed in the Calkin algebra $\mathcal{B}^\pi = \mathcal{B}/\mathcal{K}$ and contains the Banach algebra $\mathcal{D}^\pi := \mathcal{D}/\mathcal{K}$, and $\mathcal{Z}^\pi$ is a central subalgebra of $\mathcal{D}^\pi$ and Λ^π (see (6.1)). For $A \in \mathcal{B}$, let $A^\pi := A + \mathcal{K}$.

For every $\xi \in \mathcal{M}(SO) = \mathcal{M}(SO_{p,w})$, let J_ξ^π and $\widetilde{J}_\xi^\pi$ denote, respectively, the closed two-sided ideal of Λ^π and $\mathcal{D}^\pi$ generated by the maximal ideal

$$I_\xi^\pi := \Big\{W^\pi(b) : b \in SO_{p,w},\ \xi(b) = 0\Big\}$$

of the commutative algebra $\mathcal{Z}^\pi$, and let $\Lambda_\xi^\pi := \Lambda^\pi / J_\xi^\pi$ and $\mathcal{D}_\xi^\pi := \mathcal{D}^\pi / \widetilde{J}_\xi^\pi$ be the corresponding quotient Banach algebras. Clearly, $\widetilde{J}_\xi^\pi \subset J_\xi^\pi$. Consider the cosets

$$W_\xi^\pi(a) := W^\pi(a) + J_\xi^\pi \in \Lambda_\xi^\pi, \quad \widetilde{W}_\xi^\pi(a) := W^\pi(a) + \widetilde{J}_\xi^\pi \in \mathcal{D}_\xi^\pi.$$

To study the Fredholmness of Wiener-Hopf operators $W(a)$ with symbols $a \in [SO_{p,w}, SAP_{p,w}]$ on the space $L^p(\mathbb{R}_+, w)$ we will apply the Allan-Douglas local principle. Thus, we need to study the invertibility of the cosets $W_\xi^\pi(a)$ in the quotient algebras Λ_ξ^π for every $\xi \in \mathcal{M}(SO)$.

Lemma 6.1. *If $a \in [SO_{p,w}, SAP_{p,w}]$ where $1 < p < \infty$ and $w \in A_p^0(\mathbb{R})$, then for every $\xi \in \mathbb{R}$ the following three assertions are equivalent:*

(i) *the coset $W_\xi^\pi(a)$ is invertible in the quotient algebra Λ_ξ^π,*

(ii) *the coset $\widetilde{W}_\xi^\pi(a)$ is invertible in the quotient algebra $\mathcal{D}_\xi^\pi$,*

(iii) $a(\xi) \neq 0$.

Proof. Given $a \in [SO_{p,w}, SAP_{p,w}]$ and $\xi \in \mathbb{R}$, let us show that

$$W^\pi(a) - W^\pi(a(\xi)) = W^\pi(a - a(\xi)) \in \widetilde{J}_\xi^\pi, \tag{6.2}$$

where the ideal $\widetilde{J}_\xi^\pi \subset J_\xi^\pi$ is represented in the form

$$\widetilde{J}_\xi^\pi = \mathrm{clos}_{\mathcal{D}^\pi}\Big\{\sum\nolimits_{i=1}^n A_i^\pi B_i^\pi : A_i^\pi \in \mathcal{D}^\pi,\ B_i^\pi \in I_\xi^\pi,\ n \in \mathbb{N}\Big\}. \tag{6.3}$$

Because the matrix function $\widetilde{a} := a - a(\xi)$ belongs to $[SO_{p,w}, SAP_{p,w}]$ along with a, we conclude that $W^\pi(\widetilde{a}) \in \mathcal{D}^\pi$. Moreover, $\widetilde{a}(\xi) = 0$. Then for any $\varepsilon > 0$ there are a function $\widetilde{a}_\varepsilon \in [SO_{p,w}, SAP_{p,w}]$ and a number $\delta > 0$ such that $\|\widetilde{a} - \widetilde{a}_\varepsilon\|_{M_{p,w}} < \varepsilon$ and $\widetilde{a}_\varepsilon(x) = 0$ for all $x \in (\xi - \delta, \xi + \delta)$ (see, e.g., the proof of [20, Theorem 3.7]). Choosing now a function $b_\xi \in C_{p,w}(\overline{\mathbb{R}})$ such that $b_\xi(\xi) = 0$ and $b_\xi(x) = 1$ for $|x - \xi| \geq \delta$, we deduce that $\widetilde{a}_\varepsilon = \widetilde{a}_\varepsilon b_\xi$ and $W^\pi(b_\xi) \in I_\xi^\pi$. Hence, by (6.1), $W^\pi(\widetilde{a}_\varepsilon) = W^\pi(\widetilde{a}_\varepsilon) W^\pi(b_\xi)$, and therefore, by (6.3), the cosets $W^\pi(\widetilde{a}_\varepsilon)$ and $W^\pi(\widetilde{a})$ belong to the ideal $\widetilde{J}_\xi^\pi$.

Since $W(a(\xi)) = \chi_+ \mathcal{F}^{-1} a(\xi) \mathcal{F} = a(\xi) I$, we have $W^\pi(a(\xi)) = [a(\xi)I]^\pi$. Hence we infer from (6.2) that for every $\xi \in \mathbb{R}$,

$$\widetilde{W}_\xi^\pi(a) = [a(\xi)I]^\pi + \widetilde{J}_\xi^\pi, \quad W_\xi^\pi(a) = [a(\xi)I]^\pi + J_\xi^\pi,$$

which immediately implies the assertion of the lemma. $\square$

Given $p \in (1, \infty)$, for every $k > 0$ we consider the isometric dilation operators

$$V_k \in \mathcal{B}(L^p(\mathbb{R})), \quad (V_k f)(x) = k^{1/p} f(kx) \quad (x \in \mathbb{R}), \tag{6.4}$$

and the limit operators $A_h = \operatorname*{s-lim}_{m \to \infty} \big(V_{h_m} A V_{h_m}^{-1}\big)$ of operators $A \in \mathcal{B}(L^p(\mathbb{R}))$ related to a sequence $h = \{h_m\} \subset \mathbb{R}_+$ tending to $+\infty$. By [3] and [5], for every function

$a \in SO$ and every sequence $h \subset \mathbb{R}_+$ tending to $+\infty$ there is a subsequence g of h such that there exists the limit operator $(aI)_g := \operatorname*{s-lim}_{m\to\infty} \big(V_{h_m} a V_{h_m}^{-1}\big) = a_g I$ and a_g is a constant, which equals $\xi(a)$ for some $\xi \in \mathcal{M}_\infty(SO)$ by Proposition 2.1.

Let V_k $(k > 0)$ denote the dilation operators (6.4) on $L^p(\mathbb{R})$ and on $L^p(\mathbb{R}_+)$.

Lemma 6.2. *If $p \in (1,\infty)$, $w \in A_p^0(\mathbb{R})$, $b \in C_{p,w}(\overline{\mathbb{R}})$, $\xi \in \mathcal{M}_\infty(SO)$, and the coset $W_\xi^\pi(b)$ is invertible in the quotient Banach algebra Λ_ξ^π, then the Wiener-Hopf operator $W\big(b(-\infty)\chi_- + b(+\infty)\chi_+\big)$, where $\chi_\pm$ are the characteristic functions of $\mathbb{R}_\pm$, is invertible on the space $L^p(\mathbb{R}_+)$, that is, for every $x \in \overline{\mathbb{R}}$,*

$$b(+\infty)\,\frac{1-\coth[\pi(x+i/p)]}{2} + b(-\infty)\,\frac{1+\coth[\pi(x+i/p)]}{2} \neq 0. \tag{6.5}$$

Proof. As the coset $W_\xi^\pi(b) = W^\pi(b) + J_\xi^\pi$ is invertible in the quotient algebra Λ_ξ^π, there exist operators $B, C_1, C_2 \in \Lambda$ such that $C_1^\pi, C_2^\pi \in J_\xi^\pi$ and

$$W(b)B = I + C_1, \quad BW(b) = I + C_2. \tag{6.6}$$

Clearly, the operators C_1, C_2 are limits in $\mathcal{B}(L^p(\mathbb{R}_+, w))$ of sequences of operators of the form $\sum_i D_i W(c_i) + K$, where i runs through a finite set, $D_i \in \Lambda$, $K \in \mathcal{K}$, $c_i \in SO_{p,w}$ and $c_i(\xi) = 0$. Passing in (6.6) to the operators acting on the space $L^p(\mathbb{R}_+)$, we obtain

$$\widehat{W}(b)\widehat{B} = I + \widehat{C}_1, \quad \widehat{B}\,\widehat{W}(b) = I + \widehat{C}_2, \tag{6.7}$$

where $\widehat{W}(b) := wW(b)w^{-1}I$, $\widehat{B} := wBw^{-1}I$, and $\widehat{C} := wCw^{-1}I$ for $C \in \{C_1, C_2\}$. By Proposition 2.1, for every $\xi \in \mathcal{M}_\infty(SO)$ and any at most countable set $\{c_i\} \subset SO_{p,w}$ there is a sequence $\{k_n\} \subset \mathbb{R}_+$ such that

$$\lim_{n\to\infty} k_n = 0, \quad c_i(\xi) := \xi(c_i) = \lim_{n\to\infty} c_i(1/k_n). \tag{6.8}$$

Since $w \in A_p^0(\mathbb{R})$, we may without loss of generality assume due to [24] that w is continuous on $\mathbb{R}$ (see [20, p. 90]), and therefore both indices of powerlikeness of w at the point 0 equal zero. Then we infer from [21, Theorem 4.3] and (6.8) that

$$\operatorname*{s-lim}_{n\to\infty} \big(V_{k_n} wW(b)w^{-1}V_{k_n}^{-1}\big) = W\big(b(-\infty)\chi_- + b(+\infty)\chi_+\big), \tag{6.9}$$

$$\operatorname*{s-lim}_{n\to\infty} \big(V_{k_n} wW(c_i)w^{-1}V_{k_n}^{-1}\big) = W\big(c_i(\xi)\big) = 0. \tag{6.10}$$

Similarly to [17], if $K \in \mathcal{K}(L^p(\mathbb{R}_+, w))$, then the operator $wKw^{-1}I \in \mathcal{K}(L^p(\mathbb{R}_+))$ can be approximated in $\mathcal{B}(L^p(\mathbb{R}_+))$ by the operators T of the form

$$(T\varphi)(x) = \sum\nolimits_{j=1}^{n} a_j(x)\left(\frac{i-x}{i+x}\,S_{\mathbb{R}_+} - S_{\mathbb{R}_+}\,\frac{i-y}{i+y}\,I\right)(b_j\varphi) \tag{6.11}$$

where a_j and b_j are continuous functions on $\mathbb{R}_+$ with compact support on $[0,\infty)$, and the operator $S_{\mathbb{R}_+}$ is given by (1.1) with $\mathbb{R}$ replaced by $\mathbb{R}_+$ (see, e.g., [19, Lemma 10.1]). Because $V_k S_{\mathbb{R}_+} V_k^{-1} = S_{\mathbb{R}_+}$, we deduce from (6.11) that for every $K \in \mathcal{K}(L^p(\mathbb{R}_+, w))$ and every sequence $\{k_n\} \subset \mathbb{R}$ such that $k_n \to 0$ as $n \to \infty$,

$$\operatorname*{s-lim}_{n\to\infty} \big(V_{k_n} wKw^{-1}V_{k_n}^{-1}\big) = 0. \tag{6.12}$$

Since the operators C_1, C_2 are approximated in $\mathcal{B}(L^p(\mathbb{R}_+, w))$ by the operators of the form $\sum_i D_i W(c_i) + K$, we infer from (6.10) and (6.12) that for every $\xi \in \mathcal{M}_\infty(SO)$ there is a sequence $\{k_n\} \subset \mathbb{R}_+$ such that $\lim_{n\to\infty} k_n = 0$ and

$$\operatorname*{s-lim}_{n\to\infty} \big(V_{k_n} \widehat{C}_1 V_{k_n}^{-1}\big) = \operatorname*{s-lim}_{n\to\infty} \big(V_{k_n} \widehat{C}_2\, V_{k_n}^{-1}\big) = 0. \tag{6.13}$$

By (6.7), for every $n \in \mathbb{N}$, we obtain

$$I = \big[V_{k_n} \widehat{B}\, V_{k_n}^{-1}\big]\big[V_{k_n} \widehat{W}(b)\, V_{k_n}^{-1}\big] - \big[V_{k_n} \widehat{C}_2\, V_{k_n}^{-1}\big]. \tag{6.14}$$

Since $\|V_{k_n} \widehat{B}\, V_{k_n}^{-1}\|_{\mathcal{B}(L^p(\mathbb{R}_+))} = \|\widehat{B}\|_{\mathcal{B}(L^p(\mathbb{R}_+))} = \|B\|_{\mathcal{B}(L^p(\mathbb{R}_+,w))}$, we deduce from (6.14) that for every $f \in L^p(\mathbb{R}_+)$ and all $n \in \mathbb{N}$,

$$\|f\|_{L^p(\mathbb{R}_+)} \le \|B\|_{\mathcal{B}(L^p(\mathbb{R}_+,w))} \big\|V_{k_n} \widehat{W}(b)\, V_{k_n}^{-1} f\big\|_{L^p(\mathbb{R}_+)} + \big\|V_{k_n} \widehat{C}_2\, V_{k_n}^{-1} f\big\|_{L^p(\mathbb{R}_+)} \tag{6.15}$$

Applying (6.9) and (6.13) and passing in (6.15) to the limit as $n \to \infty$, we get

$$\|f\|_{L^p(\mathbb{R}_+)} \le \|B\|_{\mathcal{B}(L^p(\mathbb{R}_+,w))} \big\|W\big(b(-\infty)\chi_- + b(+\infty)\chi_+\big) f\big\|_{L^p(\mathbb{R}_+)}. \tag{6.16}$$

Analogously, since $I^* = (\widehat{B})^*(\widehat{W}(b))^* - (\widehat{C}_1)^*$ due to (6.7), and hence

$$I^* = \big[(V_{k_n}^{-1})^*(\widehat{B})^*(V_{k_n})^*\big]\big[(V_{k_n}^{-1})^*(\widehat{W}(b))^*(V_{k_n})^*\big] - \big[(V_{k_n}^{-1})^*(\widehat{C}_1)^*(V_{k_n})^*\big],$$

we infer that

$$\|f\|_{L^q(\mathbb{R}_+)} \le \|B\|_{\mathcal{B}(L^p(\mathbb{R}_+,w))} \big\|\big[W\big(b(-\infty)\chi_- + b(+\infty)\chi_+\big)\big]^* f\big\|_{L^q(\mathbb{R}_+)}. \tag{6.17}$$

Finally, (6.16) and (6.17) imply that the operator $W\big(b(-\infty)\chi_- + b(+\infty)\chi_+\big)$ is invertible on the space $L^p(\mathbb{R}_+)$, and

$$\big\|\big(W\big(b(-\infty)\chi_- + b(+\infty)\chi_+\big)\big)^{-1}\big\|_{\mathcal{B}(L^p(\mathbb{R}_+))} \le \|B\|_{\mathcal{B}(L^p(\mathbb{R}_+,w))}. \tag{6.18}$$

Obviously, in (6.18) we can replace $\|B\|_{\mathcal{B}(L^p(\mathbb{R}_+,w))}$ by $\|B_\xi^\pi\|_{\Lambda_\xi^\pi}$.

By [10, Proposition 2.8], the operator $W\big(b(-\infty)\chi_- + b(+\infty)\chi_+\big) \in \mathcal{B}(L^p(\mathbb{R}_+))$ is invertible on the space $L^p(\mathbb{R}_+)$ if and only if it is Fredholm of index zero. The latter coincides with (6.5) in virtue of [6, Theorem 17.7]. □

Theorem 6.3. *Let $p \in (1,\infty)$, $w \in A_p^0(\mathbb{R})$, let $a \in SAP_{p,w}$ with almost periodic representatives $a_\pm \in AP_{p,w}$ at $\pm\infty$ and let $\xi \in \mathcal{M}_\infty(SO)$. If the operators $W(a_\pm)$ are invertible on the space $L^p(\mathbb{R}_+, w)$, then the following assertions are equivalent:*

(i) *the coset $W_\xi^\pi(a)$ is invertible in the quotient algebra Λ_ξ^π;*

(ii) *the coset $\widetilde{W_\xi^\pi}(a)$ is invertible in the quotient algebra $\mathcal{D}_\xi^\pi$;*

(iii) *for every $x \in \mathbb{R}$,*

$$\mathbf{d}(a_+)\frac{1 - \coth[\pi(x + i/p)]}{2} + \mathbf{d}(a_-)\frac{1 + \coth[\pi(x + i/p)]}{2} \ne 0. \tag{6.19}$$

Proof. (iii)⇒(ii). If the Wiener-Hopf operators $W(a_\pm)$ are invertible on the space $L^p(\mathbb{R}_+, w)$, then, by Theorem 2.6, the almost periodic functions $a_\pm$ are invertible in $AP_{p,w}$ and $\mathbf{d}(a_\pm) \ne 0$. Hence, there exists a function $b_0 \in C_{p,w}(\dot{\mathbb{R}})$ with

$b_0(\infty)=0$ such that the function $a+b_0$ belongs to $GSAP_{p,w}$. By [20, Theorem 4.7 and Remark 4.9], the Wiener-Hopf operator $W(a+b_0)$ is Fredholm on the space $L^p(\mathbb{R}_+,w)$. Moreover, in virtue of Lemma 5.1, every regularizer $(W(a+b_0))^{(-1)}$ of the operator $W(a+b_0)$ belongs to the Banach algebra $W(SAP_{p,w})$, and therefore the coset $W^\pi(a+b_0)=W(a+b_0)+\mathcal{K}$ is invertible in the quotient algebra $\mathcal{D}^\pi$. Consequently, the coset $\widetilde{W}_\xi^\pi(a+b_0)=W^\pi(a+b_0)+\widetilde{J}_\xi^\pi$ is invertible in the quotient algebra $\mathcal{D}_\xi^\pi$. It remains to observe that $\widetilde{W}_\xi^\pi(a+b_0)=\widetilde{W}_\xi^\pi(a)$.

(ii)⇒(i) is evident because $\widetilde{J}_\xi^\pi\subset J_\xi^\pi$.

(i)⇒(iii). First assume that $a_\pm\in APW_{p,w}$. Since the operators $W(a_\pm)$ are invertible on the space $L^p(\mathbb{R}_+,w)$, we infer from Theorem 2.5 that the functions $a_\pm\in APW_{p,w}$ admit canonical right $APW_{p,w}$ factorizations, which can be written in the form $a_\pm=a_\pm^-\mathbf{d}(a_\pm)a_\pm^+$, where $a_\pm^-$ and $a_\pm^+$ satisfy (5.2).

Constructing now the functions $b\in GC_{p,w}(\overline{\mathbb{R}})$ and $g^\pm\in GSAP_{p,w}\cap GD_{p,w,\pm}$ such that $b(\pm\infty)=\mathbf{d}(a_\pm)$ and $a=g^-bg^+$ as in the proof of Lemma 5.1, and applying (5.3), we infer that

$$W_\xi^\pi(a)=W_\xi^\pi(g^-)\,W_\xi^\pi(b)\,W_\xi^\pi(g^+),\tag{6.20}$$

where the cosets $W_\xi^\pi(g^\pm)$ are invertible and $(W_\xi^\pi(g^\pm))^{-1}=W_\xi^\pi((g^\pm)^{-1})$. Hence, by (6.20), the coset $W_\xi^\pi(b)=W^\pi(b)+J_\xi^\pi$ is invertible in the quotient algebra Λ_ξ^π along with the coset $W_\xi^\pi(a)=W^\pi(a)+J_\xi^\pi$. By Lemma 6.2, the invertibility of the coset $W_\xi^\pi(b)$ in the algebra Λ_ξ^π implies (6.5), which is equivalent to (6.19) because $b(\pm\infty)=\mathbf{d}(a_\pm)$.

Assume now that $a_\pm\in AP_{p,w}$. Since the invertibility of the operators $W(a_\pm)$ on the space $L^p(\mathbb{R}_+,w)$ and the invertibility of the coset $W_\xi^\pi(a)=W^\pi(a)+J_\xi^\pi$ in the quotient algebra Λ_ξ^π are stable with respect to small perturbations of the symbol $a\in SAP_{p,w}$ in the norm $\|\cdot\|_{M_{p,w}}$, we can find a function $\widehat{a}\in SAP_{p,w}$ such that its almost periodic representatives $\widehat{a}_\pm$ at $\pm\infty$ belong to $APW_{p,w}$, the coset $W_\xi^\pi(\widehat{a})$ is invertible in the quotient algebra Λ_ξ^π, the operators $W(\widehat{a}_\pm)$ are invertible on the space $L^p(\mathbb{R}_+,w)$, and $\mathbf{d}(\widehat{a}_\pm)=\mathbf{d}(a_\pm)$. Then we infer from the part already proved in the case $a_\pm\in APW_{p,w}$ that (6.19) holds for any $a\in SAP_{p,w}$. □

Applying Lemma 6.1 and Theorem 6.3, we get a Fredholm criterion for $W(a)$.

Theorem 6.4. *Let $1<p<\infty$, $w\in A_p^0(\mathbb{R})$ and $a\in[SO_{p,w},SAP_{p,w}]$. The Wiener-Hopf operator $W(a)$ is Fredholm on the space $L^p(\mathbb{R}_+,w)$ if and only if the following three conditions hold:*

- (i) *$a(x)\neq 0$ for all $x\in\mathbb{R}$;*
- (ii) *for every $\xi\in\mathcal{M}_\infty(SO)$, the operators $W(a_{\xi,\pm})$ with almost periodic symbols $a_{\xi,\pm}=\nu_{\xi,\pm}a\in AP_{p,w}$ are invertible on the space $L^p(\mathbb{R}_+,w)$ (equivalently, the functions $a_{\xi,\pm}$ are invertible in $AP_{p,w}$ and $\kappa(a_{\xi,\pm})=0$);*
- (iii) *for every $\xi\in\mathcal{M}_\infty(SO)$, the number $\eta_\xi:=\mathbf{d}^{-1}(a_{\xi,+})\mathbf{d}(a_{\xi,-})$ satisfies the condition*

$$\frac{1}{p}+\frac{1}{2\pi}\arg\eta_\xi\notin\mathbb{Z}.\tag{6.21}$$

If $W(a)$ is Fredholm, then all its regularizers belong to the Banach algebra $\mathcal{D} = \operatorname{alg} W([SO_{p,w}, SAP_{p,w}])$.

Proof. Suppose conditions (i)–(iii) are fulfilled. According to the Allan-Douglas local principle, the operator $W(a)$ with symbol $a \in [SO_{p,w}, SAP_{p,w}]$ is Fredholm on the space $L^p(\mathbb{R}_+, w)$ if for every $\xi \in \mathcal{M}(SO) = \mathbb{R} \cup \mathcal{M}_\infty(SO)$ the coset $W_\xi^\pi(a)$ is invertible in the quotient algebra $\mathcal{D}_\xi^\pi$.

First, let $\xi \in \mathbb{R}$. Then $a(\xi) \neq 0$ by condition (i), which implies due to Lemma 6.1 that the coset $\widetilde{W}_\xi^\pi(a)$ is invertible in the quotient algebra $\mathcal{D}_\xi^\pi$.

Suppose now that $\xi \in \mathcal{M}_\infty(SO)$. By (2.2), for every $\xi \in \mathcal{M}_\infty(SO)$ there is a non-unique function $a_\xi \in SAP_{p,w}$ with uniquely determined almost periodic representatives $a_{\xi,\pm} \in AP_{p,w}$ of a_ξ at $\pm\infty$ such that $\beta_\xi(a - a_\xi) = 0$ and therefore, by (2.3), $\nu_{\xi,\pm}(a - a_\xi) = 0$. Hence from (2.5) it follows that the coset $W^\pi(a - a_\xi)$ belongs to the ideals J_ξ^π and $\widetilde{J}_\xi^\pi$. Consequently,

$$W_\xi^\pi(a) = W_\xi^\pi(a_\xi), \quad \widetilde{W}_\xi^\pi(a) = \widetilde{W}_\xi^\pi(a_\xi). \tag{6.22}$$

Since the Wiener-Hopf operators $W(a_{\xi,\pm})$ are invertible on the space $L^p(\mathbb{R}_+, w)$ according to condition (ii) of the theorem, it follows that $a_{\xi,\pm} \in GAP_{p,w}$, $\kappa(a_{\xi,\pm}) = 0$ and $\mathbf{d}(a_{\xi,\pm}) \neq 0$. It is easily seen that, under the condition $\mathbf{d}(a_{\xi,\pm}) \neq 0$, the relation (6.21) is equivalent to (6.19) with $a_\pm$ replaced by $a_{\xi,\pm}$. Hence condition (iii) of the theorem implies by Theorem 6.3 that the coset $\widetilde{W}_\xi^\pi(a_\xi) = \widetilde{W}_\xi^\pi(a)$ is invertible it the quotient algebra $\mathcal{D}_\xi^\pi$ for every $\xi \in \mathcal{M}_\infty(SO)$ too.

Finally, applying the Allan-Douglas local principle to the Banach algebra $\mathcal{D}^\pi$, we infer that the Wiener-Hopf operator $W(a)$ with symbol $a \in [SO_{p,w}, SAP_{p,w}]$ is Fredholm on the space $L^p(\mathbb{R}_+, w)$, and all its regularizers belong to the Banach algebra $\mathcal{D}$, that is, the algebra $\mathcal{D}^\pi$ is inverse closed.

Conversely, if the operator $W(a)$ is Fredholm on the space $L^p(\mathbb{R}_+, w)$, then Corollary 4.5 implies that for every $\xi \in \mathcal{M}_\infty(SO)$ the operators $W(a_{\xi,\pm})$ are invertible on the space $L^p(\mathbb{R}_+, w)$. This gives part (ii) of the theorem and implies that $a_{\xi,\pm} \in GAP_{p,w}$. By the Allan-Douglas local principle applied to the Banach algebra Λ^π, the Fredholmness of the operator $W(a)$ on the space $L^p(\mathbb{R}_+, w)$ is equivalent to the invertibility of the cosets $W_\xi^\pi(a)$ in the quotient algebras Λ_ξ^π for all $\xi \in \mathcal{M}(SO)$. By Lemma 6.1 the invertibility of the cosets $W_\xi^\pi(a) = [a(\xi)I]^\pi + J_\xi^\pi$ for all $\xi \in \mathbb{R}$ implies part (i) of the theorem. On the other hand, if $\xi \in \mathcal{M}_\infty(SO)$, then there is a function $a_\xi \in SAP_{p,w}$ such that $W_\xi^\pi(a) = W_\xi^\pi(a_\xi)$ (see (6.22)). Since the operators $W(a_{\xi,\pm})$ are invertible on the space $L^p(\mathbb{R}_+, w)$, from Theorem 6.3 it follows that the invertibility of the coset $W_\xi^\pi(a)$ for $\xi \in \mathcal{M}_\infty(SO)$ is equivalent to the condition

$$\mathbf{d}(a_{\xi,+})\frac{1 - \coth[\pi(x + i/p)]}{2} + \mathbf{d}(a_{\xi,-})\frac{1 + \coth[\pi(x + i/p)]}{2} \neq 0 \quad \text{for all } x \in \mathbb{R},$$

which in its turn is equivalent to condition (iii) of the theorem. □

7. Matrix case

Given $1 < p < \infty$, $w \in A_p^0(\mathbb{R})$ and $N \in \mathbb{N}$, in this section we study the Fredholmness of Wiener-Hopf operators $W(a)$ with matrix symbols $a \in [SO_{p,w}, SAP_{p,w}]_{N\times N}$ on the space $L_N^p(\mathbb{R}_+, w)$ under the condition that for every $\xi \in \mathcal{M}_\infty(SO)$ the matrix functions $a_{\xi,\pm} = \nu_{\xi,\pm} a$ admit right $AP_{p,w}$ factorizations. We save the notation Λ_ξ^π and $\mathcal{D}_\xi^\pi$ for corresponding quotient algebras in the case $N > 1$.

According to [20, Section 5], a matrix function $a \in [AP_{p,w}]_{N\times N}$ is said to admit a *right $AP_{p,w}$ factorization* if it can be represented in the form

$$a = a^- \operatorname{diag}\{e_{\lambda_1}, \dots, e_{\lambda_N}\}\, a^+$$

where $a^\pm \in G[AP_{p,w}^\pm]_{N\times N}$ and $\kappa(a) := (\lambda_1, \dots, \lambda_N) \subset \mathbb{R}$. A right $AP_{p,w}$ factorization with $\kappa(a) = (0, \dots, 0)$ is referred to as a *canonical right $AP_{p,w}$ factorization.* If $a \in [AP_{p,w}]_{N\times N}$ admits a canonical right $AP_{p,w}$ factorization, then the *geometric mean* $\mathbf{d}(a) = M(a^-)M(a^+) \in \mathbb{C}^{N\times N}$ is independent of the particular choice of the canonical right $AP_{p,w}$ factorization of a.

The next result is a matrix version of Theorem 6.3.

Theorem 7.1. *Let $a \in [SAP_{p,w}]_{N\times N}$ where $1 < p < \infty$, $w \in A_p^0(\mathbb{R})$ and $N \in \mathbb{N}$, and let the almost periodic representatives $a_\pm \in [AP_{p,w}]_{N\times N}$ of a admit canonical right $AP_{p,w}$ factorizations. Then for every $\xi \in \mathcal{M}_\infty(SO)$ the following three assertions are equivalent:*

(i) *the coset $W_\xi^\pi(a) = W^\pi(a) + J_\xi^\pi$ is invertible in the quotient algebra Λ_ξ^π,*
(ii) *the coset $\widetilde{W_\xi^\pi}(a) = W^\pi(a) + \widetilde{J_\xi^\pi}$ is invertible in the quotient algebra $\mathcal{D}_\xi^\pi$,*
(iii) *for every $x \in \mathbb{R}$,*

$$\det\left[\mathbf{d}(a_+)\frac{1-\coth[\pi(x+i/p)]}{2} + \mathbf{d}(a_-)\frac{1+\coth[\pi(x+i/p)]}{2}\right] \neq 0.$$

Since the matrix functions $a_\pm \in [AP_{p,w}]_{N\times N}$ admit canonical right $AP_{p,w}$ factorizations, we conclude that $\det \mathbf{d}(a_\pm) \neq 0$ and the Wiener-Hopf operators $W(a_\pm)$ are invertible on the space $L_N^p(\mathbb{R}_+, w)$. Consequently, by analogy with Lemma 5.1, the regularizers of the Fredholm Wiener-Hopf operator $W(a)$ with symbol $a = a_-u_- + a_+u_+ + a_0 \in [SAP_{p,w}]_{N\times N}$ belong to the Banach algebra $\operatorname{alg} W([SAP_{p,w}]_{N\times N}) \subset \mathcal{B}(L_N^p(\mathbb{R}_+, w))$. Arguing similarly to the proof of Theorem 6.3 and applying the matrix version of Lemma 6.2, we obtain Theorem 7.1.

Theorem 7.2. *Let $1<p<\infty$, $w\in A_p^0(\mathbb{R})$, $N\in\mathbb{N}$, and let $a\in[SO_{p,w}, SAP_{p,w}]_{N\times N}$. If for every $\xi \in \mathcal{M}_\infty(SO)$ the matrix functions $a_{\xi,\pm} = \nu_{\xi,\pm} a \in [AP_{p,w}]_{N\times N}$ admit right $AP_{p,w}$ factorizations, then the Wiener-Hopf operator $W(a)$ is Fredholm on the space $L_N^p(\mathbb{R}_+, w)$ if and only if the following three conditions are satisfied:*

(i) $\det a(x) \neq 0$ *for all $x \in \mathbb{R}$;*
(ii) *for every $\xi \in \mathcal{M}_\infty(SO)$, $\kappa(a_{\xi,\pm}) = (0, \dots, 0)$;*

(iii) *for every* $\xi \in \mathcal{M}_\infty(SO)$ *and all* $j = 1, 2, \dots, N$, *the eigenvalues* $\eta_{\xi,j}$ *of the matrix* $\mathbf{d}^{-1}(a_{\xi,+})\mathbf{d}(a_{\xi,-})$ *satisfy the condition*

$$\frac{1}{p} + \frac{1}{2\pi} \arg \eta_{\xi,j} \notin \mathbb{Z}. \tag{7.1}$$

If $W(a)$ *is Fredholm, then all its regularizers belong to the Banach algebra* $\mathcal{D} = \operatorname{alg} W([SO_{p,w}, SAP_{p,w}]_{N\times N})$.

Proof. Necessity. Let the operator $W(a)$ be Fredholm on the space $L^p_N(\mathbb{R}_+, w)$. Then, by the Allan-Douglas local principle, $\det a(x) \neq 0$ for all $x \in \mathbb{R}$ and for every $\xi \in \mathcal{M}_\infty(SO)$ the coset $W^\pi_\xi(a) = W^\pi(a) + J^\pi_\xi$ is invertible in the quotient algebra Λ^π_ξ. By Corollary 4.5, for every $\xi \in \mathcal{M}_\infty(SO)$ the operators $W(a_{\xi,\pm})$ are invertible on the space $L^p_N(\mathbb{R}_+, w)$. Since all the matrix functions $a_{\xi,\pm} = \nu_{\xi,\pm} a \in [AP_{p,w}]_{N\times N}$ admit right $AP_{p,w}$ factorizations, from Theorem 2.6 it follows that $\kappa(a_{\xi,\pm}) = (0, \dots, 0)$, and therefore the right $AP_{p,w}$ factorizations of $a_{\xi,\pm}$ are canonical. Hence, taking an $a_\xi \in [SAP_{p,w}]_{N\times N}$ such that $W^\pi_\xi(a_\xi) = W^\pi_\xi(a)$, we infer from Theorem 7.1 that for every $\xi \in \mathcal{M}_\infty(SO)$ and all $x \in \mathbb{R}$,

$$\det\left[\mathbf{d}(a_{\xi,+})\frac{1 - \coth[\pi(x + i/p)]}{2} + \mathbf{d}(a_{\xi,-})\frac{1 + \coth[\pi(x + i/p)]}{2}\right] \neq 0. \tag{7.2}$$

Since $\det \mathbf{d}(a_{\xi,\pm}) \neq 0$, we conclude that (7.2) holds for all $x \in \overline{\mathbb{R}}$, which is equivalent to condition (iii) of Theorem 7.2.

Sufficiency. Suppose all the conditions of Theorem 7.2 are fulfilled. Then for every $\xi \in \mathcal{M}_\infty(SO)$ the matrix functions $a_{\xi,\pm} \in [AP_{p,w}]_{N\times N}$ admit canonical right $AP_{p,w}$ factorizations, and hence, taking a matrix function $a_\xi \in [SAP_{p,w}]_{N\times N}$ such that $\widetilde{W}^\pi_\xi(a_\xi) = \widetilde{W}^\pi_\xi(a)$ and applying Theorem 7.1, we infer from condition (iii) of Theorem 7.2, which is equivalent to (7.2) in view of $\det \mathbf{d}(a_{\xi,\pm}) \neq 0$, that the coset $\widetilde{W}^\pi_\xi(a) = W^\pi(a) + \widetilde{J}^\pi_\xi$ is invertible in the quotient algebra $\mathcal{D}^\pi_\xi$. On the other hand, condition (i) implies that the cosets $\widetilde{W}^\pi_\xi(a) = [a(\xi)I]^\pi + \widetilde{J}^\pi_\xi$ are invertible in $\mathcal{D}^\pi_\xi$ for all $\xi \in \mathbb{R}$. Hence, by the Allan-Douglas local principle, the Wiener-Hopf operator $W(a)$ with a matrix symbol $a \in [SO_{p,w}, SAP_{p,w}]_{N\times N}$ is Fredholm on the space $L^p_N(\mathbb{R}_+, w)$ and all its regularizers belong to the Banach algebra $\mathcal{D}$. □

Applying [6, Corollary 19.11] or Theorem 2.7 with $w = 1$, we immediately deduce the following corollary from Theorem 7.2.

Theorem 7.3. *If* $1 < p < \infty$, $N \in \mathbb{N}$, $a \in [SO_p, SAP_p]_{N\times N}$, *and for every* $\xi \in \mathcal{M}_\infty(SO)$ *the matrix functions* $a_{\xi,\pm} = \nu_{\xi,\pm} a$ *are in* $[APW]_{N\times N}$, *then the Wiener-Hopf operator* $W(a)$ *is Fredholm on the space* $L^p_N(\mathbb{R}_+)$ *if and only if the following three conditions are satisfied:*

(i) $a \in G[SO_p, SAP_p]_{N\times N}$ *(equivalently,* $a^{-1} \in L^\infty_{N\times N}(\mathbb{R})$*);*

(ii) *for every* $\xi \in \mathcal{M}_\infty(SO)$, *the matrix functions* $a_{\xi,\pm}$ *admit canonical right APW factorizations;*

(iii) *for every* $\xi \in \mathcal{M}_\infty(SO)$ *and every* $j = 1, 2, \dots, N$, *the eigenvalues* $\eta_{\xi,j}$ *of the matrix* $\mathbf{d}^{-1}(a_{\xi,+})\mathbf{d}(a_{\xi,-})$ *satisfy* (7.1).

References

[1] M.A. Bastos, A. Bravo, and Yu.I. Karlovich, *Convolution type operators with symbols generated by slowly oscillating and piecewise continuous matrix functions.* Operator Theory: Advances and Applications **147** (2004), 151–174.

[2] M.A. Bastos, A. Bravo, and Yu.I. Karlovich, *Symbol calculus and Fredholmness for a Banach algebra of convolution type operators with slowly oscillating and piecewise continuous data.* Math. Nachrichten **269–270** (2004), 11–38.

[3] M.A. Bastos, Yu.I. Karlovich, and B. Silbermann, *Toeplitz operators with symbols generated by slowly oscillating and semi-almost periodic matrix functions.* Proc. London Math. Soc. **89** (2004), 697–737.

[4] A. Böttcher and Yu.I. Karlovich, *Carleson Curves, Muckenhoupt Weights, and Toeplitz Operators.* Progress in Mathematics **154**, Birkhäuser, Basel, 1997.

[5] A. Böttcher, Yu.I. Karlovich, and V.S. Rabinovich, *The method of limit operators for one-dimensional integrals with slowly oscillating data.* J. Operator Theory **43** (2000), 171–198.

[6] A. Böttcher, Yu.I. Karlovich, and I.M. Spitkovsky, *Convolution Operators and Factorization of Almost Periodic Matrix Functions.* Operator Theory: Advances and Applications **131**, Birkhäuser, Basel, 2002.

[7] A. Böttcher and B. Silbermann, *Analysis of Toeplitz Operators*, 2nd edition, Springer, Berlin, 2006.

[8] A. Böttcher and I.M. Spitkovsky, *Wiener-Hopf integral operators with PC symbols on spaces with Muckenhoupt weight.* Revista Matemática Iberoamericana **9** (1993), 257–279.

[9] A. Böttcher and I.M. Spitkovsky, *Pseudodifferential operators with heavy spectrum.* Integral Equations and Operator Theory **19** (1994), 251–269.

[10] R.V. Duduchava, *Integral Equations with Fixed Singularities.* B.G. Teubner Verlagsgesellschaft, Leipzig, 1979.

[11] R.V. Duduchava, *On algebras generated by convolutions and discontinuous functions.* Integral Equations and Operator Theory **10** (1987), 505–530.

[12] R.V. Duduchava and A.I. Saginashvili, *Convolution integral equations on a half-line with semi-almost-periodic presymbols.* Differential Equations **17** (1981), 207–216.

[13] J.B. Garnett, *Bounded Analytic Functions.* Academic Press, New York, 1981.

[14] I.M. Gelfand, D.A. Raikov, and G.E. Shilov, *Commutative Normed Rings.* Fizmatgiz, Moscow, 1960 [Russian]. English transl.: Chelsea, New York, 1964.

[15] I. Gohberg and I. Feldman, *Wiener-Hopf integro-difference equations.* Soviet Math. Dokl. **9** (1968), 1312–1316.

[16] I. Gohberg and I.A. Feldman, *Convolution Equations and Projection Methods for Their Solutions.* Transl. of Math. Monographs **41**, Amer. Math. Soc., Providence, R.I., 1974. Russian original: Nauka, Moscow, 1971.

[17] I. Gohberg and N. Krupnik, *Singular integral operators with piecewise continuous coefficients and their symbols.* Math. USSR – Izv. **5** (1971) 955–979.

[18] R. Hunt, B. Muckenhoupt, and R. Wheeden, *Weighted norm inequalities for the conjugate function and Hilbert transform.* Trans. Amer. Math. Soc. **176** (1973), 227–251.

[19] Yu.I. Karlovich, *An algebra of pseudodifferential operators with slowly oscillating symbols.* Proc. London Math. Soc. (3) **92** (2006), 713–761.

[20] Yu.I. Karlovich and J. Loreto Hernández, *Wiener-Hopf operators with matrix semi-almost periodic symbols on weighted Lebesgue spaces.* Integral Equations and Operator Theory **62** (2008), 85–128.

[21] Yu.I. Karlovich and J. Loreto Hernández, *Wiener-Hopf operators with slowly oscillating matrix symbols on weighted Lebesgue spaces.* Integral Equations and Operator Theory **64** (2009), 203–237.

[22] Yu.I. Karlovich and E. Ramírez de Arellano, *Singular integral operators with fixed singularities on weighted Lebesgue spaces.* Integral Equations and Operator Theory **48** (2004), 331–363.

[23] M.A. Krasnoselskii, P.P. Zabreiko, E.I. Pustylnik, and P.E. Sobolevskii, *Integral Operators in Spaces of Summable Functions.* Noordhoff I.P., Leyden, 1976. Russian original: Nauka, Moscow, 1966.

[24] V. Petkova, *Symbole d'un multiplicateur sur* $L^2_\omega(\mathbb{R})$. Bull. Sci. Math. **128** (2004), 391–415.

[25] S.C. Power, *Fredholm Toeplitz operators and slow oscillation.* Can. J. Math. **32** (1980), 1058–1071.

[26] V. Rabinovich, S. Roch, and B. Silbermann, *Limit Operators and Their Applications in Operator Theory,* Birkhäuser, Basel, 2004.

[27] S. Roch and B. Silbermann, *Algebras of Convolution Operators and Their Image in the Calkin Algebra.* Report R-Math-05/90, Akad. Wiss. DDR, Karl-Weierstrass-Institut f. Mathematik, Berlin, 1990.

[28] D. Sarason, *Toeplitz operators with semi-almost periodic symbols.* Duke Math. J. **44** (1977) 357–364.

[29] R. Schneider, *Integral equations with piecewise continuous coefficients in* L^p *spaces with weight.* J. Integral Equations **9** (1985), 135–152.

Yu.I. Karlovich
Facultad de Ciencias
Universidad Autónoma del Estado de Morelos
Av. Universidad 1001, Col. Chamilpa,
C.P. 62209
Cuernavaca, Morelos, México
e-mail: `karlovich@uaem.mx`

J. Loreto Hernández
Instituto de Matemáticas
Universidad Nacional Autónoma de México
Av. Universidad 1001, Col. Chamilpa,
C.P. 62210
Cuernavaca, Morelos, México
e-mail: `juan@matcuer.unam.mx`

Operator Theory:
Advances and Applications, Vol. 210, 147–174

On the Ill-posed Hyperbolic Systems with a Multiplicity Change Point of Not Less Than the Third Order

Valeri V. Kucherenko and Andriy Kryvko

Dedicated to 60th birthday of Nikolai Vasilevski

Abstract. Hyperbolic systems with noninvolutive multiple characteristics are considered and an example of ill-posed Cauchy problem is proposed.

Mathematics Subject Classification (2000). 34E20; 35L40.

Keywords. Hyperbolic system, Cauchy problem, multiplicity change point, pseudodifferential operator, leading symbol, Poisson bracket, Hamiltonian system.

1. Main results

In this paper we are concerned with nostrongly hyperbolic systems, i.e., systems such that the well-posedness of Cauchy problem depends on the lowest terms.

We assume that the leading symbol of a real first-order hyperbolic system

$$\partial_t u + \widehat{A} u = 0 \tag{1.1}$$

has the form

$$A(x', \eta, \xi) = \eta I + D(x', \xi) \left\| b_{ij}^0 (x', \xi) \right\| D^{-1}(x', \xi), \tag{1.2}$$

where $x' = (t, x)$, $\xi' = (\eta, \xi)$, I is the identity $((r+2) \times (r+2))$-matrix, the symbols $D, D^{-1} \in S^0 \left(\mathbb{R}_{x'}^{n+1} \times \mathbb{R}_\xi^n \right)$, $b_{ij}^0 (x', \xi) \in S^1 \left(\mathbb{R}_{x'}^{n+1} \times \mathbb{R}_\xi^n \right)$, and

$$\left\| b_{ij}^0 (x', \xi) \right\| = \begin{pmatrix} \lambda_1 I_r & 0 & 0 \\ 0 & \lambda_2 & d \\ 0 & 0 & \lambda_3 \end{pmatrix}, \tag{1.3}$$

where I_r is the identity $(r \times r)$-matrix and it holds that $\operatorname{Im} \lambda_j \equiv 0, j = 1, 2, 3$.

This research was partially supported by project of CONACYT Mexico and by project of SIP – IPN.

The substitution $u = D(x', -i\partial_x)\,v$ reduces the leading symbol of system (1.1) to the triangular form. It follows from the analysis of pseudodifferential operators (PDO) that the complete matrix symbol A_c of the hyperbolic system for the function v has the form

$$A_c := \eta I + \left\|b^0_{ij}(x', \xi)\right\| + \sum_{k=1}^{\infty} b^k(x', \xi), \tag{1.4}$$

where $b^k(x', \xi) = \left\|b^k_{ij}(x', \xi)\right\|$, and $b^k(x', \xi) \in S^{1-k}\left(\mathbb{R}^n_x \times [0,T] \times \mathbb{R}^n_\xi\right)$, $k \in \mathbb{N}$. In what follows the element $b^1_{r+2,r+1}$ of the matrix symbol $b^1(x', \xi)$ plays an important role.

We assume that the set

$$\Sigma_{123} := \{x', \xi : \lambda_1 = \lambda_2 = \lambda_3\},$$

is not empty and it is a smooth C^∞-manifold of dimension $(2n-1)$ in $\mathbb{R}^{2n+1}_{x',\xi} \backslash \xi = 0$, the sets

$$\Sigma_{ij} := \{x', \xi : \lambda_i = \lambda_j, 1 \le i, j \le 3, i \ne j\},$$

are C^∞-manifolds of dimension $2n$ in $\mathbb{R}^{2n}_{x,\xi} \times [0,T] \setminus \{\xi = 0\}$, and the Poisson brackets in variables $(x'; \xi')$ satisfy conditions

$$\{\eta + \lambda_i, \eta + \lambda_j\}\,|_{\Sigma_{ij} \cap \{|\xi|=1\}} \ne 0, \tag{1.5}$$

for $1 \le i, j \le 3, i \ne j$ (i.e., $\Sigma_{123} \subset \Sigma_{ij}$). *In what follows we use the abbreviations*

$$[i,j] := \{\eta + \lambda_i, \eta + \lambda_j\}. \tag{1.6}$$

Also suppose that the restriction $d|_{\Sigma_{123} \cap \{|\xi|=1\}}$ is not identically zero.

An example of a real first-order hyperbolic system with a leading symbol reducible to form (1.3) provides a linearization of hydrodynamic system for ideal compressible fluids. It was proved in [12] that the characteristic roots $\eta = -\lambda(x', \xi)$ satisfy the equation of the form

$$\begin{aligned} &\det\left(\eta I + \left\|b^0_{ij}(x', \xi)\right\|\right) \\ &\quad = (\eta + (u_0, \xi))^2\,(\eta + (u_0, \xi) - c(x')\,|\xi|)\,(\eta + (u_0, \xi) + c(x')\,|\xi|) = 0. \end{aligned}$$

Evidently at the points x' such that $c(x') = 0$ the characteristic roots $\lambda_\pm = (u_0, \xi) \pm c\,|\xi|$ coincide with the characteristic root $\lambda = (u_0, \xi)$. Therefore, at such points the multiplicity is equal to three. A simple calculus shows that at the points $x : c(\rho_0(x)) = 0$, there are three eigenvectors and one adjoint vector, i.e., a Jordan block of second order arise. Evidently in this case the set of multiplicity $\Sigma \subset \mathbb{R}^{2n+1}_{x',\xi}$ is such that

$$\Sigma = \{x', \xi : c(x') = 0\}. \tag{1.7}$$

A differential system with the complete matrix symbol of the form

$$A_c = \eta I + \left\|b^0_{ij}(x', \xi)\right\| + \left\|b^1_{ij}(x', \xi)\right\| \tag{1.8}$$

is a basic example of the systems with multiplicity of third order, such that the well-posedness of Cauchy problem with the initial data at $t = 0$, depends on the lowest terms. Here we set $n = 2, r = 1$; the elements b^0_{ij} have form (1.3) with $d = \xi_2$; the elements $b^1_{ij} \in S^0\left(\mathbb{R}^n_x \times [0,T] \times \mathbb{R}^n_\xi\right)$, $b^1_{ij} \neq 0$ for $i \neq j$ and $b^1_{ii} = 0$. We define the characteristic roots λ_j as

$$\lambda_1 = 3\xi_1 + 4x_1\xi_2;\ \lambda_2 = \xi_1 + x_1\xi_2;\ \lambda_3 = \xi_1 + 5x_1\xi_2 - \frac{9}{2}t\xi_2. \tag{1.9}$$

In the basic example it holds that $\Sigma_{123} = \left\{(x',\xi) : x_1 = \frac{9}{8}t, \xi_1 = -\frac{27}{16}\xi_2 t\right\}$, and evidently conditions (1.5) hold true.

The ill-posedness of Cauchy problem (1.1) when $\partial_t \widehat{A} \equiv 0$ can occur because the closure of the operator $\widehat{A}$ in a Sobolev space has the sequence of spectral points σ_m such that $\operatorname{Re}\sigma_m \longrightarrow -\infty$. Such an example of real hyperbolic system with three involuting characteristic roots $\{\eta + \lambda_i, \eta + \lambda_j\}\,|_{\Sigma_{ij}\cap\{|\xi|=1\}} \equiv 0$ was provided in [1].

It is well known that when the symbol of system (1.1) has form (1.3) with a Jordan block of second order and constant multiplicity: $\lambda_2 \equiv \lambda_3, \Sigma_{12} = \varnothing$, the well-posedness of Cauchy problem depends on the lower-order term $b^1_{r+2,r+1}$ [11], [4]. That is, if $db^1_{r+2,r+1}(x',\xi) \in R^1_+$ (here $R^1_+ = \{y : y \geq 0\}$), for all $(x',\xi) \in R^{2n+1}_{x',\xi}$ it holds that the Cauchy problem for system (1.1) is well posed in the Sobolev spaces. Whereas, if at some point $(t_0, x_0, \xi_0), t_0 \geq 0, \xi_0 \neq 0$ it holds that $db^1_{r+2,r+1}(t_0, x_0, \xi_0) \notin R^1_+$, then the Cauchy problem for system (1.1) is ill posed in all Sobolev spaces. In the case of variable multiplicity of second order, that is $\Sigma_{12} = \varnothing$, $\Sigma_{13} = \varnothing$, $\Sigma_{32} \neq \varnothing$ and condition (1.5) holds for $i = 2, j = 3$, the Cauchy problem is well posed [4] in the Sobolev spaces for all lower-order terms. In the present paper we show that the additional multiplicity $\Sigma_{12} \neq \varnothing$, $\Sigma_{32} \neq \varnothing$, $\Sigma_{123} \neq \varnothing$ implies the ill-posednes of Cauchy problem for some lower-order terms.

We investigate the well-posedness of system (1.1) through a formally asymptotic solution (FAS) of the Cauchy problem with oscillating initial data (the formal definition of FAS is presented below).

Suppose now that the wave front of the initial data of the Cauchy problem is concentrated on the Lagrangian manifold Λ_0, such that

$$\rho\left(\Lambda_0|_{|\xi|=1}, \Sigma_{ij}|_{|\xi|=1,t=0}\right) > 0, \text{ for } i,j = 1,2,3, i \neq j. \tag{1.10}$$

Let $\pi_{x,t}$ and $\pi_{x,\eta}$ be the orthogonal projections of the space $R^{2n+2}_{x'\xi'}$ onto the planes $\{\xi' = 0\}$ and $\{\xi = 0, t = 0\}$, respectively, let $g^{t,\tau}_k : R^{2n}_{x,\xi} \to R^{2n}_{x,\xi}$ be the one parameter family of diffeomorphisms generated by the Hamiltonian flow of the Hamiltonian $\lambda_k(x,t,\xi)$ and $g^{\tau,\tau}_k = I$. Following the papers [3], [5], [7], [9], we consider the set of trajectories $g^{(t,0)}_j(x,\xi)$; $(x,\xi) \in \Lambda_0, j = 1,2,3$, in the extended phase space $R^{2n+1}_{x',\xi}$ and their ramifications [3]. Ramification arises when the trajectory $g^{(t,0)}_j(x,\xi), \xi \neq 0$, in a moment of time t_j passes through the point (t_j, x_j, ξ_j) of the manifold $\Sigma_{jk}, j \neq k$ ($\Sigma_{jk} \subset R^{2n+1}_{x',\xi}$) and generates a new branch – the trajectory $g^{(t,t_j)}_k(x_j,\xi_j), t \geq t_j$. Moreover, when these branches $g^{(t,t_j)}_k(x_j,\xi_j)$ in moments of time t_k passes through the points (t_k, x_k, ξ_k) of the manifold $\Sigma_{km}, k \neq m$ they generate new branches $g_{\lambda_m}^{(t,t_k)}(x_k,\xi_k)$, $t \geq t_k, k \neq m$ and so on. The union of the initial trajectory $g_j^{(t,0)}(x,\xi)$ with the sequence of all branches generated by it during the period of time $[0,T]$ is called the complete ramification of the trajectory $g_j^{(t,0)}(x,\xi)$ in $[0,T]$.

Condition 1.1. *Suppose that for all the trajectories* $g_j^{(t,0)}(x_0,\xi_0), |\xi_0| > 0, j = 1,2,3,$ *in a finite period of time:* $t,\tau \in [0,T]$ *it holds that*

$$\inf_{t,\tau\in[0,T]} \left|\xi \circ g_j^{(t,\tau)}(x_0,\xi_0)\right| \geq c(T)\,|\xi_0|\,, c(T) > 0, \text{ for all } (x_0,\xi_0) \in \mathbb{R}^{2n}_{x,\xi}. \tag{1.11}$$

Conditions (1.5) imply that the functions $(\lambda_j - \lambda_k)(t, g_j^{(t,0)}(x,\xi)), \xi \neq 0, j \neq k$ are strictly monotone. Hence, if the distance between the complete ramification in $[0,T]$ of this trajectory and the set Σ_{123} is positive, then the trajectory and its branches can intersect only once every surface Σ_{jk} in the time interval $[0,T]$. In this case every trajectory $g_j^{(t,\tau)}(x,\xi), |\xi| > 0$ can generate only a finite quantity of branches in $[0,T]$.

When Σ_{123} is not empty the complete ramifications of the trajectories located in some vicinity of Σ_{123} may contain periodic cycles or stable and nonstable motions that have the pattern of "logarithmic spiral", and the Cauchy problem becomes ill posed under some conditions on lower-order terms. This effect does not exist in the case of multiplicity of second order, that is when $\Sigma_{123} = \varnothing$.

Now we introduce conditions on the roots $\lambda_j, j = 1,2,3,$ that provide the pattern of "logarithmic spiral".

Define the sets $\Delta_{ij}^{\pm}$, as

$$\Delta_{ij}^{+} := \{(x',\xi) : 0 \leq \lambda_i - \lambda_j\},\ \Delta_{ij}^{-} := \{(x',\xi) : \lambda_i - \lambda_j \leq 0\},$$

and let $\Delta_{ij}^{\pm int}$ be the set of the internal points.

Let $x_0' = (0,x_0), |\xi_0| > 0$ and assume that the point (x_0',ξ_0) belongs to Σ_{123}. Suppose that in some vicinity

$$U_0^{\varepsilon} := \{(x',\xi) : 0 \leq t < \varepsilon, |x_0 - x| < \varepsilon, |\xi_0 - \xi| < \varepsilon\} \tag{1.12}$$

it holds that

$$\Sigma_{32} \cap U_0^{\varepsilon} \subset \left\{\Delta_{12}^{+} \cap \Delta_{31}^{-}\right\} \cup \left\{\Delta_{12}^{-} \cap \Delta_{31}^{+}\right\}, \tag{1.13}$$

and suppose that

$$[1,2]_{U_0^{\varepsilon}} > 0, [3,2]_{U_0^{\varepsilon}} < 0, [3,1]_{U_0^{\varepsilon}} > 0. \tag{1.14}$$

It is easy to show that basic example (1.8), (1.9) satisfies conditions (1.13), (1.14). For the basic example ($n = 2$) in Figure 1 we present the projection of the manifolds Σ_{jk} and the complete ramification of the trajectory $g_2^{(t,0)}(x,\xi), (0,x;\xi) \in \Delta_{12}^{-} \cap \Delta_{31}^{-} \cap U_0^{\varepsilon}$ at $R^2_{x,\xi} \times \{t : t = 0\}$. In Figure 1 the complete ramification contains a piecewise smooth trajectory of converging "logarithmic spiral" form (the dotted lines describe the additional branches that belong to the complete ramification, but are not included in the emphasized trajectory of "logarithmic spiral" form).

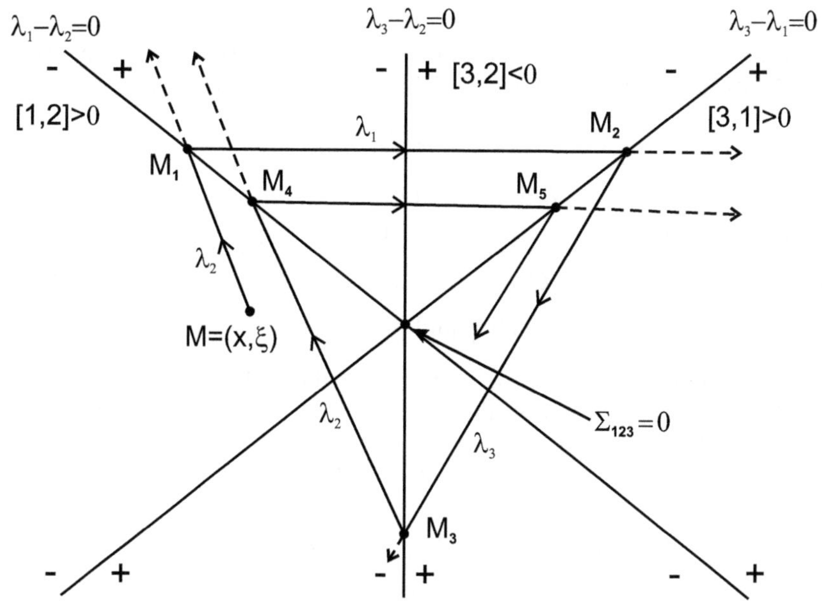

FIGURE 1

In some cases the complete ramification can contain a trajectory of diverging "logarithmic spiral" form or a periodic trajectory.

Now consider the Cauchy problem for hyperbolic system (1.1) with highly oscillating initial data of the form

$$u(0,x,h) = D\,(0,x,-i\partial_x)\,\{\phi\,(x)\,e_2 \exp\left\{\frac{i}{h}\,(x,\overline{\xi})\right\}\}, \overline{\xi} \neq 0; h \in (0,1], \tag{1.15}$$

$$e_2 := (0,1,0)^t, \text{ here } t \text{ stands for the transposition,}$$

$$\left(0 \times \text{supp } \phi \times \overline{\xi}\right) \in \{\Delta_{12}^- \cap \Delta_{31}^- \cap U_0^{\varepsilon_0}\} \setminus \Sigma_{123}, \varepsilon_0 < \varepsilon. \tag{1.16}$$

Here $U_0^{\varepsilon_0}$ is the vicinity of the form (1.12) at $\varepsilon = \varepsilon_0$; $(\text{supp } \phi)^\mu$ is the μ vicinity of the set supp ϕ and let Λ_0 be a initial manifold defined as

$$\Lambda_0 := \left\{(x,\xi) : x \in (\text{supp } \phi)^\mu, \xi = \overline{\xi}, \overline{\xi} \neq 0\right\}. \tag{1.17}$$

Evidently for a sufficiently small value μ it holds that

$$\rho\left(\Lambda_0, \Sigma_{ij}\big|_{,t=0}\right) > 0, j = 1,2,3, \text{ for } i \neq j, \tag{1.18}$$

$$\Lambda_0 \in \{\Delta_{12}^- \cap \Delta_{31}^- \cap U_0^{\varepsilon_0}\} \setminus \Sigma_{123}. \tag{1.19}$$

Definition 1.2. A function $v(x,t,h) \in C^{\infty}\left(C_0^{\infty}\left(\mathbb{R}_x^n\right),[0,T]\right)$, which depends on the parameter $h \in (0,1]$, is called the formally asymptotic solution of Cauchy problem (1.1), (1.15) with the precision $O(h^M)$ (FAS/$O(h^M)$) if it holds that

$$u(0,x,h) - v(x,0,h) = O_{in}(h^M), \tag{1.20}$$

$$\partial_t v + \widehat{A} v = O_{rh}(h^M). \tag{1.21}$$

The functions $O_{in}(h^M) \in C_0^{\infty}\left(\mathbb{R}_x^n\right)$, $O_{rh}(h^M) \in C^{\infty}\left(C_0^{\infty}\left(\mathbb{R}_x^n\right),[0,T]\right)$ are such that $\sup_{\mathbb{R}_{x'}^{n+1}}\left|\partial_{x'}^{\beta} O_{in}(h^M)\right| \le C_{\beta} h^{M-|\beta|}$; $\sup_{\mathbb{R}_x^n \times [0,T]}\left|\partial_{x'}^{\beta} O_{rh}(h^M)\right| \le C_{\beta} h^{M-|\beta|}$, $C_{\beta} < +\infty$, where $\partial_{x'}^{\beta} = \partial_t^{\beta_0}\partial_{x_1}^{\beta_1}\dots\partial_{x_n}^{\beta_n}$ and $|\beta| = \beta_0 + \dots + \beta_n$.

Definition 1.3. A function $w(x,t,h) \in C^{\infty}(\Omega)$, where $\Omega \in \mathbb{R}_x^n \times [0,T]$, is called the formal asymptotic solution of differential system (1.1) in the domain Ω with the precision $O(h^M)$ (FAS/$\Omega, O(h^M)$) if for every point $(\overline{t},\overline{x}) \in \Omega_0$ in a sphere $U(\overline{t},\overline{x},h^{1/2-\delta/2})$ of radius $h^{1/2-\delta/2}, 0 < \delta < 1/4$, centered at the point $(\overline{t},\overline{x})$, it holds

$$(\partial_t + \widehat{A}) w(x,t,h)\,\varphi(h) = O(h^M),$$

$$\sup_{U(\overline{t},\overline{x},h^{1/2-\delta/2})}\left|\partial_{x'}^{\beta} O(h^M)\right| \le C_{\overline{t},\overline{x}} h^{M-|\beta|}.$$

Let $b_{r+2,r+1}^1$ be the matrix element of the lower-order symbol $b^1(x,\xi)$, let σ_{32} be the function defined on Σ_{123} as

$$\sigma_{32} := \left.\frac{i b_{r+2,r+1}^1 d}{\{\eta+\lambda_3, \eta+\lambda_2\}}\right|_{\Sigma_{123}},$$

and denote by σ the function

$$\sigma = \operatorname{Re}\sigma_{32}(0,x_0,\xi_0) + 3/2.$$

Suppose that the complete ramification of the trajectory $g_2^{(t,0)}(x,\xi), (0,x;\xi) \in \Delta_{12}^{-} \cap \Delta_{31}^{-} \cap U_0^{\varepsilon_0}$, with $(0,x;\xi) \notin \Sigma_{123}$, in the interval $[0,T_1]$, $T_1 \le T$ intersects the manifold Σ_{32} at least N times. In what follows we prove that as time passes the amplitude of FAS/$O(h^M)$ ($M \gg N$) became of order $h^{\sigma N}$. Therefore, if it holds that $\sigma < 0$ and N became large as the time passes then the Cauchy problem became ill posed.

Theorem 1.4. *Assume that hyperbolic system* (1.1) *has leading symbol* (1.2) *and the substitution* $u = D(x', -i\partial_x)\,v$ *reduces it to a system with the complete symbol of form* (1.4). *Suppose that the eigenvalues* λ_j *satisfy conditions* (1.5), (1.1); $\Sigma_{123} \neq \varnothing$; *and condition* (1.11) *holds true. Let* $(x_0', \xi_0) \in \Sigma_{123}$, $x_0' = (0,x_0), |\xi_0| > 0$ *be a point such that conditions* (1.13), (1.14) *hold true and* $d(x_0',\xi_0) \neq 0$. *Assume that the lower-order terms satisfy the conditions:* $b_{ji}^1(x_0',\xi_0) \neq 0$, *for* $j \neq i$ *and* $\operatorname{Re}\sigma_{32}(0,x_0,\xi_0) + 3/2 < 0$. *In this case for all natural numbers* k, l, m *such that*

$k \geq m, l > 0$, and any constant $C > 0$ there exist a moment of time $T > 0$ and a function $u_{k,l,m}(x,t) \in C^\infty\left(C_0^\infty\left(\mathbb{R}_x^n\right), [0,T]\right)$ such that

$$\max_{0 \leq t \leq T}\left\{\|u_{k,l,m}(t)\|_{H_{k-m}(\mathbb{R}_x^n)}\right\} > C \max_{0 \leq t \leq T}\left\{\|Lu_{k,l,m}(t)\|_{H_l(\mathbb{R}_x^n)} + \|u_{k,l,m}(0)\|_{H_k(\mathbb{R}_x^n)}\right\}. \tag{1.22}$$

The value of T can be chosen arbitrary small.

2. Ramifications of trajectories

Existence of a function $u_{k,l,m}(x,t)$ that satisfies estimate (1.22) for all k, l, m such that $k \geq m, l > 0$ signifies that the Cauchy problem is ill posed in any Sobolev space. The functions $u_{k,l,m}(x,t)$ are calculated by the asymptotic methods [3], [10]. FAS of the Cauchy problem (1.1), (1.15) attached to the Lagrangian manifold generated by the complete ramification of the trajectories originated on the manifold $\Lambda_0 = \left\{(x,\xi) : x \in \operatorname{supp} \phi, \xi = \overline{\xi}\right\} \in \Delta_{12}^- \cap \Delta_{31}^- \cap U_0^{\varepsilon_0}$.

Lemma 2.1. *For every $T > 0$ and every integer number N there exists $\varepsilon_0 : 0 < \varepsilon_0 < \varepsilon/2$ and time $T_1 : 0 < T_1 < T$ such that:*

1. *For all initial data $(0,x;\xi) \in \{\Delta_{12}^- \cap \Delta_{31}^- \cap U_0^{\varepsilon_0}\} \setminus \Sigma_{123}$ the complete ramification of the trajectory $g_2^{(t,0)}(x,\xi), (0,x;\xi) \in \{\Delta_{12}^- \cap \Delta_{31}^- \cap U_0^{\varepsilon_0}\} \setminus \Sigma_{123}$ in the interval $[0,T_1]$ has at least N intersections with the manifold Σ_{32}.*
2. *This complete ramification belongs to $U_0^{\varepsilon/2}$.*
3. *The distance between Σ_{123} and the complete ramification of the trajectory $g_2^{(t,0)}(x,\xi)$ in the interval $[0,T_1]$ is positive.*

Proof. 1.) Let the initial point $(x;\xi)$ of the trajectory $g_2^{(t,0)}(x,\xi)$ be such that the point $M = (0,x;\xi) \in \{\Delta_{12}^- \cap \Delta_{31}^- \cap U_0^{\varepsilon_0}\} \setminus \Sigma_{123}$ and $\varepsilon_0 < \varepsilon$, as it is drawn in Figure 1. Also suppose that the trajectory $g_2^{(t,0)}(x,\xi)$ in a moment of time t_1 intersects Σ_{12} at the point $M_1 = (t_1, x_1, \xi_1)$, the trajectory $g_1^{(t,t_1)}(x_1,\xi_1)$ in a moment of time t_2 intersects Σ_{31} at the point $M_2 = (t_2, x_2, \xi_2)$, the trajectory $g_3^{(t,t_2)}(x_2,\xi_2)$ in a moment of time t_3 intersects Σ_{32} at the point $M_3 = (t_3, x_3, \xi_3)$, and the trajectory $g_2^{(t,t_3)}(x_3,\xi_3)$ in a moment of time t_4 intersects Σ_{12} at the point $M_4 = (t_4, x_4, \xi_4)$.

At first suppose that there exist such small ε_0 that all the points $M_j \in U_0^{\varepsilon/2}$(that is $t_j < \varepsilon/2$ for $j = 1, \ldots, 4$) and suppose that there exists $\tau(\varepsilon) : T > \tau(\varepsilon) > \max\{t_1, \ldots, t_4\}$ such that the trajectories $g_2^{(t,t_1)}(x_1,\xi_1)$, $g_1^{(t,t_2)}(x_2,\xi_2)$, $g_3^{(t,t_3)}(x_3,\xi_3)$ are defined for $t < \tau(\varepsilon)$. Below we prove this step by step calculating t_j.

Conditions (1.14) imply that in the domain $\Delta_{12}^+ \cap \Delta_{32}^-$ it holds

$$\frac{d}{dt}(\lambda_1 - \lambda_2)\left(t, g_2^{(t,t_1)}(x,\xi)\right) > 0, \frac{d}{dt}(\lambda_3 - \lambda_2)\left(t, g_2^{(t,t_1)}(x,\xi)\right) < 0$$

for $t \in [t_1,\tau(\varepsilon)]$. As the function $(\lambda_1 - \lambda_2)\left(t, g_2^{(t,t_1)}(x,\xi)\right), \xi \neq 0$ is monotone increasing and the function $(\lambda_3 - \lambda_2)\left(t, g_2^{(t,t_1)}(x,\xi)\right), \xi \neq 0$ is monotone decreasing, it holds that the trajectory $g_2^{(t,t_1)}(x_1,\xi_1), t > t_1$ belongs to the domain $\Delta_{12}^+ \cap \Delta_{32}^-$ for $t_1 \leq t$ as it holds for $t = t_1$. Using similar considerations we can obtain the remaining inclusions, i.e.,

$$g_2^{(t,t_1)}(x_1,\xi_1) \in U_0^{\varepsilon/2} \cap \Delta_{12}^+ \cap \Delta_{32}^-;\ g_1^{(t,t_2)}(x_2,\xi_2) \in U_0^{\varepsilon/2} \cap \Delta_{31}^+ \cap \Delta_{12}^+; \quad (2.1)$$
$$g_3^{(t,t_3)}(x_3,\xi_3) \in U_0^{\varepsilon/2} \cap \Delta_{31}^+ \cap \Delta_{32}^- \text{ for } t \in [t_k,\tau(\varepsilon)], k = 1,2,3.$$

2.) Evidently for every initial point $(\tau,x,\xi) \in U_0^{\varepsilon/2}$ there exists τ_0 such that the trajectories $g_j^{(t,\tau)}(x,\xi), j = 1,2,3,$ for $\tau \leq t \leq \tau_0$ exist and belong to U_0^{ε}. Consider the function $(\lambda_1 - \lambda_2)\left(t, g_2^{(t,0)}(x,\xi)\right)$. Its first derivative with respect to t is $[1,2]\left(t, g_2^{(t,0)}(x,\xi)\right)$, and the second derivative is bounded for $t \leq \tau_0$. Inequality (1.14) implies that the function $(\lambda_1 - \lambda_2)\left(t, g_2^{(t,0)}(x,\xi)\right)$ is monotone increasing for $t \leq \tau_0$. Expanding the function $(\lambda_1 - \lambda_2)\left(t, g_2^{(t,0)}(x,\xi)\right)$ in t up to $O(t^2)$ we prove that there exists $\delta_1 > 0$ such that if $|(\lambda_1 - \lambda_2)(M)| \leq \delta_1$ then the trajectory $g_2^{(t,0)}(x,\xi)$ intersects the manifold Σ_{12} in the moment of time t_1 given by formula

$$t_1 = \{|(\lambda_1 - \lambda_2)|/[1,2]\}(M) + O\left(|(\lambda_1 - \lambda_2)|^2(M)\right) \text{ and } t_1 \leq \tau_0. \quad (2.2)$$

Evidently $|(\lambda_1 - \lambda_2)(M)| \to 0$ as $\varepsilon_0 \to 0$, and $t_1 \to 0$ also. Hence $M_1 \to (0,x_0,\xi_0)$ as $\varepsilon_0 \to 0$, and therefore $|(\lambda_3 - \lambda_1)|(M_1) \to 0$ also. Thus we prove that $M_1 \in U_0^{\varepsilon/2}$ if number ε_0 is sufficiently small.

3.) In what follows we consider the value $(\lambda_3 - \lambda_1)(M_1)$ (the points M_j are defined above in 1.)) and prove that if $|(\lambda_3 - \lambda_1)(M_1)| \leq \delta_1$, $M_1 \in U_0^{\varepsilon/2}$ then it holds that

$$(\lambda_3 - \lambda_1)(M_4) = (\lambda_3 - \lambda_1)(M_1) + O(|(\lambda_3 - \lambda_1)(M_1)|^2). \quad (2.3)$$

First we suppose that the assumptions of item 1.) are valid. In this case applying the considerations of the item 2.) we obtain that

$$t_2 - t_1 = \{|(\lambda_3 - \lambda_1)/[3,1]|\}(M_1) + O\left(|(\lambda_3 - \lambda_1)|^2(M_1)\right) \leq \tau_0, \quad (2.4)$$
$$t_3 - t_2 = \{|(\lambda_3 - \lambda_2)/[3,2]|\}(M_2) + O\left(|(\lambda_3 - \lambda_2)|^2(M_2)\right) \leq \tau_0, \quad (2.5)$$
$$t_4 - t_3 = \{|(\lambda_1 - \lambda_2)/[1,2]|\}(M_3) + O\left(|(\lambda_1 - \lambda_2)|^2(M_3)\right) \leq \tau_0, \quad (2.6)$$

if $|(\lambda_3 - \lambda_1)|(M_1) \leq \delta_1$, $|\lambda_3 - \lambda_2|(M_2) \leq \delta_1$, $|\lambda_1 - \lambda_2|(M_3) \leq \delta_1$ and $M_j \in U_0^{\varepsilon/2}$. In item 2.) we have seen that $|(\lambda_3 - \lambda_1)|(M_1) \to 0$ as $\varepsilon_0 \to 0$. Therefore $t_2 - t_1 \to 0$ and $g_1^{(t_2,t_1)}(x_1,\xi_1) \to (x_0,\xi_0)$ as $\varepsilon_0 \to 0$. Thus we prove that $M_1, M_2 \in U_0^{\varepsilon/2}$ if ε_0 is sufficiently small. Similarly is obtained that $M_j \in U_0^{\varepsilon/2}$ for $j = 1,\ldots,4$. So the number τ_0 from item 2.) stands for number $\tau(\varepsilon)$ from item 1.).

Below we use the identity

$$(\lambda_1 - \lambda_2) + (\lambda_3 - \lambda_1) = (\lambda_3 - \lambda_2). \tag{2.7}$$

Now, as $(\lambda_3 - \lambda_1)(M_2) = 0$,$(\lambda_1 - \lambda_2)(M_1) = 0$, then from (2.7), (2.4) it follows that

$$\begin{aligned} |\lambda_3 - \lambda_2|(M_2) &= |\lambda_1 - \lambda_2|(M_2) \\ &= (t_2 - t_1)\,|[1,2]|\,(M_1) + O\left(|t_2 - t_1|^2\right) \\ &= \frac{|\lambda_3 - \lambda_1|(M_1)\,|[1,2]|\,(M_1)}{|[3,1]|\,(M_1)} + O\left(|(\lambda_3 - \lambda_1)|^2 (M_1)\right). \end{aligned} \tag{2.8}$$

Hence from (2.5) it follows that

$$t_3 - t_2 = \frac{|\lambda_3 - \lambda_1|(M_1)\,|[1,2]|\,(M_1)}{|[3,1]|\,(M_1)\,|[3,2]|\,(M_2)} + O\left(|(\lambda_3 - \lambda_1)|^2 (M_1)\right). \tag{2.9}$$

Equalities $(\lambda_3 - \lambda_1)(M_2) = 0$, $(\lambda_3 - \lambda_2)(M_3) = 0$ and formulas (2.7), (2.9) imply that

$$\begin{aligned} |\lambda_1 - \lambda_2|(M_3) = |\lambda_3 - \lambda_1|(M_3) &= (t_3 - t_2)\,|[3,1]|\,(M_2) + O((t_3 - t_2)^2) \\ &= \frac{|\lambda_3 - \lambda_1|(M_1)\,|[1,2]|\,(M_1)\,|[3,1]|\,(M_2)}{|[3,1]|\,(M_1)\,|[3,2]|\,(M_2)} + O\left(|(\lambda_3 - \lambda_1)|^2 (M_1)\right). \end{aligned} \tag{2.10}$$

Therefore, from formulas (2.9), (2.6) we obtain

$$t_4 - t_3 = \frac{|\lambda_3 - \lambda_1|(M_1)\,|[1,2]|\,(M_1)\,|[3,1]|\,(M_2)}{|[3,1]|\,(M_1)\,|[3,2]|\,(M_2)\,|[1,2]|\,(M_3)} + O\left(|(\lambda_3 - \lambda_1)|^2 (M_1)\right). \tag{2.11}$$

Similarly we have that $|\lambda_3 - \lambda_1|(M_4) = |\lambda_3 - \lambda_2|(M_4) = (t_4 - t_3)\,|[3,2]|\,(M_3) + O(|t_4 - t_3|^2)$, and formula (2.11) implies that

$$\begin{aligned} |\lambda_3 - \lambda_1|(M_4) = &\frac{|\lambda_3 - \lambda_1|(M_1)\,|[1,2]|\,(M_1)\,|[3,1]|\,(M_2)\,|[3,2]|\,(M_3)}{|[3,1]|\,(M_1)\,|[3,2]|\,(M_2)\,|[1,2]|\,(M_3)} \\ &+ O\left(|(\lambda_3 - \lambda_1)|^2 (M_1)\right). \end{aligned} \tag{2.12}$$

Evidently the distances have an estimate

$$\rho(M_{i+1}, M_i) \leq c_\varepsilon (t_{i+1} - t_i), \tag{2.13}$$

and we proved that $|t_{i+1} - t_i| = O(|\lambda_3 - \lambda_1|(M_1))$. Thus

$$\rho(M_1, M_i) \leq c_1\,|\lambda_3 - \lambda_1|(M_1), \text{for } i = 2, 3, 4.$$

It follows from this estimate and formula (2.12) that the formula

$$|\lambda_3 - \lambda_1|(M_4) = |\lambda_3 - \lambda_1|(M_1) + O(|(\lambda_3 - \lambda_1)(M_1)|^2) \tag{2.3}$$

holds true.

4.) Above we prove that the existence of ε_0 such that the assumptions of the item 1.) and the statements of items 2.) and 3.) hold true.

Now in what follows we prove the statements 1, 2, and 3 of the lemma. Evidently in expression (2.3) it holds that $\left|O\left((\lambda_3-\lambda_1)^2(M_1)\right)\right| \leq c_1\left|\lambda_3-\lambda_1\right|^2(M_1)$ for $M_1 \in U_0^{\varepsilon_0}$. In this case we have that

$$\left|\lambda_3-\lambda_1\right|(M_1)(1-c_1\varepsilon_0) \leq \left|\lambda_3-\lambda_1\right|(M_4) \leq \left|\lambda_3-\lambda_1\right|(M_1)(1+c_1\varepsilon_0). \quad (2.14)$$

If the point M_4 belongs to $U_0^{\varepsilon_0} \setminus \Sigma_{123}$ we can set $M_1' = M_4$ and determine the points M_2', M_3', M_4'. We set $M_5 = M_2', M_6 = M_3', M_7 = M_4' \in \Sigma_{12}$ and extend this construction assuming that $M_{3j+1} \in U_0^{\varepsilon_0} \setminus \Sigma_{123}$ for $j = 1, \ldots, N$. From (2.14) it follows that

$$\left|\lambda_3-\lambda_1\right|(M_1)(1-c_1\varepsilon_0)^j \leq \left|\lambda_3-\lambda_1\right|(M_{3j+1}) \leq \left|\lambda_3-\lambda_1\right|(M_1)(1+c_1\varepsilon_0)^j. \quad (2.15)$$

Estimates (2.13), (2.15) and formulas (2.4), (2.5), (2.6) imply that the above-mentioned constructions hold true assuming that $M_1 \in U_0^{\varepsilon_N} \setminus \Sigma_{123}$, where $\varepsilon_N(N+1)(1+c_1\varepsilon_0)^N c_2 < \varepsilon_0$ with some constant c_2 independent on N. Hence there exists $\widetilde{\varepsilon}$ such that if $(0,x;\xi) \in \{\Delta_{12}^- \cap \Delta_{31}^- \cap U_0^{\widetilde{\varepsilon}}\} \setminus \Sigma_{123})$ then $M_1 \in U_0^{\varepsilon_N} \setminus \Sigma_{123}$ and the complete ramification of the trajectory $g_2^{(t,0)}(x,\xi)$ has at least N trisections with the manifold Σ_{23}. All these intersections occur in $[0,T_1]$ where $T_1 = t_1 + (t_2 - t_1) + \cdots + (t_N - t_{N-1}) < \varepsilon_N(N+1)(1+c_1\varepsilon_0)^N c_2 < \varepsilon_0$. Therefore decreasing ε_N we obtain that $T_1 < T$.

In what follows we estimate the distance between the complete ramification and the manifold Σ_{123}. Evidently this distance can be estimated by the minimal value of the function $|\lambda_1-\lambda_2| + |\lambda_3-\lambda_2| + |\lambda_3-\lambda_1|$ on the ramification. Simple manipulations provide that this minimal value is bounded below by $c_3\left|\lambda_3-\lambda_1\right|(M_1)(1-c_1\varepsilon_0)^N$ with some constant c_3 independent on N. □

Now we consider the ramification of the manifolds generated by the complete ramification of the trajectories $g_2^{t,0}(x_0,\xi_0), (x_0,\xi_0) \in \Lambda_0$ in the interval $[0,T]$ in the extended phase space $R^{2n+2}_{x',\xi'}$, where $\xi' = (\xi,\eta)$. We proceed step by step considering a sequence of ramifications. Evidently this ramification of manifolds contains the manifold Λ_2^{n+1} obtained by the displacement of the manifold Λ_0 along the trajectories of the Hamiltonian vector field corresponding to the Hamiltonian λ_2, i.e.,

$$\begin{aligned}\Lambda_2^{n+1} := \Big\{(x',\xi') : (x,\xi) = g_2^{t,0}(x_0,\xi_0), (x_0,\xi_0) \\ \in \Lambda_0; \eta = -\lambda_2\left(t, g_2^{t,0}(x_0,\xi_0)\right), 0 \leq t \leq T\},\end{aligned} \quad (2.16)$$

where $T > \sup_{(x_0,\xi_0)\in\Lambda_0}(t_1(x_0,\xi_0))$.

Lemma 2.1 implies that the trajectories $g_2^{t,0}(x_0,\xi_0)$, $(x_0,\xi_0) \in \Lambda_0$ intersect the manifold Σ_{12} in the moment of time $t_1(x_0,\xi_0)$. Consider the set of the ramified trajectories $g_1^{t,t_1}\left(g_2^{t_1,0}(x_0,\xi_0)\right)$, which start at the points $\left(\widetilde{x},\widetilde{\xi}\right) := g_2^{t_1,0}(x_0,\xi_0)$, $(x_0,\xi_0) \in \Sigma_{21}$ in the moment of time $t_1(x_0,\xi_0), \left(\widetilde{x},\widetilde{\xi}, t_1\right) \in \Sigma_{21}$ and generate the manifold $\Lambda^{n+1}_{2\longrightarrow 1,1}$ defined as

$$\Lambda^{n+1}_{2\longrightarrow 1,1} := \Big\{(x',\xi') : (x,\xi) = g_1^{t,t_1}\left(\widetilde{x},\widetilde{\xi}\right),$$
$$\eta = -\lambda_1\left(t, g_1^{t,t_1}(\widetilde{x},\widetilde{\xi})\right), t_1(x_0,\xi_0) \le t \le T\Big\}.$$

We define manifold $\widetilde{\Sigma}_{12}$ as $\widetilde{\Sigma}_{12} := \Sigma_{12} \times \mathbb{R}^1_\eta$. The following lemma holds true.

Lemma 2.2.

1) *The function* $t_1(x_0,\xi_0)$ *is* $C^\infty(\Lambda_0)$.
2) *The set* $\Lambda_2^{n+1}\cap\widetilde{\Sigma}_{12}$ *is* n*-dimensional infinitely differentiable Lagrangian manifold in* $R^{2n+2}_{x'\xi'}$, *and* $\Lambda^{n+1}_{2\to 1,1}$ *is* $(n+1)$*-dimensional infinitely differentiable Lagrangian manifold.*

Proof. 1.) Define the function $F(t,x_0,\xi_0) := \{\lambda_1 - \lambda_2\}(M_t)$, $M_t := (t, g_2^{t,0}(x_0,\xi_0))$. Then the function $t_1(x_0,\xi_0)$ satisfies the equation $F(t_1,x_0,\xi_0) = 0$. Evidently, condition (1.5) implies that:

$$\partial_t F\left(t,x_0,\xi_0\right)\big|_{t=t_1} = \{\eta+\lambda_1, \eta+\lambda_2\}\left(M_{t_1}\right) \ne 0, \text{ at } M_{t_1} \in \Sigma_{12}.$$

Therefore, the statement of item 1.) follows from the implicit function theorem.

2.) Now we prove that $\Lambda_2^{n+1}\cap\widetilde{\Sigma}_{12}$ is n-dimensional infinitely differentiable manifold in $R^{2n+2}_{x'\xi'}$. Evidently, it holds that

$$\Lambda_2^{n+1}\cap\widetilde{\Sigma}_{12} = \left\{x',\xi' : (x,\xi) = g_2^{t_1,0}(x_0,\xi_0), t = t_1(x_0,\xi_0), \eta = -\lambda_2(x,t_1,\xi)\right\},$$

where $(x_0,\xi_0) \in \Lambda_0$.

The intersection of $(n+1)$-dimensional tangent plane l_D^{n+1} to the manifold Λ_2^{n+1} at the point $D \in \Lambda_2^{n+1}\cap\widetilde{\Sigma}_{12}$ with $(2n+1)$-dimensional tangent plane l_D^{2n+1} to the manifold $\widetilde{\Sigma}_{12}$ at the same point D has a dimension not less than n. The dimension is exactly equal to n if there exists a vector $f : f \in l_D^{n+1}$, such that $f \notin l_D^{2n+1}$. The vector $f = \{\partial_\xi\lambda_2, 1, -\partial_x\lambda_2, -\partial_t\lambda_2\} \in l_D^{n+1}$, but does not belong to l_D^{2n+1} due to condition (1.5). As it holds that for any point $D \in \Lambda_3^{n+1}\cap\widetilde{\Sigma}_{23}$ the dimension of the intersection $l_D^{n+1}\cap l_D^{2n+1}$ is equal to n, then $\Lambda_2^{n+1}\cap\widetilde{\Sigma}_{12}$ is n-dimensional infinitely differentiable manifold. In what follows we prove that $\Lambda_2^{n+1}\cap\widetilde{\Sigma}_{12}$ is the Lagrangian manifold. Let $(x,\xi) = g_2^{t_1,0}(x_0,\xi_0)$. Differential of the variable (x_0,ξ_0) is denoted by $\delta x_0, \delta\xi_0$. Then we have that $dx_i = \delta x_i + \partial_t x_i dt_1$, $d\xi_i = \delta\xi_i + \partial_t\xi_i dt_1$, where $\delta x_i = \sum_{j=1}^n \left(\partial x_i/\partial x_{0j}\right)\delta x_{0j} + \left(\partial x_i/\partial\xi_{0j}\right)\delta\xi_{0j}$, and $\delta\xi_i = \sum_{j=1}^n (\partial\xi_i/\partial x_{0j})\delta x_{0j} + (\partial\xi_i/\partial\xi_{0j})\delta\xi_{0j}$. Direct calculation provides:

$$\left(\sum\nolimits_{i=1}^n dx_i\wedge d\xi_i + dt\wedge d\eta\right)\Big|_{\Lambda_2^{n+1}\cap\widetilde{\Sigma}_{12}} = \sum\nolimits_{i=1}^n \delta x_i\wedge\delta\xi_i + \sum\nolimits_{i=1}^n (\partial_t x_i)dt_1\wedge\delta\xi_i$$
$$-\sum\nolimits_{i=1}^n (\partial_t\xi_i)dt_1\wedge\delta x_i - \sum\nolimits_{i=1}^n (\partial_{x_i}\lambda_2)dt_1\wedge\delta x_i - \sum\nolimits_{i=1}^n (\partial_{\xi_i}\lambda_2)dt_1\wedge\delta\xi_i.$$

As the diffeomorphism g_2^{t0} is a canonical one, then $\sum_{i=1}^n \delta x_i\wedge\delta\xi_i = 0$ at Λ_2^{n+1}; furthermore it holds that $\partial_t x_i = \partial_\xi\lambda_2, \partial_t\xi_i = -\partial_x\lambda_2$. Therefore, the other terms in the previous equality cancel and we obtain that the restriction of the symplectic form over the manifold $\Lambda_2^{n+1} \cap \widetilde{\Sigma}_{12}$ is equal to zero, that is the manifold $\Lambda_2^{n+1} \cap \widetilde{\Sigma}_{12}$ is the Lagrangian manifold. At any point $D \in \Lambda_2^{n+1} \cap \widetilde{\Sigma}_{12}$ the vector $\{\partial_\xi \lambda_1, 1, -\partial_x \lambda_1, -\partial_t \lambda_1\}$ does not belong to the tangent plane l_D^{2n+1} due to condition (1.5). Therefore, this vector is not tangent to the manifold $\Lambda_2^{n+1} \cap \widetilde{\Sigma}_{12}$ as well. Hence $\Lambda_{2\to 1,1}^{n+1}$ is $(n+1)$-dimensional infinitely differentiable Lagrangian manifold. □

Now we follow the considerations of Lemma 2.2. Note that the intersection of the Lagrangian manifold $\Lambda_{2\to 1,1}^{n+1}$ with Σ_{31} is n-dimensional infinitely differentiable Lagrangian manifold. Hence the manifold $\Lambda_{1\to 3,1}^{n+1}$ defined as

$$\Lambda_{1\to 3,1}^{n+1} := \{(x', \xi') : (x, \xi) = g_3^{t,t_2}(x_0, \xi_0), (x_0, t_2, \xi_0, \eta) \in \Lambda_{2\to 1,1}^{n+1} \cap \Sigma_{31}; \eta = -\lambda_3\left(t, g_3^{t,t_2}(x_0, \xi_0)\right), t_2 \le t \le T\},$$

is $(n+1)$-dimensional Lagrangian C^∞-manifold. In a similar manner $(n+1)$-dimensional Lagrangian C^∞-manifold $\Lambda_{3\to 2,1}^{n+1}$ is generated in $R_{x',\xi'}^{2n+2}$ by the Hamiltonian flow of the Hamiltonian $(\eta + \lambda_2)$ with the origin on the n-dimensional Lagrangian C^∞-manifold $\Lambda_{1\to 3,1}^{n+1} \cap \Sigma_{32}$; similarly $(n+1)$-dimensional Lagrangian C^∞-manifold $\Lambda_{2\to 1,2}^{n+1}$ is generated in $R_{x',\xi'}^{2n+2}$ by the Hamiltonian flow of the Hamiltonian $(\eta + \lambda_1)$ with the origin on the n-dimensional Lagrangian C^∞-manifold $\Lambda_{3\to 2,1}^{n+1} \cap \Sigma_{32}$, and so on. Therefore if the distance between the set Σ_{123} and the complete ramification of all trajectories $g_2^{(t,0)}(x_0, \xi_0), (x_0, \xi_0) \in \Lambda_0$ in the interval $[0, T]$ is positive, then the complete ramification is a finite sum of the $(n+1)$-dimensional Lagrangian C^∞-manifolds $\Lambda_{k\to j,i}^{n+1}$ generated in $R_{x',\xi'}^{2n+2}$ by the Hamiltonian flow of the Hamiltonian $(\eta + \lambda_j)$ with the origin on the n-dimensional Lagrangian C^∞-manifold $\Lambda_{m\to k,l}^{n+1} \cap \Sigma_{kj}$, where $i = l$ or $i = l+1$. The distance in $R_{x',\xi'}^{2n+2}$ between all the manifolds $\Lambda_{m\to k,1}^{n+1}, \Lambda_{m\to k,2}^{n+1}, \Lambda_{m\to k,3}^{n+1}, \ldots$ is positive as they start on the manifold Σ_{mk} in disjoint intervals of time.

Lemma 2.3. *For every $T > 0$ and every integer number N there exist $\varepsilon_0 : 0 < \varepsilon_0 < \varepsilon$; time $T_1 : 0 < T_1 < T$; C_0^∞-function $\phi(x)$ and a point $\overline{\xi} \neq 0$ such that:*

1. *Lagrangian manifold* (1.17) *satisfies conditions* (1.18), (1.19)
2. *The closure of the complete ramification of every trajectory $g_2^{(t,0)}(x, \xi)$, $(x, \xi) \in \Lambda_0$ in the interval $[0, T_1]$ is at positive distance from Σ_{123} and belongs to $U_0^{\varepsilon/2}$.*
3. *The complete ramification of every trajectory $g_2^{(t,0)}(x, \xi)$, $(x, \xi) \in \Lambda_0$ in the interval $[0, T_1]$ intersects the manifold Σ_{32} at least N times.*
4. *The complete ramification of every trajectory $g_2^{(t,0)}(x, \xi)$, $(x, \xi) \in \Lambda_0$ it the interval $[0, T_1]$ is a sum of finite number K of $(n+1)$-dimensional Lagrangian C^∞-manifolds $\Lambda_{m\to k.i}^{n+1}$ and $(n+1)$-dimensional Lagrangian C^∞-manifold Λ_2^{n+1}.*

The proof of Lemma 2.3 follows directly from Lemma 2.1 and Lemma 2.2.

3. Proof of the theorem

1. *We assume that on all the Lagrangian manifolds* $\Lambda_2^{n+1}, \Lambda_{m\to k,l}^{n+1}$ *it is possible to choose local coordinates* $(t, x_1, \ldots, x_n)$. *The general case is considered in the same manner employing Maslov canonical operator. Further suppose* $r = 1$ *and the substitute* $u = D(x', -i\partial_x)\, v$ *has reduced the symbol of equation* (1.1) *to form* (1.4).

To calculate FAS of Cauchy problem (1.1), (1.15) with high precision it is necessary to consider the manifolds $\Lambda_1^{n+1}, \Lambda_3^{n+1}$ in $R^{2n+2}_{x',\xi'}$ generated by the displacement of the manifold Λ_0 along the trajectories of Hamiltonian vector fields corresponding to the Hamiltonians λ_1, λ_3. To simplify the proof of Theorem 1.4 in what follows we slightly change initial data (1.15) in such a way that the manifolds $\Lambda_1^{n+1}, \Lambda_3^{n+1}$ can be neglected in the considerations. Furthermore we assume that the support of the initial data is in $h^{1/2-\delta/2}$-vicinity (for $0 < \delta < 1/2$) of a point y, and that the point M in Figure 1 is: $M := (t = 0, x = y, \xi = \overline{\xi}), M \notin \Sigma_{123}$. Recall that the trajectory $g_2^{(t,0)}(y, \overline{\xi})$ in the moment of time $t_1(y)$ intersects Σ_{12} at the point $M_1 = (t_1, x_1, \xi_1)$, the trajectory $g_1^{(t,t_1)}(x_1, \xi_1)$ in the moment of time $t_2(y)$ intersects Σ_{31} at the point $M_2 = (t_2, x_2, \xi_2)$, the trajectory $g_3^{(t,t_2)}(x_2, \xi_2)$ in the moment of time $t_3(y)$ intersects Σ_{32} at the point $M_3 = (t_3, x_3, \xi_3)$, and the trajectory $g_2^{(t,t_3)}(x_3, \xi_3)$ in the moment of time $t_4(y)$ intersects Σ_{12} at the point $M_4 = (t_4, x_4, \xi_4)$, and so on.

2. FAS/$O(h^M)$ of the Cauchy problem is calculated step by step. Consider in the space $R^{2n+1}_{x',\xi}$ all $h^{1/2-\delta}$-vicinities of the intersections: $\Lambda_2^{n+1} \cap \widetilde{\Sigma}_{21}$; $\Lambda_{2\to 1,1}^{n+1} \cap \widetilde{\Sigma}_{21}, \Lambda_{2\to 1,1}^{n+1} \cap \widetilde{\Sigma}_{31}$; $\Lambda_{1\to 3,1}^{n+1} \cap \widetilde{\Sigma}_{31}, \Lambda_{1\to 3,1}^{n+1} \cap \widetilde{\Sigma}_{32}$; etc. for $0 < \delta < 1/2$. These vicinities are denoted as $U\left(\Lambda_{m\to k,i}^{n+1} \cap \widetilde{\Sigma}_{kj}, h^{1/2-\delta}\right)$. Outside these vicinities the WKB-Maslov [7] method can be applied to construct a FAS/$\Omega, O(h^M)$ of the Cauchy problem as a sum of Maslov operators on the Lagrangian manifolds $\Lambda_{m\to k,i}^{n+1}$. However, near the points of intersections $\Lambda_{m\to k,i}^{n+1} \cap \widetilde{\Sigma}_{mj}$ and $\Lambda_{m\to k,i}^{n+1} \cap \widetilde{\Sigma}_{kj}$ the Maslov construction [7], [8] should be modified. In the vicinities $U\left(\Lambda_{2\to 1,i}^{n+1} \cap \widetilde{\Sigma}_{12}, h^{1/2-\delta}\right)$, $U\left(\Lambda_{2\to 1,i}^{n+1} \cap \widetilde{\Sigma}_{31}, h^{1/2-\delta}\right)$ symbol (1.3) does not have the Jordan block, hence the multiphasic method developed in the paper [3] can be applied. Note that in the vicinities $U\left(\Lambda_{1\to 3,i}^{n+1} \cap \widetilde{\Sigma}_{32}, h^{1/2-\delta}\right)$ symbol (1.3) has a Jordan block of the second order. Therefore it is possible to use a modification of the method developed in [10].

3. Assume that the symbol of system (1.1) is reducible to form (1.4). Denote by π_τ the projection of $R^{2n+2}_{x',\xi'}$ onto $R^{2n}_{x,\xi} \times \{t = \tau\}$, and let $(\Lambda_2^{n+1})_t := \pi_t \Lambda_2^{n+1}$. Suppose that $\pi_t \Lambda_2^{n+1} \subset \Delta_{12}^{-int} \cap \Delta_{31}^{-int}$ for $t \in [0, \overline{t}_1]$. The linear subspace of functions of the space $C^\infty\left(\Lambda_2^{n+1} \cap \{t : 0 \le t \le \overline{t}_1\}\right)$ such that their restrictions on $(\Lambda_2^{n+1})_t$ belong to $C_0^\infty\left((\Lambda_2^{n+1})_t\right)$ for $t \in [0, \overline{t}_1]$ is denoted by $\widetilde{C}_0^\infty\left(\Lambda_2^{n+1} \cap \{t : 0 \le t \le \overline{t}_1\}\right)$. Evidently $\overline{t}_1$ can be taken as $t_1(y) - h^{1/2-\delta/2}$. Furthermore, let π_x be the projection of $R^{2n+2}_{x',\xi'}$ onto R^n_x; $\pi_{x'}$ be the projection of $R^{2n+2}_{x',\xi'}$ onto $R^{n+1}_{x'}$ and let a set Ω_0

be defined as $\Omega_0 := \left[\Lambda_2^{n+1} \cap \{t : 0 \le t \le \bar{t}_1\}\right]$. Following the WKB-Maslov method [7], [8] we calculate a function

$$\varphi(h) = \sum_{j=0}^{M+n} h^j \varphi_j, \ \varphi_j \in \widetilde{C}_0^\infty((\Lambda_2^{n+1} \cap \{t : 0 \le t \le \bar{t}_1\}) \tag{3.1}$$

such that the Maslov operator $K_{\Lambda_2^{n+1}}\varphi(h)$ provides a FAS$/\pi_{x'}\Omega_0, O(h^M)$. The vector functions φ_j are calculated from a recurrent systems of differential equations on the manifold $\Lambda_2^{n+1} \cap \{t : 0 \le t \le \bar{t}_1\}$ [8]. The function $\varphi_0 \in \widetilde{C}_0^\infty(\Lambda_2^{n+1})$ is a solution of so-called transport equation [3], [8] with the initial data $\left(\varphi_0|_{(\Lambda_2^{n+1})_0}\right) \circ \pi_x^{-1} = \phi(x)\, e_2$ (here the function $\phi(x)\, e_2$ is the amplitude in data (1.15)). This solution is unique and [8] it holds that

$$\begin{aligned}\varphi_0(t, x_0) = \phi(x_0)\, e_2|_{\Lambda_2^{n+1}} \exp\Bigg\{ \int_0^t \Bigg[\frac{1}{2} \sum_{k=1}^{n} \partial^2_{x_k \xi_k} \lambda_2(\tau, x, \xi) \\ -i \left(b^1_{22} - \frac{db^1_{23}}{\lambda_2 - \lambda_3} \right) (\tau, x, \xi) \Bigg] \circ g_2^{(\tau,0)}(x_0, \bar{\xi}) d\tau \Bigg\}.\end{aligned} \tag{3.2}$$

Direct calculations imply that the function $K_{\Lambda_2^{n+1}}\varphi(h)$ satisfies perturbed initial data (1.15)

$$K_{\Lambda_2^{n+1}}\varphi(h)\Big|_{t=0} = \phi_0(x, h) \exp\left\{ \frac{i}{h}(x, \bar{\xi}) \right\}, \text{ where}$$

$$\phi_0(x, h) = \phi(x)\, e_2 + \sum_{j=1}^{M+n} h^j \psi_j; \ \psi_j \in C_0^\infty(\mathbb{R}^n), \text{supp } \psi_j \subseteq \text{supp } \phi. \tag{3.3}$$

In what follows we use corrected initial amplitude (3.3).

4. In the vicinity $U\left(\Lambda_2^{n+1} \cap \widetilde{\Sigma}_{12}, h^{1/2-\delta}\right)$ it holds that $\lambda_2 \neq \lambda_3$ and $\lambda_1 = \lambda_2$ on Σ_{12}. Hence symbol (1.3) does not have the Jordan block. Therefore the formal asymptotic solution in the domain $\pi_{x'} U\left(\Lambda_2^{n+1} \cap \widetilde{\Sigma}_{12}, h^{1/2-\delta}\right)$ can be calculated following the paper [3], [16]. This formal asymptotic solution in the intersection $U\left(\Omega_0, h^{1/2-\delta/2}\right) \cap \pi_{x'} U\left(\Lambda_2^{n+1} \cap \widetilde{\Sigma}_{12}, h^{1/2-\delta}\right)$ (here the set Ω_0 is defined in item 3.) is equal to the FAS$/\pi_{x'}\Omega_0, O(h^M)$, obtained above in item 3., with a precision $O(h^M)$. Furthermore, applying the stationary phase method and its estimations [15] to the formal asymptotic solution in the domain $\pi_{x'} U\left(\Lambda_2^{n+1} \cap \widetilde{\Sigma}_{12}, h^{1/2-\delta}\right)$ [3], we obtain that this formal asymptotic solution in the domain

$$\begin{aligned}&\pi_{x'} \left[\Lambda_2^{n+1} \cap \left\{t : t_1(y) - h^{1/2-\delta/2} \le t \le t_1(y) + h^{1/2-\delta/2}\right\}\right] \\ &\cup \pi_{x'} \left[\Lambda_{2\to1,1}^{n+1} \cap \left\{t : t_1(y) - h^{1/2-\delta/2} \le t \le t_1(y) + h^{1/2-\delta/2}\right\}\right]\end{aligned}$$

can be presented, with the precision $O(h^M)$, as a sum of canonical Maslov operators on the manifolds $\Lambda_2^{n+1} \cap \Delta_{12}^+ \cap \Delta_{31}^-$ and $\Lambda_{2\to1,1}^{n+1} \cap \Delta_{12}^+ \cap \Delta_{31}^-$. Therefore, this formally asymptotic solution can be extended. As a result of the extension the function FAS/$\pi_{x'}\ \Omega_1, O(h^M)$ is obtained by WKB-Maslov method, here

$$\Omega_1 := \left[\Lambda_2^{n+1} \cap \left\{t : t_1(y) + h^{1/2-\delta/2} \le t \le T\right\}\right]$$
$$\cup\, \pi_{x'} \left[\Lambda_{2\to 1,1}^{n+1} \cap \left\{t : t_1(y) + h^{1/2-\delta/2} \le t \le t_2(y) - h^{1/2-\delta/2}\right\}\right].$$

The function FAS/$\pi_{x'}\ \Omega_1, O(h^M)$ is the sum of following canonical operators

$$K_{\Lambda_2^{n+1}} \varphi^{(2)}(h) + K_{\Lambda_{2\to 1,1}^{n+1}} \varphi^{(2\to 1,-)}(h). \tag{3.4}$$

Here: a) it holds that the amplitude

$$\varphi^{(2)}(h) = \sum_{j=0}^{M+n} h^j \varphi_j^{(2)};\ \varphi_0^{(2)} = \varphi_0,$$

where φ_0 were constructed in item 3. and

$$\varphi_j^{(2)} \in \widetilde{C}_0^\infty \left(\Lambda_2^{n+1} \cap \{t : t_1(y) + h^{1/2-\delta/2} \le t \le T\}\right).$$

Here $\sup\left|\partial_{x'}^\beta \varphi_j^{(2)}\right| \le c_{j,\beta} h^{-j(1/2+\delta)-|\beta|\delta}$ for $j \ge 2$, and for $j = 1$ it holds that $\sup\left|\partial_{x'}^\beta \varphi_1^{(2)}\right| \le c_{j,\beta} \left|\ln h\right|^{m(\beta)}$.

b) the amplitude $\varphi^{(2\to 1,-)}(h)$ has the form

$$\varphi^{(2\to 1,-)}(h) = \sqrt{h} \sum_{j=0}^{M+n} h^j \varphi_j^{(2\to 1.-)}, \tag{3.5}$$

where

$$\varphi_j^{(2\to 1,-)} \in \widetilde{C}_0^\infty \left[\Lambda_{2\to 1,1}^{n+1} \cap \left\{t : t_1(y) + h^{1/2-\delta/2} \le t \le t_2(y) - h^{1/2-\delta/2}\right\}\right];$$

the function $\varphi_0^{(2\to 1,-)}$ does not depend on h and

$$\sup\left|\partial_{x'}^\beta \varphi_j^{(2\to 1,-)}\right| \le c_{j,\beta} h^{-j(1/2+\delta)-|\beta|\delta}, \quad \text{for } j \ge 1.$$

Now we determine the function $\varphi_0^{(2\to 1,-)}$ explicitly. Let vector e_1 be $e_1 = T(x',\xi)(1,0,0)^t$. From [3] it follows that with the precision up to the factor $e^{i\beta}$, $\left|e^{i\beta}\right| = 1$, it holds that

$$\varphi_0^{(2\to 1,-)} = e_1|_{\Lambda_{2\to 1,1}^{n+1}} \times e^{i\beta} \varphi_0(t_1(x_0), x_0) b_{12}^1 \left(t_1(x_0), g_2^{t_1(x_0),0}(x_0,\overline{\xi})\right) \tag{3.6}$$
$$\times \frac{\sqrt{\left|D\{x \circ g_1^{t,t_1(x_0)} \circ g_2^{t_1(x_0),0}(x_0,\overline{\xi})\}/Dx_0\right|}}{\sqrt{\left|D\{x \circ g_1^{t,\tau} \circ g_2^{\tau,0}(x_0,\overline{\xi})\}/Dx_0\right|}\Big|_{\tau=t_1(x_0)}} \times \frac{1}{\sqrt{|[1,2]| \circ g_2^{t_1(x_0),0}(x_0,\overline{\xi})}}$$
$$\exp\left\{\int_{t_1(x_0)}^t \left[\frac{1}{2}\sum_{k=1}^{k=n} \partial_{x_k\xi_k}^2 \lambda_1(\tau,x,\xi) - i b_{11}^1(\tau,x,\xi)\right] \circ g_1^{\tau,t_1(x_0)} \circ g_2^{t_1(x_0),0}(x_0,\overline{\xi}) d\tau\right\}.$$

5. In the vicinity $U\left(\Lambda^{n+1}_{2\to1,1}\cap\widetilde{\Sigma}_{13},h^{1/2-\delta}\right)$ it holds that $\lambda_2\neq\lambda_3$ and $\lambda_1=\lambda_3$ on Σ_{13}. Hence symbol (1.3) does not have the Jordan block. Therefore the formal asymptotic solution in the domain $\pi_{x'}U\left(\Lambda^{n+1}_{2\to1,1}\cap\widetilde{\Sigma}_{13},h^{1/2-\delta}\right)$ can be calculated following the paper [3]. This formal asymptotic solution in the intersection $\pi_{x'}\Omega_1\cap\pi_{x'}U\left(\Lambda^{n+1}_{2\to1,1}\cap\widetilde{\Sigma}_{13},h^{1/2-\delta}\right)$ is equal to the FAS/$\pi_{x'}\Omega_1,O(h^M)$ obtained above in item 4., with the precision $O(h^M)$. Thus the considerations of item 4. are valid and the function FAS/$\pi_{x'}\Omega_2,O(h^M)$, where

$$\Omega_2:=\left[\Lambda^{n+1}_{2\to1,1}\cap\{t:t_2(y)+h^{1/2-\delta/2}\le t\le T\}\right]$$
$$\cup\,\pi_{x'}\left[\Lambda^{n+1}_{1\to3,1}\cap\{t:t_2(y)+h^{1/2-\delta/2}\le t\le t_3(y)-h^{1/2-\delta/2}\}\right]$$

is a sum of the following canonical operators

$$K_{\Lambda^{n+1}_{2\to1,1}}\varphi^{(2\to1,+)}(h)+K_{\Lambda^{n+1}_{1\to3,1}}\varphi^{(1\to3,-)}(h). \tag{3.7}$$

The amplitudes $\varphi^{(2\to1,+)}(h)$ and $\varphi^{(1\to3,-)}(h)$ in formula (3.7) are similar to the amplitudes $\varphi^{(2)}(h)$ and $\varphi^{(2\to1,-)}(h)$ in formula (3.4) respectively. Hence it holds that

$$\varphi^{(1\to3,-)}(h)=h\sum_{j=0}^{M+n}h^j\varphi_j^{(1\to3,-)},$$

where

$$\varphi_j^{(1\to3,-)}\in\widetilde{C}_0^\infty\left[\Lambda^{n+1}_{1\to3,1}\cap\{t:t_2(y)+h^{1/2-\delta/2}\le t\le t_3(y)-h^{1/2-\delta/2}\}\right];$$

$\varphi_0^{(1\to3,-)}$ does not depend on h and

$$\sup\left|\partial^\beta_{x'}\varphi_j^{(1\to3,-)}\right|\le c_{j,\beta}h^{-j(1/2+\delta)-|\beta|\delta},\quad\text{for } j\ge1.$$

The function $\varphi_0^{(1\to3,-)}$ has a representation with the precision up to the factor $e^{i\theta},|e^{i\theta}|=1$ similar to (3.6). Let e_3 be equal to $T(x',\xi)\left(0,\frac{d}{\lambda_3-\lambda_2},1\right)^t$, and let e_1^* be equal to $(T^*)^{-1}(x',\xi)(1,0,0)^t$, then

$$\varphi_0^{(1\to3.-)}=e^{i\theta}(\varphi_0^{(2\to1,-)},e_1^*|_{\Lambda^{n+1}_{2\to1,1}})(t_2(x_0),x_0) \tag{3.8}$$
$$\times\, e_3|_{\Lambda^{n+1}_{1\to3,1}}\times b^1_{31}\left(t_2(x_0),g_1^{t_2(x_0),t_1(x_0)}\circ g_2^{t_1(x_0),0}(x_0,\overline{\xi})\right)$$
$$\times\frac{\sqrt{\left|D\left\{x\circ g_3^{t,t_2(x_0)}\circ g_1^{t_2(x_0),t_1(x_0)}\circ g_2^{t_1(x_0),0}(x_0,\overline{\xi})\right\}/Dx_0\right|}}{\sqrt{\left|D\left\{x\circ g_3^{t,\tau}\circ g_1^{\tau,t_1(x_0)}\circ g_2^{t_1(x_0),0}(x_0,\overline{\xi})\right\}/Dx_0\right|}\Big|_{\tau=t_2(x_0)}}$$

$$\times \frac{1}{\sqrt{\|[1,2]\| \circ g_1^{t_2(x_0),t_1(x_0)} \circ g_2^{t_1(x_0),0}(x_0,\overline{\xi})}} \exp\left\{ \int_{t_2(x_0)}^{t} \left[\frac{1}{2} \sum_{k=1}^{n} \partial^2_{x_k \xi_k} \lambda_3(\tau,x,\xi) \right.\right.$$

$$\left.\left. - i \left(b^1_{32} \frac{\lambda_3 - \lambda_2}{d} + b^1_{33} \right) (\tau, x, \xi) \right] \circ g_3^{\tau,t_2(x_0)} \circ g_1^{t_2(x_0),t_1(x_0)} \circ g_2^{t_1(x_0),0}(x_0,\overline{\xi}) d\tau \right\}.$$

The application of formulas (3.2), (3.6) and (3.8) yields an estimate

$$\left| \varphi_0^{(1\to 3,-)}(t,x_0) \right| \geq |\phi(x_0)| \, e^{-ct}. \tag{3.9}$$

6. Now we extend the function FAS/$\pi_{x'}$ $\Omega_2, O(h^M)$ through Σ_{32} to obtain a formal asymptotic solution in $\pi_{x'} U \left(\Lambda^{n+1}_{1\to 3,1} \cap \widetilde{\Sigma}_{12}, h^{1/2-\delta} \right)$. In this vicinity it holds that $\lambda_1 \neq \lambda_2, \lambda_1 \neq \lambda_3$, and $\lambda_2 = \lambda_3$ on $\widetilde{\Sigma}_{32}$. Hence symbol (1.4) has a Jordan block on $\widetilde{\Sigma}_{32}$. To calculate the FAS we modify the representation of the asymptotic solution.

a) Toward this end it is convenient to modify principal symbol (1.4). For that we introduce an operator $U_2(t,\tau)$ that is the solution of the Cauchy problem

$$ih\partial_t U_2(t,\tau) = \{\lambda_2(x,t,-ih\partial_x) + \lambda_2^*(x,t,-ih\partial_x)\} U_2(t,\tau); \; U_2(\tau,\tau) = I, t \geqq \tau. \tag{3.10}$$

Evidently, the operator $U_2(\tau,t)$ such that

$$-ih\partial_\tau U_2(\tau,t) = \{\lambda_2(x,\tau,-ih\partial_x) + \lambda_2^*(x,\tau,-ih\partial_x)\} U_2(\tau,t); \; U_2(t,t) = I, t \geqq \tau,$$

is the inverse operator of $U_2(t,\tau)$. In what follows we use the asymptotic expansions $U_2^{as}(t,\tau)$ and $U_2^{as}(\tau,t)$ of these operators with respect to the small parameter h [8]. The FAS solution u of Cauchy problem (1.1), (1.15) is presented in the form $u = U_2^{as}(t,0)v$. Hence the function v satisfies with the precision $O\left(h^M\right)$ the equation

$$(-ih) \frac{\partial v}{\partial t} + \widehat{B}^U \left(x', -ih\frac{\partial}{\partial x}, h \right) v = 0. \tag{3.11}$$

with the initial data of form (1.15). Following the papers [13], [14] we obtain a matrix-valued symbol $B^U(x',\xi,h)$ of the h^{-1}-P.D.O. $\widehat{B}^U$ (actually the operator $\widehat{B}^U$ is obtained by commutation of a P.D.O. with a Fourier integral operator [9])

$$B^U(x',\xi,h) = B_0^U(x',\xi) + hB_1^U(x',\xi) + \cdots + h^j B_j^U(x',\xi) + \cdots . \tag{3.12}$$

It holds that $\left|\partial^\alpha_{x'} \partial^\beta_\xi B^U_j(x',\xi)\right| \leq c_{\alpha\beta}(\gamma)(1+|\xi|)$ for $|\xi| \geq \gamma > 0$. Consider the first two terms in formula (3.12). The notation $\widetilde{f}(x',\xi)$ signifies that the functions are valued at the argument $\left\{t, g_2^{t,0}(x,\xi)\right\}$. For example $\widetilde{\lambda}_j(t,x,\xi) := \lambda_j\left(t, g_2^{t,0}(x,\xi)\right)$. Then it holds that

$$B_0^U(x',\xi) := \left\| \widetilde{b}^0_{ij}(x',\xi) \right\| - \widetilde{\lambda}_2 I;$$

$$(B_1^U(x',\xi))_{ij} = \left(\widetilde{b}^1(x',\xi,h) \right)_{ij},$$

for $i \neq j$ and $(i,j) \neq (r+1, r+2)$. We set

$$B_*^U(x',\xi) := B_1^U(x',\xi) + \cdots + h^{j-1}B_j^U(x',\xi) + \cdots .$$

The substitution $v = U_2^{as}(t,0)u$ generates a canonical transformation G of the phase space: $G(t,x,\xi,,\eta) := \left\{t, g_2^{t,0}(x,\xi), \eta - \lambda_2\left(t, g_2^{t,0}(x,\xi)\right)\right\}$. In what follows we use new Hamiltonians $\lambda_k' := \widetilde{\lambda}_k - \widetilde{\lambda}_2, (\lambda_2' = 0)$ and as above introduce the sets $\Sigma_{ij}', \Sigma_{123}'$. As the transformation G is canonical in the space $R_{x',\xi'}^{2n+2}$, then we have that

$$\begin{aligned} &\{\eta + \lambda_i', \eta + \lambda_j'\}\,|_{\Sigma_{ij}' \cap \{|\xi|=1\}} \\ &\quad = \{\eta + \lambda_i, \eta + \lambda_j\}\,|_{\Sigma_{ij} \cap \{|\xi|=1\}} \neq 0. \end{aligned} \tag{3.13}$$

Furthermore it holds

$$\partial_t \lambda_j'\big|_{\Sigma_{2j}' \cap \{|\xi|=1\}} = \{\eta + \lambda_j, \eta + \lambda_2\}|_{\Sigma_{2j} \cap \{|\xi|=1\}} \neq 0. \tag{3.14}$$

Now we can construct manifolds similar to $\Lambda_{i \to j,k}^{n+1}$ using the Hamiltonians λ_j' and initial manifold $g_2^{0,-t}\Lambda_0$. In the notation of such manifolds similar to $\Lambda_{i \to j,k}^{n+1}$ we include the symbol $'$, i.e., the notation $\Lambda_{i \to j,k}^{n+1,\prime}$ is used. Evidently $\Lambda_{i \to j,k}^{n+1,\prime} = G^{-1}\Lambda_{i \to j,k}^{n+1}$.

b) A spacial choice of the local coordinates on the manifolds $\Lambda_{m \to k,i}^{n+1,\prime}$ is needed. First for simplicity sake assume that there exists a vicinity U_3 of the point $q \in \Lambda_{1 \to 3,1}^{n+1,\prime} \cap \Sigma_{32}'$ on the Lagrangian manifold $\Lambda_{1 \to 3,1}^{n+1}$ with two sets of local coordinates: (t,x) and (x,η). That is $\left.\frac{D(t,x)}{D(\eta,x)}\right|_{U_3} \neq 0$. Therefore the projections $\pi_{t,x} : R_{x',\xi'}^{2n+2} \to R_{x'}^{n+1}$, $\pi_{x,\eta} : R_{x',\xi'}^{2n+2} \to R_{x,\eta}^{n+1}$ are coordinate diffeomorphisms in this vicinity. (In addition II is proved that general case can be reduced to this one.) Consider the Jacobian determinant $J_3 := |D(x,\eta)/D(x_0,t)|$ on the manifold $\Lambda_{1 \to 3,1}^{n+1,\prime}$, and let $S_3(x,t)$ be the phase function in U_3 such that

$$dS_3(x,t) = \left.\left(\sum_{j=1}^{n} \xi_j dx_j - \lambda_3' dt\right)\right|_{\Lambda_{1 \to 3,1}^{n+1,\prime}} . \tag{3.15}$$

Evidently, it holds that $\partial_t S_3(x,t) = \eta\big|_{U_3} = -\lambda_3'|_{U_3}$ and the equation of intersection $U_3 \cap \Sigma_{32}'$ in the local coordinates (x,η) has the form $\eta = 0$. We suppose that the vicinity U_3 has the diameter $h^{1/2-\delta}$. Define in U_3 the Legendre transformation of the phase function

$$\widehat{S}_3(x,\eta) = S_3(x,t) - t\eta, \eta \circ \pi_{x,t}^{-1} = \partial_t S_3(x,t)$$

and assume that the vector amplitude function φ has form $\varphi = \omega(\eta,x,h)\chi(x,\eta)$. Here the function $\omega(\eta,x,h) \in C^\infty(\mathbb{R}_x^n \times \mathbb{C}\backslash\{\eta = 0\})$ is analytic with respect to the variable η in the vicinity of the point $\eta = 0$ cut by some ray radiated from the point $\eta = 0$ and $\chi(x,\eta) \in C_0^\infty(\pi_{x,\eta}U_3)$. The function $\omega(\eta,x,h)$ is calculated below. In what follows we perform the integration over the contour C in the complex plane η. Here the contour C belongs to the real axis except outside a small vicinity of the point $\eta = 0$. The contour C bypasses from below the point $\eta = 0$ by the circle $r = h^{1/2+\varepsilon}$; and over the contour C it holds that $|\mathrm{Im}\,\eta|_C \le h^{\varepsilon+1/2}$. By definition $\arg\eta = -\pi$ if $\eta < 0$ and $\arg\eta = 0$ if $\eta > 0$. Evidently, over this contour C it holds that $|h/\eta|_C \le h^{1/2-\varepsilon}$. Hence over the contour C this ratio is a small parameter. In order to apply deformations to the contour in the complex plane η we use almost analytic expansion of amplitude, that is for example, the function $\chi(x,\eta)$ at the point $\eta_1 + i\eta_2$ is defined as a sum: $\chi(x,\eta) = \sum_{k=0}^{N} \frac{(i\eta_2)^k}{k!} \partial_{\eta_1}^k \chi(x,\eta_1)$. Using the Green's formula we obtain that a deformation of the contour provides an error of order $h^{N\mu}$ if during the deformation it holds that $|\eta_2| < h^{\mu}$ and if during the process of deformation the amplitude φ is bounded. In the map U_3 over the functions $\varphi = \omega(\eta,x,h)\chi(x,\eta)$ the modified Maslov operator $K(U_3)$ is defined as

$$K(U_3)(\varphi\circ\pi_{x,\eta}) := \frac{\exp\{\frac{i\pi}{4} - i\frac{\pi}{2}\widehat{\gamma}(q)\}}{\sqrt{2\pi h}} U_2^{as}(t,0)\int_C \exp\left\{\frac{i}{h}[t\eta + \widehat{S}_3(x,\eta)]\right\}\frac{\varphi}{\sqrt{J_3}}d\eta, \tag{3.16}$$

here $\widehat{\gamma}(q)$ is an integer number connected with the Maslov index [8] on $\Lambda^{n+1,\prime}_{1\to3,1}$. Detailed analysis of asymptotic of integral (3.16) as $h \to 0$, implies that the contribution of the saddle points in the asymptotic of integral (3.16) gives rise to a wave with a wave front on the manifold $\Lambda^{n+1,\prime}_{1\to3,1}$; the contribution of the singular point $\eta = 0$ in the asymptotic of the integral (3.16) gives rise to a wave with a wave front on the manifold $\Lambda^{n+1,\prime}_{3\to2,1}$ [15].

c) In what follows we determine the equation for the amplitude φ in the domain $\pi_{x,\eta}U_3$. The manifold $\Sigma'_{32} \cap U_0^{\varepsilon_0}$ can be presented in the form

$$\Sigma'_{32} \cap U_0^{\varepsilon_0} = \left\{(x',\xi) : t = f(x,\xi), f \in S^0\right\}. \tag{3.17}$$

In the map U_3 it holds that $\left|(t - f(x,\xi))|_{U_3}\right| \le h^{(1/2-\varepsilon)}$. Therefore, with the precision $O(h^M)$ it is possible to replace all the symbols by their Taylor expansions with respect to $(t - f(x,\xi))$ up to $O\left(|t - f(x,\xi|^{3M}\right)$. Applying the Fourier transform with respect to the variable t to equation (3.11) we obtain an equation in "x,η"-representation for the vector function

$$y := (2\pi\sqrt{h})^{-1}\int \exp(-\frac{i}{h}t\eta)v(t,x,h)dt.$$

In this "x,η"-representation, to h^{-1}-P.D.O.'s $\widehat{\lambda'_j}, \widehat{d}, \left(\widehat{B}_*^U\right)_{kl}$ with symbols $\lambda'_j, \widetilde{d}$, $\left(B_*^U\right)_{kl}$ correspond h^{-1}-P.D.O.'s $\lambda_j^\eta := \widehat{\lambda'}_j(ih\partial_\eta, x, -ih\partial_x, h)$, d^η and $(B_*^U)_{kl}^\eta$, respectively. These operators are differential operators with respect to the derivative ∂_η. With this notation we obtain that

$$(\eta + \lambda_1^\eta)\,y_j = -h\sum_{k=1}^{r+2}(B_*^U)_{j,k}^\eta y_k, \quad j = 1,\dots,r;$$

$$\left\{1 + \frac{h}{\eta}(B_*^U)_{r+1,r+1}^\eta\right\} y_{r+1} + \frac{1}{\eta}d^\eta y_{r+2} = \frac{-h}{\eta}\sum_{k=1,k\neq r+1}^{r+2}\left(B_*^U\right)_{r+1,k}^\eta y_k;$$

$$(\eta + \lambda_3^\eta)\, y_{r+2} = -h \sum_{k=1}^{r+2} \left(B_*^U\right)^\eta_{r+2,k} y_k. \tag{3.18}$$

We consider system (3.18) over the contour C and search the solution of system (3.18) in a class of functions Θ defined over the contour C such that the functions $(B_*^U)^\eta_{r+1,r+1}\Theta\big|_C$ are bounded over the contour C. Therefore, due to the estimate $|h/\eta|_C \leq h^{1/2-\varepsilon}$, the operator $\left\{1 + \frac{h}{\eta}(B_*^U)^\eta_{r+1,r+1}\right\}^{-1}$ exists as a formal series in $(-1)^j \left(\frac{h}{\eta}(B_*^U)^\eta_{r+1,r+1}\right)^j$ (the functions of the class Θ are defined below). Hence in system (3.18) we can express the component y_{r+1} through the other components and reduce equations (3.18) to some equations with respect to $y_1, \ldots, y_r, y_{r+2}$:

$$y_{r+1} = -\left\{1 + \frac{h}{\eta}(B_*^U)^\eta_{r+1,r+1}\right\}^{-1} \left\{\frac{1}{\eta}d^\eta + \frac{h}{\eta} \sum_{k\neq r+1,k=1}^{r+2} (B_*^U)^\eta_{r+1,k} y_k\right\} \tag{3.19}$$

$$(\eta + \lambda_1^\eta) y_j = -h \sum_{k=1}^{r} (B_*^U)^\eta_{j,k} y_k + h \sum_{k=1}^{r} (B_*^U)^\eta_{j,r+1} \tag{3.20}$$
$$\times \left\{1 + \frac{h}{\eta}(B_*^U)^\eta_{r+1,r+1}\right\}^{-1} \left(\frac{h}{\eta}\right) (B_*^U)^\eta_{r+1,k} y_k + \Psi_j(y_{r+2}); \quad j = 1, \ldots, r,$$

$$(\eta + \lambda_3^\eta) y_{r+2} - h (B_*^U)^\eta_{r+2,r+1} \left\{1 + \frac{h}{\eta}(B_*^U)^\eta_{r+1,r+1}\right\}^{-1} \tag{3.21}$$
$$\times \left\{\frac{1}{\eta}d^\eta + \frac{h}{\eta}(B_*^U)^\eta_{r+1,r+2}\right\} y_{r+2} + h (B_*^U)^\eta_{r+2,r+2} y_{r+2} = \Phi_{r+2}(y_1,\ldots,y_r).$$

Here the functions Ψ_j; $j = 1, \ldots, r$; Φ_{r+2} have the form:

$$\Psi_j = h (B_*^U)^\eta_{j,r+1} \left\{1 + \frac{h}{\eta}(B_*^U)^\eta_{r+1,r+1}\right\}^{-1} \tag{3.22}$$
$$\times \left\{\frac{1}{\eta}d^\eta + \frac{h}{\eta}(B_*^U)^\eta_{r+1,r+2}\right\} y_{r+2} - h (B_*^U)^\eta_{j,r+2} y_{r+2}$$

$$\Phi_{r+2} = -h \sum_{k=1}^{r+1} (B_*^U)^\eta_{r+2,k} y_k + h (B_*^U)^\eta_{r+2,r+1} \tag{3.23}$$
$$\times \left\{1 + \frac{h}{\eta}(B_*^U)^\eta_{r+1,r+1}\right\}^{-1} \left(\frac{h}{\eta}\right) \sum_{k=1,k\neq r+1}^{r+2} (B_*^U)^\eta_{r+1,k} y_k.$$

Thus it is necessary to solve system of equations (3.20), (3.21) in the class Θ with the precision $O(h^{M+1})$. We do it applying the WKB method to system (3.20), (3.21) for $\eta \in C$. Over the contour C it holds $|h/\eta|_C \leq h^{1/2-\varepsilon}$; and system (3.20), (3.21) over C does not have multiplicity points and Jordan block. This system has singularity in the coefficients, but it is proposed to apply the standard WKB method to obtain formal asymptotic solution of the system [10].

The above-defined phase function $\widehat{S}_3(x,\eta)$ satisfies the equation

$$\eta + \lambda_3'\left(-\partial_\eta \widehat{S}_3, x, \partial_x \widehat{S}_3\right) = 0.$$

It follows from (3.15) and (3.17) that

$$\begin{aligned} \widehat{S}_3(x,0) &= S_3(f(x,\xi),x)|_{U_3 \cap\, \Sigma_{32}'}, \\ \partial_\eta \widehat{S}_3(x,0) &= -f(x,\xi)\circ \pi_{x,\eta}^{-1}(x,0) \end{aligned} \tag{3.24}$$

$$\widehat{S}_3(x,\eta) = \widehat{S}_3(x,0) - \eta f(x,\xi)\circ \pi_{x,\eta}^{-1}(x,0) - -\frac{D(t,x)}{D(\eta,x)}(x,0)\frac{\eta^2}{2} + O(\eta^3). \tag{3.25}$$

We calculate the WKB solution in the form

$$y = \exp\left\{\frac{i}{h}\widehat{S}_3(x,\eta)\right\}\sum_{l=0}^{3M+1} \rho_l(x,\eta,h)h^l. \tag{3.26}$$

Applying the commutation formula of h^{-1}-P.D.O. with exponent $\exp\left\{\frac{i}{h}\widehat{S}_3(x,\eta)\right\}$ we obtain the expansion in an asymptotic series with respect to the small parameter h [8]. In the expansion we consider (h/η) as a new variable and include this term in the transport equation. Equations for the vector amplitude functions ρ_l can be obtained in the standard manner. We have that $\rho_0 = (0,\dots,0,1)^t\psi_0/\sqrt{J_3}$ where the Jacobian determinant J is defined above. The function ψ_0 satisfies the so-called transport equation:

$$\begin{aligned} V_1\psi_0 &:= i\{\partial_t\lambda_3'\}(-\partial_\eta\widehat{S}_3, x, \partial_x\widehat{S}_3)\frac{d\psi_0}{d\eta} - \left\{1 + \frac{h}{\eta}(B_*^U)^\eta_{r+1,r+1}\right\}^{-1} \\ &\times \left\{\frac{d\widetilde{b}^1_{r+2,r+1}(-\partial_\eta\widehat{S}_3, x, \partial_x\widehat{S}_3)}{\eta} + \frac{h}{\eta}\widetilde{b}^1_{r+2,r+1}\widetilde{b}^1_{r+1,r+2}(-\partial_\eta\widehat{S}_3, x, \partial_x\widehat{S}_3)\right. \\ &\left. + \widetilde{b}^1_{r+2,r+2}(-\partial_\eta\widehat{S}_3, x, \partial_x\widehat{S}_3)\right\}\psi_0 + \Delta(-\partial_\eta\widehat{S}_3, x, \partial_x\widehat{S}_3)\psi_0 = 0. \end{aligned} \tag{3.27}$$

In formula (3.27) $\Delta(x',\xi)$ is the function

$$\Delta(x',\xi) := -\frac{i}{2}\sum_{j=1}^{n}\frac{\partial^2}{\partial_{\alpha_j}\partial_{\beta_j}}\lambda_3',$$

where $\alpha = (x,\eta), \beta = (\xi,-t)$, and

$$\frac{d}{d\eta} := \sum_{j=1}^{n} - \left\{\partial_{\xi_j}\left(\widehat{\lambda}_3 - \widehat{\lambda}_2\right)/\partial_t\left(\widehat{\lambda}_3 - \widehat{\lambda}_2\right)\right\}\partial_{x_j} + \partial_\eta.$$

The transport equation is considered over the contour C. As the diameter $C \le h^{-\delta+1/2}$ we can resolve the transport equation with the precision $O(h^N)$ using the expansion of its coefficients with respect to the argument η. Consider the equations for all the amplitudes $\rho_l = \{(0,\dots,0,1)^t\psi_l + f_l\}/\sqrt{J}$, where the functions f_l depend on the amplitudes ψ_l [8]. Let $\sigma_{32}(x)$ be the function defined as

$$\sigma_{32}(x) := \left\{ i\widehat{b}^0_{r+2,r+1}\widehat{d}/\partial_t \left(\widehat{\lambda}_3 - \widehat{\lambda}_2\right)\right\}\left(-\partial_\eta \widehat{S}, x, \nabla_x \widehat{S}\right)\Big|_{\eta=0}.$$

Direct calculations imply that the amplitudes $\rho_l(x,\eta,h)$, $l = 0,1,\dots,$ of the standard WKB approach can be presented in the form

$$\begin{aligned} \rho_0 &= \eta^{\sigma_{32}}\omega_0(x,\eta,h)c_0(x), \\ \rho_l &= \eta^{\sigma_{32}-l}\omega_l(x,\eta,h). \end{aligned} \tag{3.28}$$

Here the functions ω_l have the form

$$\omega_l = \sum_{l=0,j_1=0,\dots,j_5=0}^{3M+1} \left(\frac{h}{\eta}\right)^{j_1} h^{j_2}(h\ln\eta)^{j_3}(\eta\ln\eta)^{j_4}\eta^{j_5}c_{(j),l}(x,h) + O(h^N), \tag{3.29}$$

where $l = 1,2,\dots$; $(j) = (j_1,j_2,j_3,j_4,j_5)$; and the vector functions $c_{(j),l}(x,h)$ are polynomials in h with C^∞-coefficients in the projection of the vicinity U_3 onto the plane $\{\eta = 0, t = 0, \xi = 0\} \in R^{2n+2}_{x',\xi'}$.

The function constructed above (3.26) satisfies system (3.20), (3.21) with the precision up to a function $O(h^{M+1})$, such that

$$\left|\partial^\alpha_\eta \partial^\beta_x O(h^{M+1})\right|_C \le C_{\alpha\beta}h^{M+1-\alpha(\frac{1}{2}+\varepsilon)}\{|\ln h| + 1\}^{|\beta|}$$

d) Function (3.16) is a formal asymptotic solution in domain $\pi_{x'}U_3$. Applying a stationary phase method to integral (3.16) we obtain formal asymptotic solution in domains $(\pi_{x'}U_3)\cap\Delta^-_{32}$ and $(\pi_{x'}U_3)\cap\Delta^+_{32}$. In addition is proved a statement

Lemma 3.1. *Let $\alpha \in \mathbb{R}\setminus\{0\}$, $\sigma \in \mathbb{C}$ and γ is the contour that passes through the real axis and below the point $p = 0$ along the circumference of radius $|p| = h^{1/2+2\delta}$, $0 < \delta < 1/8$. By definition $\arg p = -\pi$ if $p < 0$ and $\arg p = 0$ if $\eta > 0$. Suppose the function $\chi(p) \in C^\infty_0(\mathbb{R}^1)$ and $\chi(p) = 1$ for $|p| \le h^{1/2-3\delta/2}$ and $\chi(p) = 0$ for $|p| \ge h^{1/2-2\delta}$. Then in the domain $h^{1/2-\delta/2} > t > h^{1/2-\delta}$ the integral*

$$I_\sigma(t) = \frac{1}{\sqrt{2\pi h}}\int_\gamma \exp\frac{i}{h}\left(pt + \alpha\frac{p^2}{2}\right)p^\sigma\chi(p)dp, \tag{3.30}$$

for $\alpha > 0$ has asymptotic

$$\begin{aligned} I_\sigma(t) = {}& \frac{1}{\sqrt{t}}\left(\frac{h}{t}\right)^{\sigma+1/2} e^{i\frac{\pi}{2}(\sigma+1)}\left(1 - e^{-i2\pi\sigma}\right)\left\{\Gamma(\sigma+1) + O\left(h^\delta\right)\right\} \\ &+ \exp\left(-\frac{i}{h}\frac{t^2}{2\alpha}\right)\sqrt{\frac{1}{\alpha}}\left|\frac{t}{\alpha}\right|^\sigma e^{-i\pi\sigma}\exp\left(\frac{i\pi}{4}\right)\left(1 + O\left(h^{1/2}\right)\right), \end{aligned} \tag{3.31}$$

and for $\alpha < 0$ it has asymptotic

$$I_\sigma(t) = \frac{1}{\sqrt{t}} e^{\frac{i\pi}{2}(1+\sigma)} \left(\frac{h}{t}\right)^{\sigma+1/2} \left(1 - e^{-i2\pi\sigma}\right) \left\{\Gamma(\sigma+1)1 + O\left(h^{1/2}\right)\right\} \quad (3.32)$$
$$+ \exp\left(-\frac{i}{h}\frac{t^2}{2\alpha}\right)\sqrt{\frac{1}{|\alpha|}}\left|\frac{t}{\alpha}\right|^{\sigma} \exp\left(-\frac{i\pi}{4}\right)\left(1 + O\left(h^\delta\right)\right).$$

In the domain $-h^{1/2-\delta} < t < -h^{1/2-\delta/2}$ the integral $I_\sigma(t)$ for $\alpha \in \{\mathbb{R}^1 \setminus 0\}$ has asymptotic

$$I_\sigma(t) = \exp\left(-\frac{i}{h}\frac{t^2}{2\alpha}\right)\sqrt{\frac{1}{|\alpha|}}\left|\frac{t}{\alpha}\right|^{\sigma} e^{-i\pi\sigma\theta(\alpha)} \exp\left(\frac{i\pi}{4}\operatorname{sign}\alpha\right)\left(1 + O\left(h^{1/2}\right)\right), \quad (3.33)$$

where $\theta(\alpha) = 1$ if $\alpha < 0$ and $\theta(\alpha) = 0$ if $\alpha > 0$.

The phase function in integral (3.16) is

$$t\eta + \widehat{S}_3(x,\eta) = (t - f(x,\xi)\circ\pi^{-1}_{x,\eta}(x,0))\eta - \frac{D(t,x)}{D(\eta,x)}(x,0)\frac{\eta^2}{2} + O(\eta^3).$$

Now we apply the statement to integral (3.16). We see that if $(t - f(x,\xi)\circ \pi^{-1}_{x,\eta}(x,0)) < 0$, then in the domain $(\pi_{x'}U_3)\cap\Delta^+_{32}$, the asymptotic of integral (3.16) is defined by it stationary point. Asymptotic solution (3.16) in the domain $(\pi_{x'}U_3)\cap\Delta^+_{32}$ must coincide with $K_{\Lambda^{n+1}_{1\to3,1}}\varphi^{(1\to3,-)}(h)$ up to $O(h^M)$. This condition assigns initial data for amplitudes ρ_l. The asymptotic expansion of integral (3.16) in case $(t - f(x,\xi)\circ\pi^{-1}_{x,\eta}(x,0)) > 0$, that is in the domain $(\pi_{x'}U_3)\cap\Delta^-_{32}$, has the form

$$K_{\Lambda^{n+1}_{3\to2,1}}\varphi^{(3\to2,+)}(h) + K_{\Lambda^{n+1}_{1\to3,1}}\varphi^{(1\to3,+)}(h).$$

The manifold $\Lambda^{n+1}_{1\to3,1}$ does not intersect Σ_{12} but manifold $\Lambda^{n+1}_{3\to2,1}$ intersects. Hence consider the term $K_{\Lambda^{n+1}_{3\to2,1}}\varphi^{(3\to2,+)}(h)$, which contributes into the asymptotic in the domain $\Delta^+_{12}\cap\Delta^-_{32}$. It follows from the formulas (3.9), (3.28) and formula (3.31) that the amplitude $\varphi^{(3\to2,+)}(h)$ satisfies an estimate

$$\left|\varphi^{(3\to2,+)}(h)\right| \geq Ah^{\overline{\sigma}_{32}+3/2}\left|\phi(x_0)\right|e^{-ct} \quad (3.34)$$

on such subdomain D of $\Lambda^{n+1}_{3\to2,1}\setminus\{U(\Sigma_{12},h^{1/2-\delta/2})\cup U(\Sigma_{32},h^{1/2-\delta/2})\}$ that it holds that $|\phi(x_0)|_D \geq 1/2$. Here $\overline{\sigma}_{32} = \min_{D\cap\Sigma_{123}}\operatorname{Re}\sigma_{32}$. Therefore after one turn around the set Σ_{123} the amplitude grows up in $h^{\overline{\sigma}_{32}+3/2}$ times. The trajectories make N turn around the set Σ_{123}.

For times $t \in [t_{4N}(y) - h^{1/2-\delta},\, t_{4N}(y) - h^{1/2-\delta} + h^{1/2-\delta/2}]$ the formal asymptotic solution $w(t,x,h)$ of the Cauchy problem (1.1), (1.15) will be a sum of some finite functions. Then finite function with the support in the set

$$\Omega := \{\Delta^-_{12}\cap\Delta^-_{31}\cap U_0^{\varepsilon_0}\}\setminus\{U(\Sigma_{12},h^{1/2-\delta})\cup U(\Sigma_{32},h^{1/2-\delta})\} \quad (3.35)$$

satisfies an estimate

$$\left\{\|w(t,.,h)\|_{H_k(\Omega_t)}\right\} \geq Ah^{(\overline{\sigma}_{32}+3/2)N-k}e^{-ct}. \tag{3.36}$$

Here constants A, c do not depend on parameter h; Ω_t is the cross-section of domain Ω (3.35) with a hyperplane $\mathbb{R}^{2n}_{x,\xi} \times \{\tau : \tau = t\}$. The statement of the theorem is a direct consequence of estimate (3.36) and condition $(\overline{\sigma}_{32} + 3/2) < 0$. The theorem is proved.

4. Addition

I.) Proof of Lemma (3.1).

A) *First we consider the case* $\alpha > 0$. In this case integral (3.30) has a singular point $p = 0$ and a stationary point $p = -t/\alpha$. Consider the real part of the phase function in integral (3.30):

$$\operatorname{Re}\left\{i\left(pt + \alpha\frac{p^2}{2}\right)\right\} = -p_2\left(t + \alpha p_1\right), \text{ where } p = p_1 + ip_2.$$

Note that for $(t + \alpha p_1) > 0$ the contour γ can be moved upward and for $(t + \alpha p_1) < 0$ it is possible to move the contour γ downward. In the proses of the contour deformation we change the truncated function $\chi(p)$ for its analytic continuation $\chi(p_1 + ip_2) := \sum_{k=0}^{M} \frac{(ip_2)^k}{k!}\partial^k_{p_1}\chi(p_1)$.

1. Let t be in the interval: $-h^{1/2-\delta} < t < -h^{1/2-\delta/2}$ and $p_1 < -t/\alpha$. The contour γ will be deformed downward. As

$$\int\int_{p_1\leq 0;p_2\leq 0}\left|\exp\frac{i}{h}\left(pt + \alpha\frac{p^2}{2}\right)\frac{\partial}{\partial\overline{p}}\chi(p_1 + ip_2)\right| dp_1dp_2 \leq C_M h^{M(1/2+\delta/2)},$$

then with a precision $O(h^\infty)$ integral (3.30) equal to an integral from the same integrand by the contour presented in Figure 2.

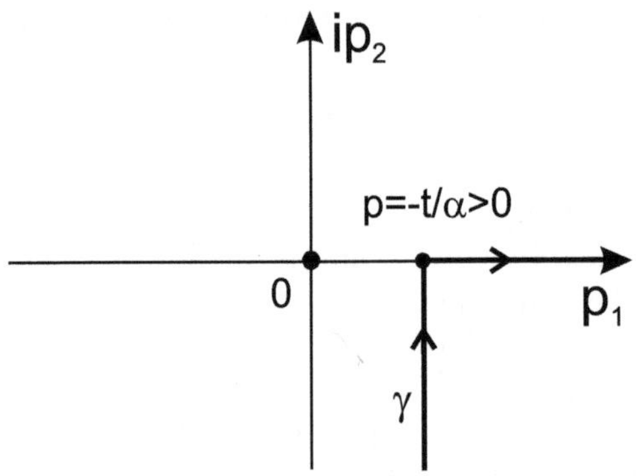

FIGURE 2

Hence for such values t only the stationary point contributes in the asymptotic solution and the singular point $p = 0$ does not contribute. Applying the stationary phase method obtain formula (3.33).

2. Now suppose that $t > 0$. Then for $p_1 > -t/\alpha$ the contour γ will be moved upward, and for $p_1 < -t/\alpha$ the contour γ will be moved downward. So let t be in the interval: $h^{1/2-\delta/2} > t > h^{1/2-\delta}$. Hence with similar arguments we obtain that with a precision $O(h^\infty)$ integral (3.30) is equal to an integral from the same integrand by a contour $\gamma_1 \cup \gamma_2$ presented in Figure 3.

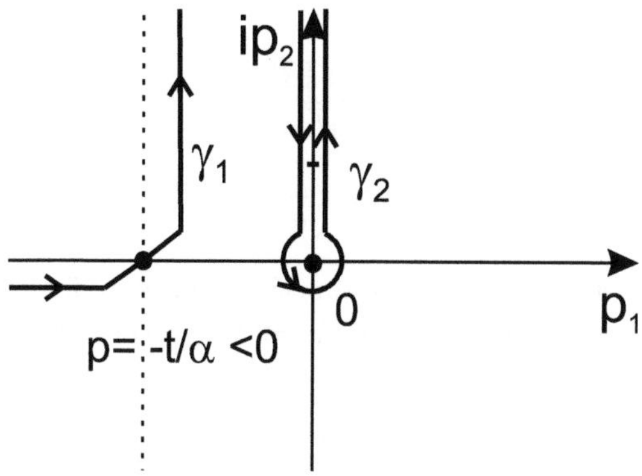

FIGURE 3

The asymptotic of integral along the contour γ_1 is calculated by the stationary phase method. Consider integral along the contour γ_2.

$$I_\sigma^{pol}(t) := \frac{1}{\sqrt{2\pi h}} \int_{\gamma_2} \exp \frac{i}{h}\left(pt + \alpha\frac{p^2}{2}\right) p^\sigma dp. \tag{4.1}$$

Introducing the substitution $p = st$ one obtains that

$$I_\sigma^{pol}(t) = \frac{t^{\sigma+1}}{\sqrt{2\pi h}} \int_{\gamma_2} \exp\left(\frac{it^2}{h}\left(s + \alpha\frac{s^2}{2}\right)\right) s^\sigma ds \tag{4.2}$$

$$= \frac{t^{\sigma+1}}{\sqrt{2\pi h}} \sum_{k=0}^{\infty} \int_{\gamma_2 \cap \{|s|\le 1\}} \exp\left(\frac{it^2}{h}s\right) \frac{s^{\sigma+2k}}{k!}\left(\frac{it^2\alpha}{h2}\right)^k ds + O(h^\infty)$$

$$\sim \frac{t^{\sigma+1}}{\sqrt{2\pi h}} \frac{h}{t^2} \sum_{k=0}^{\infty} \frac{1}{k!}\left(\frac{h}{t^2}\right)^{\sigma+k} \left(\frac{i\alpha}{2}\right)^k \int_{\gamma_2} \exp(i\tau)\, \tau^{\sigma+2k} d\tau. \tag{4.3}$$

Consider integrals in formula (4.3). If $\operatorname{Re}\sigma + 2k > -1$ then

$$\int_{\gamma_2} \exp(i\tau)\, \tau^{\sigma+2k} d\tau = e^{i(\frac{\pi}{2}\sigma + \frac{\pi}{2} + \pi k)}(1 - e^{-i2\pi\sigma})\Gamma(\sigma + 2k + 1). \tag{4.4}$$

The analytic continuation in variable σ provides formula (4.4) for all $\sigma \in \mathbb{C}$. Substituting expression (4.4) into formula (4.3) we get asymptotic formula (3.31).

B) The proof of the case $\alpha < 0$ is similar to the case of $\alpha > 0$, presented above.

II.) Now consider local coordinates on the Lagrangian manifold, i.e., the item 6.b in the proof of the theorem. Suppose that in the map $U_3 \in \Lambda^{n+1,\prime}_{1\to 3,1}$ exist local coordinates $(t, x_1, \dots, x_n)$ but at the point $q \in U_3$ it holds $\frac{\partial \eta}{\partial t}(q) = 0$. In such case we must modify formula (3.16). For this aim define a Hamiltonian $H := \frac{1}{2}\xi^2$; the flow g^τ in the phase space $\mathbb{R}^{2n}_{x\xi}$ generated by the Hamiltonian vector file with the Hamiltonian $H : g^\tau(x,\xi) = (x + \tau\xi, \xi)$; an operator $\widehat{H} = -\frac{1}{2}\sum_{k=1}^{k=n}\frac{\partial^2}{\partial x_k^2}$ and change formula (3.16) for the next one

$$K(U_3)(\varphi\circ\pi_{x,\eta}) := \frac{\exp\{\frac{i\pi}{4} - i\frac{\pi}{2}\widehat{\gamma}(q)\}}{\sqrt{2\pi h}} e^{-\frac{i}{h}\tau\widehat{H}} U_2^{as}(t,0) \times \int_C \exp\left\{\frac{i}{h}[t\eta + \widehat{S}_3^\tau(x,\eta)]\right\}\frac{\varphi}{\sqrt{J_3^\tau}}d\eta \tag{4.5}$$

where the function $\widehat{S}_3^\tau(x,\eta)$ is the phase function on a manifold U_3^τ obtained from U_3 by canonical transformations $U_3^\tau := g^\tau U_3$, and J_3^τ is the corresponding Jacobian on U_3^τ. It holds the lemma.

Lemma 4.1. *Let Λ^{n+1} be a $(n+1)$-dimensional C^∞ Lagrangian manifold in $\mathbb{R}^{2n+2}_{x'\xi'}$ and let a Hamiltonian H be equal to $\frac{1}{2}\xi^2$. Suppose q is any point at Λ^{n+1} and (t,x) be the local coordinate in some vicinity $U_q \in \Lambda^{n+1}$ of the point q, i.e., the projection $\pi_{t,x}$ is the coordinate diffeomorphism at U_q. Also suppose that $\eta|_{U_q} = \lambda(x',\xi)|_{U_q}$ where $\lambda(x',\xi) \in S^1\left(\mathbb{R}^{n+1}_{x'} \times \mathbb{R}^n_\xi\right)$, $\partial_t\lambda(x',\xi)|_{U_q} \neq 0$ and $\xi|_{U_q} \neq 0$. If $(\partial_t\eta)(\pi_{t,x}q) = 0$ then there exist value $\tau \geq 0$ and such vicinity $U_q' \subset U_q$ of the point q that in the vicinity $g^\tau(U_q')$ there are two sets of local coordinates: the set (t,x) and the set (η,x).*

Proof. Let $S(t,x)$ be the phase function in $\pi_{x'}U_q$, i.e., $\xi'|_{Uq} = \partial_{x'}S(x')$. Then $\partial_t S + \lambda(x', \partial_x S(t,x)) = 0$ and

$$\partial_t^2 S = -\partial_t\lambda(x',\partial_x S(t,x)) - \sum_{j=1}^{j=n}(\partial_{\xi_j}\lambda)(x',\partial_x S(t,x))\frac{\partial^2}{\partial_{x_j}\partial_t}S(t,x). \tag{4.6}$$

As $\partial_t\eta(q) = \partial_t^2 S(q) = 0$ and $\partial_t\lambda(x',\partial_x S(t,x)) \neq 0$ in $\pi_{x'}U_q$, then it follows from formula (4.6) that

$$\partial_t\xi(q) = \nabla_x\eta(q) = \nabla_x\partial_t S(q) \neq 0. \tag{4.7}$$

Let us $(t,\widetilde{x};\widetilde{\eta},\widetilde{\xi}) := g^\tau(x',\xi') = (t, x+\tau\xi;\eta,\xi)$; hence for a small τ local coordinates in $g^\tau(U_q')$, for some $U_q' \subset U_q$ are $(t,\widetilde{x})$. Evidently for a small τ it holds $x = \widetilde{x} - \tau\xi(t,\widetilde{x}) - \tau\xi_0 + O(\tau^2), \widetilde{\eta}(t,\widetilde{x}) = \eta(t, \widetilde{x} - \tau\xi(t,\widetilde{x}) - \tau\xi_0 + O(\tau^2))$ where $\partial_t^{k_0}\dots\partial_{x_n}^{k_n}O(\tau^2)$ have the same order τ^2. Therefore

$$\partial_t\widetilde{\eta}(t,\widetilde{x}) = \partial_t\eta(t,\widetilde{x} - \tau(\xi(t,\widetilde{x}) + \xi_0)) + \tau(\nabla_x\eta(t,\widetilde{x}), \partial_t\xi(t,\widetilde{x})) + O(\tau^2). \tag{4.8}$$

By the condition of the lemma $\partial_t\eta(\pi_{t,x}q) = 0$. Suppose $(t,\widetilde{x}) = \pi_{t,x}g^\tau L_{\xi_0}q$, then $\partial_t\eta(t,\widetilde{x}-\tau(\xi(t,\widetilde{x}))) = O(\tau^2)$ and it follows from formulas (4.7), (4.8) that $\partial_t\widetilde{\eta}(t,\widetilde{x}) = \tau(\nabla_x\eta(t,\widetilde{x}),\partial_t\xi(t,\widetilde{x})) + O(\tau^2) \neq 0$ for sufficiently small value τ. Hence there exist small vicinity U'_q and the value τ such that in the vicinity $g^\tau(U'_q)$ there exist two sets of local coordinates the set (t,x) and the set (η,x). $\square$

In general case always exist local coordinates $(t,x^I,\xi_{\overline{I}})$, $I\cup\overline{I}=(1,2,\ldots,n)$, $I\cap\overline{I}=\varnothing$. In this case representation (4.5) must be change for the next one

$$K(U_3)(\varphi\circ\pi_{x,\eta}) := \frac{\exp\{\frac{i\pi}{4}-i\frac{\pi}{2}\widehat{\gamma}(q)\}}{(2\pi h)^{(|\overline{I}|+1)/2}}\int\exp\{\frac{i}{h}x^{\overline{I}}p_{\overline{I}}\}e^{-\frac{i}{h}\tau\widehat{H}_I}U_2^{as}(t,0)$$
$$\times\int_C\exp\left\{\frac{i}{h}[t\eta+\widehat{S}^\tau_{3,I}(x^I,\xi_{\overline{I}},\eta)]\right\}\frac{\varphi}{\sqrt{J^\tau_{3,I}}}d\eta d\xi_{\overline{I}}.$$

Here $\widehat{H}_I = -\frac{1}{2}\sum_{k\in I}\frac{\partial^2}{\partial x_k^2} - \frac{1}{2}\sum_{k\in\overline{I}}\frac{\partial^2}{\partial\xi_k^2}$; $d\widehat{S}^\tau_{3,I}(x^I,\xi_{\overline{I}},\eta) = \xi_I dx^I - x^{\overline{I}}d\xi_{\overline{I}} - td\eta$; $J^\tau_{3,I} = D(\eta,x^I,\xi_{\overline{I}})/D\mu$ and $D\mu = dtdx_1^0\ldots dx_n^0$.

Acknowledgment

This research was partially supported by project of CONACYT Mexico and by project of SIP – IPN.

References

[1] Enrico Bernardi and Antonio Bove, *On a class of strongly hyperbolic systems* Arkiv för Matematik **43, N 1** (2005), 113–131

[2] A.V. Krivko, V.V. Kucherenko, *On Real Hyperbolic Systems with Characteristics of Variable Multiplicity* Doklady Mathematics **75, N.1** (2007), 83–86

[3] V.V. Kucherenko, *Asymptotics of the solution of the system $A(x,-ih\partial/\partial x)$ as $h\to 0$ in the case of characteristics of variable multiplicity* Izv. AN SSSR. Ser. Mat. **38 (3)** (1974) 625–662

[4] V.Ya. Ivrii, and V.M. Petkov, *Necessary conditions for the Cauchy problem for non-strictly hyperbolic equations to be well posed* Russ. Math. Surv. **29 (5)** (1974) 3–70

[5] R.B. Melrose and G.A. Uhlmann, *Lagrangian intersection and the Cauchy problem* Comm. Pure Appl. Math. **32** (1979) 483–519

[6] A. Kryvko and V.V. Kucherenko, *Stability of stationary solutions of nonlinear hyperbolic systems with multiple characteristics* Proceedings of ISAAC Congress 2005 (2009)

[7] V.P. Maslov, *Perturbation Theory and Asymptotic Methods* Izdat. Moskov. Gos. Univ., Moscow, 1965 (Dunod, Paris, 1972)

[8] V.P. Maslov and M.V. Fedoruk, *Semi-Classical Approximation in Quantum Mechanics* Nauka, Moscow,1976 (Reidel, Dordrecht, 1981)

[9] L. Hörmander, *The Analysis of Linear Partial Differential Operators, II* Springer Verlag, Berlin, 1983

[10] V.V. Kucherenko and Yu. Osipov, *The Cauchy problem for nonstrictly hyperbolic equations* Math. USSR Sbornik **48(1)** (1984) 81–109

[11] H. Flaschka and G. Strang, *The correctness of the Cauchy problem* Advances in Math. **6(3)** (1971) 347–379

[12] R. Courant and D. Hilbert, *Methods of Mathematical Physics* John Wiley and Sons, New York, 1962

[13] M.V. Karasev, *Weyl and ordered calculus of noncommuting operators* Math. Notes **26(6)** (1979) 885–907

[14] M.V. Karasev and V.P. Maslov, *Nonlinear Poisson Brackets Geometry and Quantization* AMS, 1993

[15] M.V. Fedoruk, *Metod Perevala* Nauka, Moscow, 1977

[16] S.Yu. Dobrokhotov, *Resonances in multi-frequency averaging theory of nonlinear partial differential equations* Singular limits of dispersive waves, NATO ASI Series, Ser. B, **320** (1994) 203–217

Valeri V. Kucherenko
Instituto Politecnico Nacional – ESFM
Av. IPN S/N
U.P.A.L.M
Edif. 9, cp 07738
Mexico, D.F., Mexico
e-mail: `valeri@esfm.ipn.mx`

Andriy Kryvko
Instituto Politecnico Nacional – ESIME Zacatenco
Av. IPN S/N
U.P.A.L.M
Edif. 5, cp 07738
Mexico, D.F., Mexico
e-mail: `kryvko@gmail.com`

Operator Theory:
Advances and Applications, Vol. 210, 175–188

On C^*-Algebras of Super Toeplitz Operators with Radial Symbols

M. Loaiza and A. Sánchez-Nungaray

To Nikolai Vasilevski on the occasion of his 60th birthday

Abstract. We study Toeplitz operators with radial symbols acting on the Bergman space of the super unit disk. We prove that, generalizing the classical case, every super Toeplitz operator with radial symbol is diagonal. This fact implies that the algebra generated by all super Toeplitz operators with radial symbols is commutative.

Mathematics Subject Classification (2000). Primary 47B35; Secondary 47L80, 32A36, 58A50.

Keywords. Toeplitz operators, commutative C^*-algebras, Bergman spaces, supermanifolds and graded manifolds.

1. Introduction

The existence of commutative algebras of Toeplitz operators is not usual, for example there are not non trivial commutative algebras of Toeplitz operators acting on the Hardy space. For the case of the Bergman space of the unit disk the situation is different, there are many commutative algebras of such kind of operators. In [5] S. Grudsky, R. Quiroga and N. Vasilevski classified the algebras of symbols such that the corresponding algebras of Toeplitz operators, acting on the Bergman space of the unit disk are commutative. Families of symbols that produce commutative Toeplitz algebras are invariant under the action of certain kind of commutative groups of Möbius transformations (see [5]). A classical example of this kind of algebras is the following: Consider the algebra $\mathcal{R}$ of all radial functions defined on the unit disk. The C^*-algebra generated by all Toeplitz operators which symbol is in $\mathcal{R}$ is commutative. Since the unit circle acts on the unit disk as a group of rotations, the algebra of symbols $\mathcal{R}$ consists precisely of the functions $f : D \to \mathbb{C}$ which are invariant under the action of the unit circle. Some recent works related to this topic in other domains such that the unit sphere, Reinhart domains and the unit ball are the following [8, 9, 10, 11].

Recently, Toeplitz operators acting on super Bergman spaces have been object of study. Specially because they are close related to physical objects. In [7] Super Toeplitz operators on the unit disk were studied. In this article the authors find an explicit representation for super Toeplitz operators in terms of classical Toeplitz operators. In [12] is proved that the algebra of Toeplitz operators with invariant symbols under the action of the unit supercircle on the super-sphere is a commutative algebra.

The aim of this article is to study super Toeplitz operators with radial symbols acting on the super Bergman space of the unit disk, where by a radial function we mean that it is invariant under the action of the unit supercircle. We show that the algebra generated by all super Toepliz operators with radial symbols is commutative. Finding an orthonormal basis of eigenvectors of the super Toeplitz operator we give explicit conditions for boundeness and compactness of such kind of operators.

The article is organized as follows. In Section 2 we give some results from [7]. The representation of a Super Toeplitz operator in terms of classical Toeplitz operators is given (Theorem 2.5). In Section 3 we study the super unit circle, we prove that it is a subgroup of $SU(1,1|1)$, the group of Möbius transformations of the super unit disk. Since a radial function is invariant under the action of the circle (i.e., $f(z) = f(e^{it}z), t \in [0, 2\pi)$) we define, in Section 4, a radial function as an invariant function under the action of the super circle and find an explicit form for this kind of functions (Theorem 4.1). Section 5 is devoted to the study of super Toeplitz operators with radial symbols. We prove that a super Toeplitz operator with radial symbol is diagonal and therefore they generate a commutative $C*$-algebra. The last section is devoted to the study of a commutative algebra of super Toeplitz operators which include as a special case radial (classical) functions as symbols.

2. Super Toeplitz operators on the unit disk

In [7] Super Toeplitz operators on the unit disk were studied. We write here the main results down for the unit disk. Let $\mathcal{O}(\mathbb{B})$ denote the algebra of all holomorphic functions $\psi(z)$ on the open unit disk

$$\mathbb{B} := \{z \in \mathbb{C} : \ |z| < 1\}.$$

Let Λ_1 denote the complex Grassmann algebra with generator ζ, satisfying the relation $\zeta^2 = 0$. Thus

$$\Lambda_1 = \mathbb{C}\langle 1, \zeta \rangle.$$

The tensor product algebra

$$\mathcal{O}(\mathbb{B}^{1|1}) := \mathcal{O}(\mathbb{B}) \otimes \Lambda_1 = \mathcal{O}(\mathbb{B})\langle 1, \zeta \rangle$$

consists of all "super-holomorphic" functions

$$\Psi = \psi_0 + \zeta\psi_1$$

with $\psi_0, \psi_1 \in \mathcal{O}(\mathbb{B})$. We sometimes write

$$\Psi(z, \zeta) = \psi_0(z) + \zeta\psi_1(z)$$

for all $z \in \mathbb{B}$.

Definition 2.1. For $\nu > 1$, the *weighted Bergman space*

$$H^2_\nu(\mathbb{B}) := \mathcal{O}(\mathbb{B}) \cap L^2(\mathbb{B}, d\mu_\nu)$$

consists of all holomorphic functions on $\mathbb{B}$ which are square-integrable for the probability measure

$$d\mu_\nu(z) = \frac{\nu - 1}{\pi}(1 - |z|^2)^{\nu-2}dz. \tag{2.1}$$

Here dz denotes Lebesgue measure on $\mathbb{C}$.

It is well known [HKZ] that $H^2_\nu(\mathbb{B})$ has the reproducing kernel

$$K_\nu(z, w) = (1 - z\overline{w})^{-\nu}$$

for all $z, w \in \mathbb{B}$. Let $\Lambda_1^{\mathbb{C}}$ denote the complex Grassmann algebra with 2 generators $\zeta, \overline{\zeta}$ satisfying

$$\zeta^2 = \overline{\zeta}^2 = 0, \quad \zeta\overline{\zeta} = -\overline{\zeta}\zeta.$$

Thus

$$\Lambda_1^{\mathbb{C}} = \mathbb{C}\langle 1, \zeta, \overline{\zeta}, \overline{\zeta}\,\zeta\rangle = \Lambda_1\langle 1, \overline{\zeta}\rangle.$$

Let $\mathcal{C}(\overline{\mathbb{B}})$ denote the algebra of continuous functions on $\overline{\mathbb{B}}$. The tensor product

$$\mathcal{C}(\overline{\mathbb{B}}^{1|1}) := \mathcal{C}(\overline{\mathbb{B}}) \otimes \Lambda_1^{\mathbb{C}} = \mathcal{C}(\overline{\mathbb{B}})\langle 1, \zeta, \overline{\zeta}, \overline{\zeta}\zeta\rangle$$

consists of all "continuous super-functions"

$$F = f_{00} + \overline{\zeta}\, f_{10} + \zeta\, f_{01} + \overline{\zeta}\zeta\, f_{11}\,, \tag{2.2}$$

where $f_{00}, f_{10}, f_{01}, f_{11} \in \mathcal{C}(\overline{\mathbb{B}})$. The involution on $\mathcal{C}(\overline{\mathbb{B}}^{1|1})$ is given by

$$\overline{F} = \overline{f}_{00} + \zeta\,\overline{f}_{10} + \overline{\zeta}\,\overline{f}_{01} + \overline{\zeta}\zeta\,\overline{f}_{11}$$

where $\overline{f}(z) := \overline{f(z)}$ (pointwise conjugation).

$\mathcal{C}(\overline{\mathbb{B}}^{1|1})$ contains $\mathcal{O}(\mathbb{B}^{1|1})$ as a subalgebra, and for $\Psi = \psi_0 + \zeta\,\psi_1 \in \mathcal{O}(\mathbb{B}^{1|1})$ we have

$$\overline{\Psi}\Psi = \overline{\psi}_0\psi_0 + \zeta\,\overline{\psi}_0\,\psi_1 + \overline{\zeta}\,\overline{\psi}_1\,\psi_0 + \overline{\zeta}\zeta\,\overline{\psi}_1\,\psi_1.$$

Given a super-function $F \in \mathcal{C}(\overline{\mathbb{B}}^{1|1})$, we define its *Berezin integral*

$$\int\limits_{\mathbb{C}^{0|1}} d\zeta\; F := f_{11} \in \mathcal{C}(\overline{\mathbb{B}})$$

and

$$\int\limits_{\mathbb{B}^{1|1}} dz\, d\zeta\; F(z, \zeta) := \int\limits_{\mathbb{B}} dz \int\limits_{\mathbb{C}^{0|1}} d\zeta\; F(z, \zeta) = \int\limits_{\mathbb{B}} dz\; f_{11}(z). \tag{2.3}$$

Thus the "fermionic integration" is determined by the rules

$$\int_{\mathbb{C}^{0|1}} d\zeta \cdot \zeta = \int_{\mathbb{C}^{0|1}} d\zeta \cdot \overline{\zeta} = \int_{\mathbb{C}^{0|1}} d\zeta \cdot 1 = 0\,, \quad \int_{\mathbb{C}^{0|1}} d\zeta \cdot \overline{\zeta}\zeta = 1.$$

As an example, we have

$$\int_{\mathbb{B}^{1|1}} dz\, d\zeta\ \overline{F}(z,\zeta)\, F(z,\zeta)$$

$$= \int_{\mathbb{B}} dz(\overline{f_{00}(z)}\, f_{11}(z) + \overline{f_{11}(z)}\, f_{00}(z) - \overline{f_{10}(z)}\, f_{10}(z) + \overline{f_{01}(z)}\, f_{01}(z))$$

which shows that the (unweighted) Berezin integral is not positive. For $\Psi = \psi_0 + \zeta\psi_1 \in \mathcal{O}(\mathbb{B}^{1|1})$, it follows that

$$\int_{\mathbb{B}^{1|1}} dz\, d\zeta\ \overline{\Psi}(z,\zeta)\, \Psi(z,\zeta) = \int_{\mathbb{B}^{1}} dz\ \overline{\psi_1(z)}\psi_1(z)$$

is positive, but not positive definite since the ψ_0 term is not present.

Definition 2.2. For any parameter $\nu > 1$ the (weighted) *super-Bergman space*

$$H^2_\nu(\mathbb{B}^{1|1}) \subset \mathcal{O}(\mathbb{B}^{1|1})$$

consists of all super-holomorphic functions $\Psi(z,\zeta)$ which satisfy the square-integrability condition

$$(\Psi|\Psi)_\nu := \frac{1}{\pi} \int_{\mathbb{B}^{1|1}} dz\, d\zeta\ (1 - z\overline{z} - \zeta\overline{\zeta})^{\nu-1}\, \overline{\Psi(z,\zeta)}\, \Psi(z,\zeta) < +\infty.$$

Proposition 2.3. *For* $\Psi = \psi_0 + \zeta\,\psi_1 \in \mathcal{O}(\mathbb{B}^{1|1})$ *we have*

$$\frac{1}{\pi} \int_{\mathbb{B}^{1|1}} dz\, d\zeta\ (1 - z\overline{z} - \zeta\overline{\zeta})^{\nu-1} \overline{\Psi}(z,\zeta)\, \Psi(z,\zeta) = (\psi_0|\psi_0)_\nu + \frac{1}{\nu}\, (\psi_1|\psi_1)_{\nu+1}\,,$$

i.e., there is an orthogonal decomposition

$$H^2_\nu(\mathbb{B}^{1|1}) = H^2_\nu(\mathbb{B}) \oplus [H^2_{\nu+1}(\mathbb{B}) \otimes \Lambda^1(\mathbb{C}^1)]$$

into a sum of weighted Bergman spaces, where $\Lambda^1(\mathbb{C}^1)$ *is the one-dimensional vector space with basis vector* ζ.

Proposition 2.4. *For* $\Psi = \psi_0 + \zeta\,\psi_1 \in H^2_\nu(\mathbb{B}^{1|1})$ *we have the reproducing kernel property*

$$\Psi(z,\zeta) = \frac{1}{\pi} \int_{\mathbb{B}^{1|1}} dw\, d\omega\ (1 - w\overline{w} - \omega\overline{\omega})^{\nu-1} (1 - z\overline{w} - \zeta\overline{\omega})^{-\nu}\Psi(w,\omega),$$

i.e., $H^2_\nu(\mathbb{B}^{1|1})$ *has the reproducing kernel*

$$K_\nu(z,\zeta,w,\omega) = (1 - z\overline{w} - \zeta\overline{\omega})^{-\nu}.$$

For $F \in \mathcal{C}(\overline{\mathbb{B}}^{1|1})$, the *super-Toeplitz operator* $T_F^{(\nu)}$ on $H^2_\nu(\mathbb{B}^{1|1})$ is defined as

$$T_F^{(\nu)}\Psi = P^{(\nu)}(F\Psi),$$

where $P^{(\nu)}$ denotes the orthogonal projection onto $H^2_\nu(\mathbb{B}^{1|1})$.

Theorem 2.5. *With respect to the decomposition $\Psi = \psi_0 + \zeta\,\psi_1$, the super-Toeplitz operator $T_F^{(\nu)}$ on $H^2_\nu(\mathbb{B}^{1|1})$ is given by the block matrix*

$$T_F^{(\nu)} = \begin{pmatrix} T_\nu^\nu\left(f_{00} + \frac{1-w\overline{w}}{\nu-1}\,f_{11}\right) & T_\nu^{\nu+1}\left(\frac{1-w\overline{w}}{\nu-1}\,f_{10}\right) \\ T_{\nu+1}^\nu\left(f_{01}\right) & T_{\nu+1}^{\nu+1}\left(f_{00}\right) \end{pmatrix}. \tag{2.4}$$

Here $T_{\nu+i}^{\nu+j}(f)$, for $0 \le i,j \le 1$, denotes the Toeplitz type operator from $H^2_{\nu+j}(\mathbb{B})$ to $H^2_{\nu+i}(\mathbb{B})$ defined by

$$T_{\nu+i}^{\nu+j}(f)\,\psi := P_{\nu+i}(f\psi)$$

for $\psi \in H^2_{\nu+j}(\mathbb{B})$ and $P_{\nu+i}$ is the orthogonal projection from $L^2_{\nu+i}(\mathbb{B})$ onto $H^2_{\nu+i}(\mathbb{B})$.

3. Action of the super circle on the super disk

The Berezinian is the analogous of the determinant in super analysis. It is defined by the following formula

$$\operatorname{Ber}\begin{pmatrix} A & B \\ C & D \end{pmatrix} = \det(A - BD^{-1}C)\det D^{-1},$$

where A, D are even, D is invertible and B, C are odd. The super Lie Group $SU(1,1|1)$ is the supermanifold $SU(1,1)$ and its structure shift is generated by the 3×3 matrices (s_{ij}) and $(\overline{s}_{ij})$, where the element s_{ij} (resp. $\overline{s}_{ij}$) is even if $1 \le i,j \le 2$ or $i = j = 3$ and odd otherwise. We also require that $s^*Js = J$, where

$$J = \begin{pmatrix} 1 & 0 & 0 \\ 0 & -1 & 0 \\ 0 & 0 & -1 \end{pmatrix}$$

and that $\operatorname{Ber} s = 1$.

The super Lie Group $SU(1,1|1)$ acts on the unit super disk as follows. Given a matrix $\begin{pmatrix} a & b & \alpha \\ c & d & \beta \\ \gamma & \delta & e \end{pmatrix} \in SU(1,1|1)$ the corresponding transformation are given by the formulas

$$\begin{aligned} z &\mapsto \frac{az + b + \alpha\zeta}{cz + d + \beta\zeta} \\ \zeta &\mapsto \frac{\gamma z + \delta + e\zeta}{cz + d + \beta\zeta}. \end{aligned} \tag{3.1}$$

Consider the form

$$g = \frac{\partial^2 \log(1 - z\bar{z} - \zeta\bar{\zeta})}{\partial z \partial \bar{z}} dz d\bar{z} + \frac{\partial^2 \log(1 + z\bar{z} - \zeta\bar{\zeta})}{\partial z \partial \bar{\zeta}} dz d\bar{\zeta} + \frac{\partial^2 \log(1 + z\bar{z} - \zeta\bar{\zeta})}{\partial \zeta \partial \bar{z}} d\zeta d\bar{z} + \frac{\partial^2 \log(1 + z\bar{z} - \zeta\bar{\zeta})}{\partial \zeta \partial \bar{\zeta}} d\zeta d\bar{\zeta}$$

where

$$\frac{\partial^2 \log(1 - z\bar{z} - \zeta\bar{\zeta})}{\partial z \partial \bar{z}} = \frac{1}{(1 - z\bar{z})^2} + \frac{(1 + z\bar{z})\zeta\bar{\zeta}}{(1 - z\bar{z})^3},$$

$$\frac{\partial^2 \log(1 + z\bar{z} - \zeta\bar{\zeta})}{\partial z \partial \bar{\zeta}} = \frac{\bar{z}\zeta}{(1 - z\bar{z})^2},$$

$$\frac{\partial^2 \log(1 + z\bar{z} - \zeta\bar{\zeta})}{\partial \zeta \partial \bar{z}} = \frac{z\bar{\zeta}}{(1 - z\bar{z})^2},$$

$$\frac{\partial^2 \log(1 + z\bar{z} - \zeta\bar{\zeta})}{\partial \zeta \partial \bar{\zeta}} = \frac{1}{(1 - z\bar{z})}.$$

In [2] is proved that the form g is an invariant form under the action of the super Lie group $SU(1,1|1)$. This form produces a metric (Bergman super metric) for the unit super disk. This metric is invariant under the action of $SU(1,1|1)$.

The super circle is defined as the super manifold

$$S^{1|1} = \{(a = a_1 + ia_2, \tau = \eta_1 + i\eta_2)| \ a_1^2 + a_2^2 = 1, \ \ \eta_1 a_1 + \eta_2 a_2 = 0\}$$

where η_1, η_2 are real Grassmann variables.

The definition of $S^{1|1}$ given above is equivalent to define the super unit circle as the set of matrices

$$A = \begin{pmatrix} a & \tau \\ \tau & a \end{pmatrix},$$

such that $A^* A = I$, $\operatorname{Ber} A = 1$.

In the following theorem we see that $S^{1|1}$ can be seen as a subgroup of $SU(1,1|1)$, the super group of isometries of the super unit disk.

Theorem 3.1. *The super-circle is a super-group of isometries of the super unit disc.*

Proof. In order to see any element of $S^{1|1}$ as an element of $SU(1,1|1)$ we define the following function

$$A : S^{1|1} \longrightarrow SU(1,1|1),$$

by

$$A\left[\begin{pmatrix} a & \tau \\ \tau & a \end{pmatrix}\right] = \begin{pmatrix} a & 0 & -\tau \\ 0 & 1 & 0 \\ \tau & 0 & a \end{pmatrix}.$$

Easy calculations show that

$$A\left[\begin{pmatrix} a & \tau \\ \tau & a \end{pmatrix}\right]^* J A\left[\begin{pmatrix} a & \tau \\ \tau & a \end{pmatrix}\right] = J$$

Furthermore

$$\operatorname{Ber}\begin{pmatrix} a & 0 & \tau \\ 0 & 1 & 0 \\ \tau & 0 & a \end{pmatrix} = \det\left(\begin{pmatrix} a & 0 \\ 0 & 1 \end{pmatrix} - \begin{pmatrix} \tau \\ 0 \end{pmatrix} \cdot (a)^{-1} \cdot \begin{pmatrix} -\tau & 0 \end{pmatrix}\right) \cdot \det(a)^{-1}$$
$$= \det\begin{pmatrix} a & 0 \\ 0 & 1 \end{pmatrix} a^{-1} = 1.$$

Implying that the map A is well defined. Easy calculations show that A is a homomorphism of Lie super groups. Then $S^{1|1}$ is a super-subgroup of $SU(1,1|1)$ which is a group of isometries of the super unit disk. □

Since the action of the group $SU(1,1|1)$ on the super-manifold $\mathbb{B}^{1|1}$ is given by formulas (3.1), the corresponding action of $S^{1|1}$ is given by

$$z \mapsto w = az - \tau\zeta, \tag{3.2}$$
$$\zeta \mapsto \eta = \tau z + a\zeta. \tag{3.3}$$

4. Radial functions: $S^{1|1}$-invariant functions

In this section we find the explicit form of all functions, defined on the super unit disk, which are invariant under the action of $S^{1|1}$.

Theorem 4.1. *Let f be a smooth function defined on the super unit disk. If f is invariant under the action of $S^{1|1}$. Then f has the form*

$$f(z,\zeta) = f_0(r) + f_1(r)z\bar{\zeta} + f_1(r)\bar{z}\zeta - \frac{f_0'(r)}{2r}\bar{\zeta}\zeta, \tag{4.1}$$

where $r = |z|$ and f_0 and f_1 are radial functions.

Proof. An element $f \in C^\infty(\mathbb{B}^{1|1})$ is of the form

$$f(z,\zeta) = f_{00}(z) + f_{10}(z)\bar{\zeta} + f_{01}(z)\zeta + f_{11}(z)\bar{\zeta}\zeta, \tag{4.2}$$

where $f_{ij} \in C^\infty(\mathbb{B})$.

If f is an invariant function under the action of $S^{1|1}$, f holds the following equation

$$f(z,\zeta) = f(az - \tau\zeta, \tau z + a\zeta), \tag{4.3}$$

for each $(a,\tau) \in S^{1|1}$.

We start finding all functions which are invariant under the action of elements of the supercircle which have the form $(a,0)$. Let f be a function having this property. Then

$$f_{00}(z)+f_{10}(z)\bar{\zeta}+f_{01}(z)\zeta+f_{11}(z)\bar{\zeta}\zeta = f_{00}(az)+f_{01}(az)a\zeta+f_{10}(az)\bar{a}\bar{\zeta}+f_{11}(az)\bar{\zeta}\zeta.$$

Since a is any complex number in the unit circle, last equation implies that the functions f_{00} and f_{11} are radial functions. Moreover the function $f_{10}(z)$ can be written as $z\tilde{f}_{10}$ where $\tilde{f}_{10}$ is a radial function and $f_{01}(z) = \bar{z}\tilde{f}_{01}$, being $\tilde{f}_{01}$ also a radial function.

$$f(z,\zeta) = f_{00}(r) + z\tilde{f}_{10}(r)\bar{\zeta} + \bar{z}\tilde{f}_{01}(r)\zeta + f_{11}(r)\bar{\zeta}\zeta. \tag{4.4}$$

An invariant function under the action of $S^{1|1}$ must be invariant under the action of any element of the form $(1,\tau)$. Consider the special case for the variable $\tau = i\theta$, where θ is a real Grassmann variable. Under the transformation given by formula (3.2) we get the new variables

$$w = z - i\theta\zeta \quad \text{and} \qquad \eta = iz\theta + \zeta.$$

We note that

$$\begin{aligned} s := (w\bar{w})^{1/2} &= ((z - i\theta\zeta)(\bar{z} - i\theta\bar{\zeta}))^{1/2} \\ &= (z\bar{z} - i\theta\zeta\bar{z} - i\theta\bar{z}\zeta)^{1/2} \\ &= r - \tfrac{i\theta(\bar{z}\zeta + z\bar{\zeta})}{2r}. \end{aligned}$$

For any differentiable function $h(x,y)$ defined on the plane $\mathbb{R}^2$, the superposition $h(s)$ is defined by the formula

$$h(s) = h(r) - \frac{i\theta(\bar{z}\zeta + z\bar{\zeta})}{2r} h'(r). \tag{4.5}$$

We will use the same notation for the superposition of a function and the original function. Then

$$h(s)\bar{w}\eta = \left(h(r) - \frac{i\theta(\bar{z}\zeta + z\bar{\zeta})}{2r} h'(r) \right) (\bar{z} - i\theta\bar{\zeta})(iz\theta + \zeta) = h(r)(ir^2\theta + \bar{z}\zeta), \tag{4.6}$$

and

$$h(s)w\bar{\eta} = h(r)(z\bar{\zeta} - ir^2\theta). \tag{4.7}$$

Analogously

$$h(s)\bar{\eta}\eta = \left(h(r) - \frac{i\theta(\bar{z}\zeta + z\bar{\zeta})}{2r} h'(r) \right) (-i\bar{z}\theta + \bar{\zeta})(iz\theta + \zeta) = h(r)(-i\theta(\bar{z}\zeta + iz\bar{\zeta}) + \bar{\zeta}\zeta). \tag{4.8}$$

As a consequence of equations (4.6)–(4.8), we have

$$\begin{aligned} f(w,\eta) &= f_{00}(s) + \tilde{f}_{01}(s)\bar{w}\eta + \tilde{f}_{10}(s)w\bar{\eta} + f_{11}(s)\bar{\eta}\eta \\ &= f_{00}(r) - \frac{i\theta(\bar{z}\zeta + z\bar{\zeta})}{2r} f'_{00}(r) + \tilde{f}_{01}(r)(ir^2\theta + \bar{z}\zeta) + \tilde{f}_{10}(r)(z\bar{\zeta} - ir^2\theta) \\ &\quad + f_{11}(r)(-i\theta(\bar{z}\zeta + z\bar{\zeta}) + \bar{\zeta}\zeta). \end{aligned}$$

Therefore, a function f is invariant under the action of the unit supercircle if this function satisfies Equation (4.4) and the following condition:

$$f_{11}(r) = \frac{-f'_{00}(r)}{2r} \text{ and } \tilde{f}_{10}(r) = \tilde{f}_{01}(r). \tag{4.9}$$

□

5. Super Toeplitz operators with radial symbols

Generalizing the characterization of radial complex-valued functions we define a radial super function as follows. A radial super function is an invariant function under the action of $S^{1|1}$. By Theorem 4.1 radial super functions have the following form

$$f(z) = f_0(r) + f_1(r)\overline{z}\zeta + f_1(r)z\overline{\zeta} - \frac{f_0'(r)}{2r}\overline{\zeta}\zeta. \tag{5.1}$$

For a radial super function f consider the super Toeplitz operator T_f acting on $H^2_\nu(\mathbb{B}^{1|1})$. The main result of this paper is the following theorem.

Theorem 5.1. *A super Toeplitz operator with radial symbol is a diagonal operator.*

Proof. Consider the base $\{z^n, z^n\zeta | n \geq 0\}$ for $H^2_\nu(\mathbb{B}^{1|1})$. Let f be a radial super function and $n \geq 1$. Then

$$\begin{aligned}
&T^\nu_\nu(f_0 - \frac{1-w\overline{w}}{\nu-1}\frac{f_0'}{2r})(z^n)\\
&= \frac{\nu-1}{\pi}\int_{\mathbb{B}} dw(1-w\overline{w})^{\nu-2}(1-z\overline{w})^{-\nu}f_0(r)w^n\\
&\quad - \int_{\mathbb{B}} dw(1-w\overline{w})^{\nu-2}(1-z\overline{w})^{-\nu}(1-w\overline{w})\frac{f_0'(r)}{2r}w^n\\
&= \frac{\nu-1}{\pi}\int_{\mathbb{B}} dw(1-w\overline{w})^{\nu-2}\left(\sum_{k=0}^{\infty}\frac{\Gamma(k+\nu)}{k!\Gamma(\nu)}z^k\overline{w}^k\right)f_0(r)w^n\\
&\quad - \int_{\mathbb{B}} dw(1-w\overline{w})^{\nu-2}\left(\sum_{k=0}^{\infty}\frac{\Gamma(k+\nu)}{k!\Gamma(\nu)}z^k\overline{w}^k\right)(1-w\overline{w})\frac{f_0'(r)}{2r}w^n.
\end{aligned}$$

Calculating the first integral of last equation we get

$$\begin{aligned}
&\frac{\nu-1}{\pi}\int_{\mathbb{B}} dw(1-w\overline{w})^{\nu-2}\left(\sum_{k=0}^{\infty}\frac{\Gamma(k+\nu)}{k!\Gamma(\nu)}z^k\overline{w}^k\right)f_0(r)w^n\\
&= \frac{\Gamma(n+\nu)}{n!\Gamma(\nu)}\frac{\nu-1}{\pi}z^n\int_{\mathbb{B}} dw(1-w\overline{w})^{\nu-2}f_0(r)|w|^{2n}\\
&= \frac{\Gamma(n+\nu)}{n!\Gamma(\nu)}\frac{\nu-1}{\pi}z^n\int_0^1\int_0^{2\pi} drdt(1-r^2)^{\nu-2}r^{2n}f_0(r)r\\
&= 2\frac{\Gamma(n+\nu)}{n!\Gamma(\nu)}(\nu-1)z^n\int_0^1 dr(1-r^2)^{\nu-2}r^{2n+1}f_0(r)\\
&= \frac{\Gamma(n+\nu)}{n!\Gamma(\nu)}z^n\left[\int_0^1(1-r^2)^{\nu-1}\frac{f_0'(r)}{2r}r^{2n+1} + 2n\int_0^1 f_0(r)\frac{r^{2n-1}}{2}dr\right].
\end{aligned}$$

The first integral of last equation can be written in the following form

$$\int_0^1 dr(1-r^2)^{\nu-2}r^{2n+1}f_0(r) = \frac{1}{2(\nu-1)}\int(1-r^2)^{\nu-1}[f_0'(r)r^{2n} + f_0(r)2nr^{2n-1}]dr.$$

On the other hand

$$\int_{\mathbb{B}} dw(1-w\overline{w})^{\nu-2}\Big(\sum_{k=0}^{\infty}\frac{\Gamma(k+\nu)}{k!\Gamma(\nu)}z^k\overline{w}^k\Big)(1-w\overline{w})\frac{f_0'(r)}{2r}w^n$$

$$=\frac{\Gamma(n+\nu)}{n!\Gamma(\nu)}2z^n\int_0^1\frac{(1-r^2)^{\nu-1}r^{2n+1}f_0'(r)}{2r}dr.$$

Then

$$T_\nu^\nu(f_{00}+\frac{1-\omega\overline{\omega}}{\nu-1}f_{11})(z^n)=\frac{\Gamma(n+\nu)}{n!\Gamma(\nu)}z^n2\int_0^1(1-r^2)^{\nu-1}r^{2n}f_0'(r)dr.$$

$$\begin{aligned}T_{\nu+1}^\nu(f_1(r)\overline{z})(z^n)&=\frac{\nu}{\pi}\int_{\mathbb{B}}dw(1-w\overline{w})^{\nu-1}(1-z\overline{w})^{-\nu-1}f_1(r)\overline{w}w^n\\&=\frac{\nu}{\pi}\int_{\mathbb{B}}dw(1-w\overline{w})^{\nu-1}[\sum_{k=0}^{\infty}\frac{\Gamma(k+\nu+1)}{n!\Gamma(\nu+1)}z^k\overline{w}^k]f_1(r)\overline{w}w^n\\&=\frac{\nu}{\pi}\int_{\mathbb{B}}dw(1-w\overline{w})^{\nu-1}\frac{\Gamma(n-1+\nu+1)}{(n-1)!\Gamma(\nu+1)}z^{n-1}|w|^{2n}f_1(r)\\&=\frac{\Gamma(n+\nu)}{(n-1)!\Gamma(\nu+1)}\frac{\nu}{\pi}z^{n-1}\int_{\mathbb{B}}dw(1-w\overline{w})^{\nu-1}|w|^{2n}f_1(r)\\&=\frac{\Gamma(n+\nu)}{(n-1)!\Gamma(\nu+1)}\frac{\nu}{\pi}z^{n-1}2\pi\int_0^1(1-r^2)^{\nu-1}r^{2n+1}f_1(r)dr\\&=\left[\frac{2\nu\Gamma(n+\nu)}{(n-1)!\Gamma(\nu+1)}\int_0^1(1-r^2)^{\nu-1}r^{2n+1}f_1(r)dr\right]z^{n-1}.\end{aligned}$$

Using the results above we get

$$\begin{aligned}T_f(z^n)=&\left[\frac{2n\Gamma(n+\nu)}{n!\Gamma(\nu)}\int_0^1(1-r^2)^{\nu-1}r^{2n-1}f_0(r)dr\right]z^n\\&+\left[\frac{2\nu\Gamma(n+\nu)}{(n-1)!\Gamma(\nu+1)}\int_0^1(1-r^2)^{\nu-1}r^{2n+1}f_1(r)dr\right]z^{n-1}\zeta,n\geq 1.\end{aligned}\tag{5.2}$$

The case $n=0$ easily reduces to the following formula

$$T_f(z^0)=\frac{\nu-1}{\pi}\int_{\mathbb{B}}dw(1-w\bar{w})^{\nu-2}f_0(r)-\int_{\mathbb{B}}dw(1-w\bar{w})^{\nu-1}\frac{f_0'(r)}{2r}.\tag{5.3}$$

According to formula (2.4) in order to calculate $T_f(z^n\zeta),n\geq 0$, we need to compute $T_\nu^{\nu+1}(\frac{1-w\overline{w}}{\nu-1}(f_1(r)w)(z^n\zeta)$ and $T_{\nu+1}^{\nu+1}(f_0(r))(z^n)$. Then

$$T_\nu^{\nu+1}\left(\frac{1-w\overline{w}}{\nu-1}(f_1(r)w\right)z^n$$

$$=\frac{1}{\pi}\int_{\mathbb{B}}dw(1-w\overline{w}^{\nu-2})(1-z\overline{w}))^{-\nu}(1-w\overline{w})(-f_1(r))w^{n+1}$$

$$= \frac{1}{\pi}\int_{\mathbb{B}} dw(1-w\overline{w})^{\nu-1}[\sum_{k=0}^{\infty}\frac{\Gamma(k+\nu)}{k!\Gamma(\nu)}z^k\overline{w}^k]f_1(r)w^{n+1}$$

$$= \frac{1}{\pi}\int_{\mathbb{B}} dw(1-w\overline{w})^{\nu-1}\frac{\Gamma(n+1+\nu)}{(n+1)!\Gamma(\nu)}z^{n+1}|w|^{2(n+1)}f_1(r)$$

$$= \frac{2\Gamma(n+1+\nu)}{(n+1)!\Gamma(\nu)}z^{n+1}\int_0^1 (1-r^2)^{\nu-1}r^{2(n+1)+1}f_1(r)dr. \tag{5.4}$$

The operator $T_{\nu+1}^{\nu+1}(f_0(r))$ acts on z^n in the following form

$$T_{\nu+1}^{\nu+1}(f_0(r))z^n = \frac{\nu}{\pi}\int dw(1-w\overline{w})^{\nu-1}(1-z\overline{w})^{-\nu-1}f_0(r)w^n$$

$$= \frac{\nu}{\pi}\int dw(1-w\overline{w})^{\nu-1}[\sum_{k=0}^{\infty}\frac{\Gamma(k+\nu+1)}{k!\Gamma(\nu+1)z^k\overline{w}^k}]f_0(r)w^n$$

$$= \frac{\nu}{\pi}\int dw(1-w\overline{w})^{\nu-1}\frac{\Gamma(n+\nu+1)}{n!\Gamma(\nu+1)}z^n|w|^{2n}f_0(r)$$

$$= \frac{2\nu\Gamma(n+\nu+1)}{n!\Gamma(\nu+1)}[\int_0^1 (1-r^2)^{\nu-1}r^{2n+1}f_0(r)dr]z^n. \tag{5.5}$$

Using equations (5.4) and (5.5) we get the following formula

$$T_f(z^n\zeta) = \left[\frac{2\Gamma(n+1+\nu)}{(n+1)!\Gamma(\nu)}\int_0^1 (1-r^2)^{\nu-1}r^{2(n+1)+1}f_1(r)dr\right]z^{n+1}$$
$$+\left[\frac{2\nu\Gamma(n+\nu+1)}{n!\Gamma(\nu+1)}\int_0^1 (1-r^2)^{\nu-1}r^{2n+1}f_0(r)dr\right]z^n\zeta, n\geq 0. \tag{5.6}$$

From equations (5.2), (5.3) and (5.6), the space generated by z^0, i.e., all constant functions, and the space generated by $\{z^n, z^{n-1}\zeta\}, n\geq 0$, are invariant subspaces for the operator T_f.

For $n\geq 1$ the restriction of the operator T_f to the bi-dimensional space generated by $\{z^n, z^{n-1}\zeta\}$ can be written in matrix form as follows

$$\frac{2\Gamma(n+\nu)}{(n-1)!\Gamma(\nu)}\begin{pmatrix} \int_0^1(1-r^2)^{\nu-1}r^{2n-1}f_0(r)dr & \int_0^1(1-r^2)^{\nu-1}r^{2n+1}f_1(r)dr/n \\ \int_0^1(1-r^2)^{\nu-1}r^{2n+1}f_1(r)dr & \int_0^1(1-r^2)^{\nu-1}r^{2n-1}f_0(r)dr \end{pmatrix}.$$

This 2×2 matrix has the following eigenvalues

$$\frac{2\Gamma(n+\nu)}{(n-1)!\Gamma(\nu)}\left(\int_0^1 (1-r^2)^{\nu-1}r^{2n-1}f_0(r)dr) \pm \int_0^1 (1-r^2)^{\nu-1}r^{2n+1}f_1(r)dr/\sqrt{n}\right).$$

The corresponding eigenvectors are

$$\omega_{n_1} = \frac{z^n}{\sqrt{n}} + z^{n-1}\zeta, \qquad \omega_{n_2} = -\frac{z^n}{\sqrt{n}} + z^{n-1}\zeta,$$

which norms are

$$\|\omega_{n_1}\| = \|\omega_{n_2}\| = \sqrt{\frac{(n-1)!\Gamma(\nu)}{\Gamma(n+\nu)} + \frac{n!\Gamma(\nu)}{(n+\nu)\Gamma(n+\nu)}}.$$

Let $\{1, \omega_{n_1}, \omega_{n_2} | n \geq 0\}$ be the orthonormal basis obtained normalizing all eigenvectors. Then the super Toeplitz operator with radial symbol f is the diagonal operator such that

$$T_f(w_{n_i}) = \frac{2\Gamma(n+\nu)}{(n-1)!\Gamma(\nu)} \left(\int_0^1 (1-r^2)^{\nu-1} r^{2n-1} f_0(r) dr \right. \\ \left. + (-1)^{i+1} \int_0^1 (1-r^2)^{\nu-1} r^{2n+1} f_1(r) dr/\sqrt{n} \right) w_{n_i}, \tag{5.7}$$

for $i = 1, 2, n = 1, \ldots$ □

Corollary 5.2. *The spectrum of the Toeplitz operator with radial symbol* $f(z) = f_0(r) + f_1(r)\overline{z}\zeta + f_1(r)z\overline{\zeta} - \frac{f_0'(r)}{2r}\overline{\zeta}\zeta$ *is*

$$\left\{ \frac{2\Gamma(n+\nu)}{(n-1)!\Gamma(\nu)} \left(\int_0^1 (1-r^2)^{\nu-1} r^{2n-1} f_0(r) dr \pm \int_0^1 (1-r^2)^{\nu-1} r^{2n+1} f_1(r) dr/\sqrt{n} \right), \right.$$

$$\left. n = 1, \ldots, \frac{\nu-1}{\pi} \int_{\mathbb{B}} dw (1-w\bar{w})^{\nu-2} f_0(r) - \int_{\mathbb{B}} dw (1-w\bar{w})^{\nu-1} \frac{f_0'(r)}{2r} \right\}$$

The corresponding eigenfunctions are

$$\omega_{n_1} = \frac{z^n}{\sqrt{n}} + z^{n-1}\zeta$$
$$\omega_{n_2} = -\frac{z^n}{\sqrt{n}} + z^{n-1}\zeta$$
$$\omega_0 := z^0$$

Proof. It follows from the proof of last theorem. □

Corollary 5.3. *The Toeplitz algebra generated by all super Toeplitz operators with radial symbols is commutative.*

Since the second factor of formula (5.7) has the term $1/\sqrt{n}$, if f_1 is a bounded function we have the following statements.

Corollary 5.4.

1. *The super Toeplitz operator* T_f *is bounded if and only if*

$$\left(\frac{2\Gamma(n+\nu)}{(n-1)!\Gamma(\nu)} \int_0^1 (1-r^2)^{\nu-1} r^{2n-1} f_0(r) dr \right)$$

   is a bounded sequence
2. *The super Toeplitz operator* T_f *is compact if and only if the sequence*

$$\left(\frac{2\Gamma(n+\nu)}{(n-1)!\Gamma(\nu)} \int_0^1 (1-r^2)^{\nu-1} r^{2n-1} f_0(r) dr \right),$$

   tends to zero when $n \to \infty$.

6. A generalization of the Toeplitz algebra with radial symbol (classical)

In the definition of radial function given above we exclude all complex-valued radial functions. In order to see the classical case of Toeplitz operators with radial symbols as elements of a commutative algebra of super Toeplitz we introduce the following group of elements of $SU(1|1|1)$. For two complex numbers $z = e^{it}, w = e^{is}$ in the unit circle, consider the element of $SU(1,1|1)$ given by the following matrix

$$T_{s,t} = \begin{pmatrix} e^{is} & 0 & 0 \\ 0 & e^{it} & 0 \\ 0 & 0 & e^{i(t+s)} \end{pmatrix}.$$

It is easy to prove that $T^2 = \{T_{s,t} | s, t \in [0, 2\pi)\}$ is a subgroup of $SU(1,1|1)$.

Theorem 6.1. *Let f be a bounded function on the unit super disk. If f is invariant under the action of T^2 then f has the form*

$$f(z,\zeta) = f_0(r) + f_1(r)\bar{\zeta}\zeta, \tag{6.1}$$

where f_0 and f_1 are radial functions.

Proof. The proof is analogous to the proof of Theorem 4.1. □

From Theorem 2.5 the super Toeplitz operator with symbol $f(z,\zeta) = f_0(r) + f_1(r)\bar{\zeta}\zeta$ has the form

$$\begin{pmatrix} T_\nu^\nu \left(f_0 + \frac{1-w\overline{w}}{\nu-1} f_1\right) & 0 \\ 0 & T_{\nu+1}^{\nu+1}(f_0) \end{pmatrix}. \tag{6.2}$$

Theorem 6.2. *The Toeplitz algebra generated by all super Toeplitz operators which symbols are invariant under the action of T^2 is commutative.*

Proof. Note that since f_0 and f_1 are radial functions the super Toeplitz operator with symbol $f_0(r) + f_1(r)\bar{\zeta}\zeta$ is a diagonal operator. □

References

[1] F.A. Berezin, *Introduction to Superanalysis*, Reidel, Dordrecht, 1987.

[2] D. Borthwick, S. Klimek, A. Lesniewski, M. Rinaldi, *Super Toeplitz operators and non-perturbative deformation quantization of supermanifolds*, Commun. Math. Phys. **153** (1993), 49–76.

[3] S. Grudsky, A. Karapetyants, N. Vasilevski, *Dynamics of properties of Toeplitz Operators with radial symbols*, Integral Equations and Operator Theory 50 (2004), 217–253.

[4] S. Grudsky, A. Karapetyants, N. Vasilevski, *Toeplitz operators on the unit ball in $\mathbb{C}^n$ with radial symbols*, Journal of Operator Theory 49 (2003), 325–346.

[5] S. Grudsky, R. Quiroga, N. Vasilevski, *Commutative C^*-algebras of Toeplitz operators and quantization on the unit disk*, Journal of Functional Analysis 234 (2006), 1–44.

[6] H. Hedenmalm, B. Korenblum, K. Zhu, *Theory of Bergman Spaces*, Springer-Verlag, New York, 2000.

[7] M. Loaiza, H. Upmeier, *Toeplitz C*-algebras on super Cartan domains*, Revista Matemática Complutense, vol. 21 (2), 489–518.

[8] E. Prieto-Zanabria, *Phd Dessertation* 2007, CINVESTAV, Mexico.

[9] R. Quiroga-Barranco, N. Vasilevski, *Commutative algebras of Toeplitz operators on Reinhardt domains*, Integral Equations and Operator Theory 59, 67–98, 2007.

[10] Quiroga-Barranco R., and Vasilevski N., *Commutative C*-algebra of Toeplitz operators on the unit ball, I. Bargmann type transform and spectral representations of Toeplitz operators*, Integral Equations and Operator Theory, 59(3):379–419, 2007.

[11] Quiroga-Barranco R., and Vasilevski N., *Commutative C*-algebra of Toeplitz operators on the unit ball, II. Geometry of the level sets of symbols*, Integral Equations and Operator Theory, 59(1):89–132, 2008.

[12] A. Sánchez-Nungaray, *Commutative Algebras of Toeplitz Operators on the Supersphere of dimension* (2|2) Phd Dissertation 2009, CINVESTAV, Mexico.

[13] K. Zhu, *Spaces of Holomorphic Functions in the Unit Ball*, Springer-Verlag, New York, 2005.

M. Loaiza and A. Sánchez-Nungaray
Departamento de Matemáticas
CINVESTAV
Apartado Postal 14-740 07000
México, Mexico
e-mail: `mloaiza@math.cinvestav.mx`
`arn@math.cinvestav.mx`

Operator Theory:
Advances and Applications, Vol. 210, 189–193

Universality of Some C^*-Algebra Generated by a Unitary and a Self-adjoint Idempotent

Helena Mascarenhas and Bernd Silbermann

Dedicated to Nikolai Vasilevski on the occasion of his 60th birthday

Abstract. We prove that there is essentially only one C^*-algebra generated by a unitary element u and a self-adjoint idempotent p such that

$$up = pup \text{ and } up \neq pu\,.$$

This result is related to a theorem of L. Coburn stating that there is essentially only one C^*-algebra generated by a non-unitary isometry.

Mathematics Subject Classification (2000). 46L05.

Keywords. C^*-algebra, finitely generated, universality.

1. Introduction

A theorem of L. Coburn [1] tells that the C^*-algebra generated by a non-unitary isometry v (that is $v^*v = e, vv^* \neq e$) is universal in the sense that any two C^*-algebras with this property are isometrically isomorphic. An instructive example for such a C^*-algebra is given as follows: Let $H^2 \subset L^2(\mathbb{T}), \mathbb{T} := \{z \in \mathbb{C} : |z| = 1\}$, be the familiar Hardy space and $P^+ : L^2(\mathbb{T}) \to H^2$ the Riesz projection which is known to be self-adjoint and surjective. Given $a \in L^\infty(\mathbb{T})$, define the operator

$$T(a) : H^2 \to H^2\,,\ f \mapsto P^+ af\,.$$

The operator $T(a)$ is clearly bounded, and is called a Toeplitz operator. Let $\chi_{\pm 1}$ be the functions $\chi_{\pm 1}(t) = t^{\pm 1}, t \in \mathbb{T}$. Then $T^*(\chi_1) = T(\chi_{-1})$ and $T^*(\chi_1)T(\chi_1) = I$, but $T(\chi_1)T^*(\chi_1) \neq I$. Hence, $T(\chi_1)$ is a non-unitary isometry. It is well known that the smallest C^*-algebra $\mathcal{T}(C) \subset \mathcal{B}(H^2)$ generated by $T(\chi_1)$ ($\mathcal{B}(X)$ stands for the Banach algebra of all bounded linear operators acting on the Banach space

First author partially supported by CEAF, IST, Technical Univ. of Lisbon, and "Fundação para a Ciência e a Tecnologia" through the program FCT/FEDER/POCTI/MAT/59972/2004.

X) contains all Toeplitz operators $T(a)$, with $a \in C(\mathbb{T})$, where $C(\mathbb{T})$ denotes the algebra of all continuous functions on $\mathbb{T}$. Moreover,

$$\mathcal{T}(C) = \{T(a) : a \in C(\mathbb{T})\} \dot{+} \mathcal{K}(H^2),$$

where $\mathcal{K}(H^2)$ denotes the ideal of all compact operators acting on H^2 (see for instance [3], Corollary 4.15 and Theorem 4.24).

Clearly, $\chi_1 I$ is a unitary operator on $L^2(\mathbb{T}), (\chi_1 I)^* = \chi_{-1} I$, and P^+ is a self-adjoint projection. It is easy to see that

$$\chi_1 P^+ = P^+ \chi_1 P^+ \text{ and } \chi_1 P^+ \neq P^+ \chi_1 I.$$

The smallest C^*-subalgebra $SO(C) \subset \mathcal{B}(L^2(\mathbb{T}))$ containing $\chi_1 I$ and P^+ contains all singular integral operators

$$A = aP^+ + bP^-$$

with $P^- := I - P^+$ and $a, b \in C(\mathbb{T})$. Moreover,

$$SO(C) = \{P^+ a P^+ + P^- b P^- : a, b \in C(\mathbb{T})\} \dot{+} \mathcal{K}(L^2(\mathbb{T})),$$

where $\mathcal{K}(L^2(\mathbb{T}))$ is again the ideal of all compact operators acting on $L^2(\mathbb{T})$ (see [3], Corollary 4.76; notice that this corollary is applicable in our case, and that $QC(U)$ can be identified with $\mathcal{K}(L^2(\mathbb{T}))$).

These considerations show that $\mathcal{T}(C)$ is a subalgebra of $SO(C)$. The close relationship between Toeplitz and singular integral operators gives rise to the question whether $SO(C)$ is a model for a universal C^*-algebra. We will show that this is indeed the case. Our proof is based on features known from the theory of singular integral operators with continuous coefficients. This theory is developed in many text books, and we will use mainly [2] and [3] because the approach given there is mostly convenient for us. Let us also notice that the study of finitely generated algebras is an important task. More about this topic can be found, in particular, in [4] and [5].

2. The main result and its proof

Let $\mathcal{A}$ be any C^*-algebra generated by a unitary element u (that is $uu^* = u^*u = e$) and by a self-adjoint idempotent p^+ such that

$$up^+ = p^+ up^+ \text{ und } up^+ \neq p^+ u. \tag{1}$$

Passing to adjoints and using $u^{-1} = u^*$ yields

$$p^+ u^{-1} = p^+ u^{-1} p^+, \tag{2}$$

and thus

$$u^{-1} p^- = p^- u^{-1} p^-,$$

where $p^- := e - p^+$.

We denote by $\sum$ the class of all C^*-algebras of this type.

Theorem. *Let $\mathcal{A}_1$ and $\mathcal{A}_2$ be arbitrary C^*-algebras belonging to $\sum$. Then these algebras are isometrically isomorphic. This isomorphism can be chosen so that u_1, p_1 are taken into u_2, p_2, respectively.*

Proof. Due to the GNS-construction we may assume (without loss of generality) that there is a Hilbert space $\mathcal{H}$ such that for $\mathcal{A} \in \sum$ the elements u and p^+ are operators on $\mathcal{H}$ which generate $\mathcal{A}$.

Let $q \in C(\mathbb{T})$ be an arbitrarily given quasipolynomial,

$$q(t) = \sum_{j=-k}^{k} a_j t^j \,, \; a_j \in \mathbb{C} \text{ and form } q(u) = \sum_{j=-k}^{k} a_j u^j \,.$$

We call $q(u) \in \mathcal{A}$ a quasipolynomial of u and let $\mathcal{L}_0(u)$ stand for the (non-closed) algebra of all quasipolynomials of u, and let $\mathcal{L}(u)$ be the closure of $\mathcal{L}_0(u)$ in $\mathcal{B}(\mathcal{H})$. Since u is a unitary element, we know that

$$\mathrm{sp}(u) = \mathrm{sp}(u^{-1}) = \mathbb{T} \,.$$

The general theory of commutative C^*-algebras entails that $\mathcal{L}(u)$ is isometrically isomorphic to $C(\mathbb{T})$, and the isomorphism takes u into χ_1:

$$\mathcal{L}(u) \cong C(\mathbb{T}) \,.$$

Now let $\mathcal{A} \in \sum$ be generated by u and p^+. Introduce the (dense) subalgebra Z of $\mathcal{A}$:

$$Z := \left\{ \sum_{v=1}^{M} \prod_{s=1}^{N} (a_{vs}(u)p^+ + b_{vs}(u)p^-) : a_{vs}(u), b_{vs}(u) \in \mathcal{L}_0(u), M, N \in \mathbb{N} \right\} .$$

Take an element $c \in Z$, say

$$c := \sum_{v=1}^{M} \prod_{s=1}^{N} (a_{vs}(u)p^+ + b_{vs}(u)p^-)$$

and suppose that it is invertible in $\mathcal{A}$. We would like to express its invertibility in terms which can be used for further analysis. For this aim we proceed as follows: Given $r \in \mathbb{N}$ let $\mathcal{H}^r$ stand for $\{(h_1, \ldots, h_r) : h_j \in \mathcal{H}, \, j = 1, \ldots, r\}$. This linear space becomes a Hilbert space by introducing the scalar product

$$\langle (h_1, \ldots, h_r), (g_1, \ldots, g_r) \rangle = \sum_{j=1}^{r} \langle h_j, g_j \rangle \,,$$

and $\mathcal{B}(\mathcal{H}^r)$ can be identified with $(\mathcal{B}(\mathcal{H}))^{r \times r}$ in a natural way. Then form a linear extension $\tilde{c}$ of c exactly as it is done in [2], Chapter VIII, § 10. The properties of $\tilde{c}$ are (and only these are needed):

1. $\tilde{c} \in (\mathcal{B}(\mathcal{H}))^{r \times r}$ for some $r \in \mathbb{N}$ and the entries of $\tilde{c}$ are among the elements $O, I, \, a_{11}(u)p^+ + b_{11}(u)p^-, \ldots, a_{MN}(u)p^+ + b_{MN}(u)p^-$.
2. c is invertible in $\mathcal{B}(\mathcal{H})$ (and therefore in $\mathcal{A}$ by the inverse closedness property of unital C^*-subalgebras) if and only if $\tilde{c}$ is invertible in $\mathcal{B}(\mathcal{H}^r)$.

Now observe that $\tilde{c}$ can be written by 1. as

$$\tilde{c} = A_1(u)\,\mathrm{diag}\,\{\underbrace{p^+,\ldots,p^+}_{r-\text{times}}\} + A_2(u)\,\mathrm{diag}\,\{\underbrace{p^-,\ldots,p^-}_{r-\text{times}}\},$$

where $A_1(u), A_2(u)$ are $r \times r$-matrices with entries from $\mathcal{L}_0(u)$. Now we apply the theory of [2], Chapter VIII, § 8 (the condition dim coker $(u|\mathrm{im}P^+) < \infty$, which is formulated there, can be dropped down because we are only interested in invertibility):

a) the invertibility of $\tilde{c}$ implies the invertibility of $A_1(u)$ and $A_2(u)$, that is

$$\det A_1(t) \neq 0, \det A_2(t) \neq 0 \text{ for all } t \in \mathbb{T},$$

where $A_1(t)$ and $A_2(t)$ are the matrices whose entries are the Gelfand transforms of the entries of $A_1(u)$ and $A_2(u)$, respectively.

b) Notice that the functions $A_1(t), A_2(t)$ belong to the class $W^{r\times r}$, where W is the Wiener algebra. Moreover, the invertibility of $\tilde{c}$ implies that the right canonical Wiener-Hopf factorization of $t \mapsto A_2^{-1}(t)A_1(t)$ $(\in W^{r\times r})$ exists with all partial indices equal to zero. Conversely, if this condition is fulfilled then $\tilde{c}$, and thus also c, is invertible. Notice that this fact is true independently of the choice of $\mathcal{A}$.

Now take two algebras $\mathcal{A}_1$ and $\mathcal{A}_2$ from $\sum$. The element c and the algebra Z are denoted now by c_1, c_2 and Z_1, Z_2, respectively. Further, we define a map $\psi : Z_1 \to Z_2, c_1 \mapsto c_2$ and we need to show that ψ is correctly defined. Using that c_1 is invertible if and only if c_2 is invertible (in $\mathcal{A}_i$) then by the above argument, we get

$$\mathrm{sp}(c_1) = \mathrm{sp}(c_2).$$

Because $c_1^*c_1 \in Z_1, c_2^*c_2 \in Z_2$, we therefore obtain

$$\mathrm{sp}(c_1^*c_1) = \mathrm{sp}(c_2^*c_2),$$

and using $\|c_1\|^2 = \|c_1^*c_1\| = \|c_2^*c_2\| = \|c_2\|^2$, that

$$\|c_1\| = \|c_2\|.$$

(Recall that for a self-adjoint element the norm and the spectral radius coincide.) This equality shows that ψ is correctly defined and that ψ represents an isometric isomorphism between Z_1 and Z_2. Moreover, $\psi u_1 = u_2, \psi p_1^+ = p_2^+$. The continuous extension of ψ to the whole of $\mathcal{A}_1$ provides us with the wanted isomorphism and finishes the proof. □

Remark: *This proof can easily be adapted to prove Coburns result which was originally proved by different methods.*

References

[1] L. Coburn: The C^*-algebra generated by an isometry. *Bull. Amer. Soc.* **73** (1967), 722–726.

[2] I. Gohberg, I. Feldman: Convolution Equations and Projection Methods for Their Solution. Nauka, Moskva 1971 (Russian) [Engl. translation: Transl. of Math. Monographs 41, Amer. Math. Soc., Providence, R.I., 1974].

[3] S. Prössdorf, B. Silbermann: *Numerical Analysis for Integral and Related Operator Equations.* Akademie-Verlag Berlin 1991, and Birkhäuser Verlag, Basel-Boston-Stuttgart, 1991.

[4] V. Ostrovskyi, Y. Samoilenko: *Introduction to the theory of representations of finitely presented $*$-algebras.* I. Representations by bounded operators, Harwood Academic Publishers, Amsterdam, 1999.

[5] N. Vasilevskii: C^*-algebras generated by orthogonal projections and their applications. *IEOT* **31** (1998), no. 1, 113–132.

Helena Mascarenhas
Instituto Superior Técnico
Univ. Técnica de Lisboa
Av. Rovisco Pais
1049-001 Lisbon, Portugal
e-mail: `hmasc@math.ist.utl.pt`

Bernd Silbermann
Technical University of Chemnitz
Faculty of Mathematics
D-09107 Chemnitz, Germany
e-mail: `silbermn@mathematik.tu-chemnitz.de`

Operator Theory:
Advances and Applications, Vol. 210, 195–201

Commutative Algebras of Toeplitz Operators and Lagrangian Foliations

R. Quiroga-Barranco

To Nikolai Vasilevski on the occasion of his 60th birthday

Abstract. Let D be a homogeneous bounded domain of $\mathbb{C}^n$ and $\mathcal{A}$ a set of (anti-Wick) symbols that defines a commutative algebra of Toeplitz operators on every weighted Bergman space of D. We prove that if $\mathcal{A}$ is rich enough, then it has an underlying geometric structure given by a Lagrangian foliation.

Mathematics Subject Classification (2000). 47B35, 32M10, 57R30, 53D05.

Keywords. Toeplitz operators, Lagrangian submanifolds, foliations.

1. Introduction

In recent work, Vasilevski and his collaborators discovered unexpected commutative algebras of Toeplitz operators acting on weighted Bergman spaces on the unit disk (see [7] and [8]). It was even possible to classify all such algebras as long as they are assumed commutative for every weight and suitable richness conditions hold; the latter ensure that the (anti-Wick) symbols that define the Toeplitz operators have enough infinitesimal linear independence. We refer to [3] for further details.

Latter on, it was found out that the phenomenon of the existence of nontrivial commutative algebras of Toeplitz operators extends to the unit ball $\mathbb{B}^n$ in $\mathbb{C}^n$. There exists at least $n + 2$ nonequivalent such commutative algebras which were exhibited explicitly in [5] and [6]. The discovery of such commutative algebras of Toeplitz operators on $\mathbb{B}^n$ was closely related to a profound understanding of the geometry of this domain. The description of these commutative algebras of Toeplitz operators involved Lagrangian foliations, i.e., by Lagrangian submanifolds (see Section 2 for detailed definitions), with distinguished geometric properties.

In order to completely understand the commutative algebras of Toeplitz operators on $\mathbb{B}^n$ and other domains, it is necessary to determine the general features that produce such algebras. In particular, there is the question as to whether or

Research supported by CONACYT, CONCYTEG and SNI.

not the commutative algebras of Toeplitz operators have always a geometric origin. More precisely, whether or not they are always given by a Lagrangian foliation.

The main goal of this paper is to prove that, on a homogeneous domain, commutative algebras of Toeplitz operators are always obtained from Lagrangian foliations. This is so at least when the commutativity holds on every weighted Bergman space and a suitable richness condition on the symbols holds. To state this claim we use the following notation. We denote by $C^\infty(M)$ the space of smooth complex-valued functions on a manifold M and by $C^\infty_b(M)$ the subspace of bounded functions. For a given foliation $\mathcal{F}$ of a manifold M, we will denote by $\mathcal{A}_\mathcal{F}(M)$ the vector subspace of $C^\infty(M)$ that consists of those functions which are constant along every leaf of $\mathcal{F}$. Our main result is the following.

Main Theorem. *Let D be a bounded domain in $\mathbb{C}^n$ and $\mathcal{A}$ a vector subspace of $C^\infty_b(M)$ which is closed under complex conjugation. If for every $h \in (0,1)$ the Toeplitz operator algebra $\mathcal{T}_h(\mathcal{A})$, acting on the weighted Bergman space $\mathcal{A}^2_h(D)$, is commutative and $\mathcal{A}$ satisfies the following richness condition:*

- *for some closed nowhere dense subset $S \subseteq D$ and for every $p \in D \setminus S$ there exist real-valued elements $a_1, \dots, a_n \in \mathcal{A}$ such that $da_{1p}, \dots, da_{np}$ are linearly independent over $\mathbb{R}$,*

then, there is a Lagrangian foliation $\mathcal{F}$ of $D \setminus S$ such that $\mathcal{A}|_{D\setminus S} \subseteq \mathcal{A}_\mathcal{F}(D \setminus S)$. In other words, every element of $\mathcal{A}$ is constant along the leaves of $\mathcal{F}$.

To obtain this result we prove the following characterization of spaces of functions which define commutative algebras with respect to the Poisson brackets in a symplectic manifold.

Theorem 1.1. *Let M be a $2n$-dimensional symplectic manifold and $\mathcal{A}$ a vector subspace of $C^\infty(M)$ which is closed under complex conjugation. Suppose that the following conditions are satisfied:*

1. *$\mathcal{A}$ is a commutative algebra for the Poisson brackets of M, i.e., $\{a, b\} = 0$ for every $a, b \in \mathcal{A}$, and*
2. *for every $p \in M$ there exist real-valued elements $a_1, \dots, a_n \in \mathcal{A}$ such that $da_{1p}, \dots, da_{np}$ are linearly independent over $\mathbb{R}$.*

Then, there is a Lagrangian foliation $\mathcal{F}$ of M such that $\mathcal{A} \subseteq \mathcal{A}_\mathcal{F}(M)$.

This together with Berezin's correspondence principle (see Section 3) allows us to prove the Main Theorem.

We note that it is a known fact that for a Lagrangian foliation $\mathcal{F}$ in a symplectic manifold M, the vector subspace $\mathcal{A}_\mathcal{F}(M)$ is a commutative algebra with respect to the Poisson brackets. More precisely we have the following result.

Proposition 1.2. *If $\mathcal{F}$ is a Lagrangian foliation of a symplectic manifold M, then:*

$$\{a, b\} = 0,$$

for every $a, b \in \mathcal{A}_\mathcal{F}(M)$.

Hence, Theorem 1.1 can also be thought of as a converse of Proposition 1.2.

2. Preliminaries on symplectic geometry and foliations

The goal of this section is to establish some notation and state some very well-known results on symplectic geometry and foliations. For the next remarks on symplectic geometry we refer to [4] for further details.

Let M be a symplectic manifold with symplectic form ω. Being nondegenerate, the symplectic form ω defines an isomorphism between the tangent and cotangent spaces. More precisely, we have the following elementary fact.

Lemma 2.1. *For every $p \in M$ the map:*

$$\begin{array}{rcl} T_pM & \to & T_p^*M \\ v & \to & \omega(v,\cdot), \end{array}$$

is an isomorphism of vector spaces.

This remark allows us to construct vector fields associated to 1-forms. In particular, for a complex-valued smooth function f defined over M, we define the Hamiltonian field associated to f as the smooth vector field over M that satisfies the identity:

$$df(X) = \omega(X_f, X),$$

for every vector field X over M. The Poisson brackets of two complex-valued smooth functions f, g over M is then given as the smooth function:

$$\{f,g\} = \omega(X_f, X_g) = df(X_g).$$

The following well-known result relates the Poisson brackets on smooth functions to the Lie brackets of vector fields.

Lemma 2.2. *The Poisson brackets define a Lie algebra structure on the space $C^\infty(M)$. Also, we have the identity:*

$$[X_f, X_g] = X_{\{f,g\}},$$

for every $f, g \in C^\infty(M)$. In particular, the assignment:

$$f \mapsto X_f$$

is a homomorphism of Lie algebras onto the Lie algebra of Hamiltonian fields.

Another important object in our discussion is given by the notion of a foliation $\mathcal{F}$ of a manifold M. This is given as a decomposition into connected submanifolds which is locally given by submersions. More precisely, we define a foliated chart for M as a smooth submersion $\varphi : U \subseteq M \to \mathbb{R}^k$ from an open subset of M onto an open subset of $\mathbb{R}^k$. Given two such foliated charts φ, ψ, defined on open subsets U, V respectively, we will say that they are compatible if there is a smooth diffeomorphism $\xi : \varphi(U \cap V) \to \psi(U \cap V)$ such that $\xi \circ \varphi = \psi$ on $U \cap V$. A foliated atlas for M is a family of compatible foliated charts whose domains cover M; note that we also need k above to be the same for all the foliated charts. For any such foliated atlas, we define the plaques as the connected components of the fibers of its foliated charts. With these plaques we define the following equivalence relation in M:

$$x \sim y \iff \begin{array}{l} \text{there is a sequence of plaques } (P_j)_{j=0}^{l} \\ \text{of the foliated atlas, such that } x \in P_0,\ y \in P_l, \\ \text{and } P_{j-1} \cap P_j \neq \phi \text{ for every } j = 1, \dots, l \end{array}$$

The equivalence classes of such an equivalence relation are called the leaves of the foliation, which are easily seen to be submanifolds of M. For further details on this definition and some of its consequences and properties we refer to [2].

Finally, a foliation $\mathcal{F}$ in a symplectic manifold M is called Lagrangian if its leaves are Lagrangian submanifolds of M.

3. Berezin's correspondence principle for bounded domains

In the rest of this section D denotes a homogeneous bounded domain of $\mathbb{C}^n$. We now recollect some facts on the analysis of D leading to Berezin's correspondence principle; we refer to [1] for further details.

For every $h \in (0,1)$, let us denote by $\mathcal{A}_h^2(D)$ the weighted Bergman space defined as the closed subspace of holomorphic functions in $L^2(D, d\mu_h)$. Here, $d\mu_h$ denotes the weighted volume element obtained from the Bergman kernel. If we denote by $B_D^{(h)} : L^2(D, d\mu_h) \to \mathcal{A}_h^2(D)$ the orthogonal projection, then for every $a \in L^\infty(D, d\mu_h)$ the Toeplitz operator $T_a^{(h)}$ with (anti-Wick) symbol a is given by the assignment:

$$\begin{aligned} T_a^{(h)} : \mathcal{A}_h^2(D) &\to \mathcal{A}_h^2(D) \\ \varphi &\mapsto B_D^{(h)}(a\varphi). \end{aligned}$$

For any such (anti-Wick) symbol a and its associated Toeplitz operator $T_a^{(h)}$, Berezin [1] constructed a (Wick) symbol $\widetilde{a}_h : D \times \overline{D} \to \mathbb{C}$ defined so that the relation:

$$T_a^{(h)}(\varphi)(z) = \int_D \widetilde{a}_h(z, \overline{\zeta})\varphi(\zeta)F_h(\zeta, \overline{\zeta})d\mu(\zeta),$$

holds for every $\varphi \in \mathcal{A}_h^2(D)$, where $d\mu$ is the (weightless) Bergman volume and F_h is a suitable kernel defined in terms of the Bergman kernel and depending on h. This provides the means to describe the algebra of Toeplitz operators as a suitable algebra of functions. To achieve this, one defines a $*$-product of two Wick symbols $\widetilde{a}_h, \widetilde{b}_h$ as the symbol given by:

$$(\widetilde{a}_h * \widetilde{b}_h)(z, \overline{z}) = \int_D \widetilde{a}_h(z, \overline{\zeta})\widetilde{b}_h(\zeta, \overline{z})G_h(\zeta, \overline{\zeta}, z, \overline{z})d\mu(\zeta),$$

where G_h is again some kernel defined in terms of the Bergman kernel and depending on h. We denote by $\widetilde{\mathcal{A}}_h$ the vector space of Wick symbols associated to anti-Wick symbols that belong to $C_b^\infty(D)$. Then $\widetilde{\mathcal{A}}_h$ can be considered as an algebra for the $*$-product defined above.

Berezin's correspondence principle is then stated as follows.

Theorem 3.1 (Berezin [1]). *Let D be a homogeneous bounded domain of $\mathbb{C}^n$. Then, the map given by:*

$$\begin{array}{rcl} \mathcal{T}_h(C_b^\infty(D)) & \to & \widetilde{\mathcal{A}}_h \\ T_a^{(h)} & \mapsto & \widetilde{a}_h \end{array}$$

is an isomorphism of algebras, where $\mathcal{T}_h(C_b^\infty(D))$ is the Toeplitz operator algebra defined by bounded smooth symbols. Furthermore, the following correspondence principle is satisfied:

$$(\widetilde{a}_h * \widetilde{b}_h - \widetilde{b}_h * \widetilde{a}_h)(z,\overline{z}) = ih\{a,b\}(z) + O(h^2),$$

for every $h \in (0,1)$ and every $a, b \in C_b^\infty(D)$.

4. Proofs of the main results

For the sake of completeness, we present here the proof of Proposition 1.2.

Proof of Proposition 1.2. For a given $a \in \mathcal{A}_\mathcal{F}(M)$, the condition of a being constant along the leaves of $\mathcal{F}$ implies that:

$$\omega(X_a, X) = da(X) = 0,$$

for every smooth vector field X tangent to $\mathcal{F}$. Since the foliation $\mathcal{F}$ is Lagrangian, at every $p \in M$ the space $T_p\mathcal{F}$ is a maximal isotropic subspace for ω and so the above identity shows that $(X_a)_p$ belongs to $T_p\mathcal{F}$. Hence, for every $a \in \mathcal{A}_\mathcal{F}(M)$ the vector field X_a is tangent to the foliation $\mathcal{F}$. From this we conclude that:

$$\{a, b\} = da(X_b) = 0,$$

for every $a, b \in \mathcal{A}_\mathcal{F}(M)$. □

We now prove our result on commutative Poisson algebras and Lagrangian foliations.

Proof of Theorem 1.1. Let us consider a subspace $\mathcal{A}$ of $C^\infty(M)$ as in the hypotheses of Theorem 1.1. For every $p \in M$ define the vector subspace of T_pM given by:

$$E_p = \{(X_a)_p : a \in \mathcal{A}\}.$$

We will now prove that $E = \cup_{p\in M} E_p$ is a smooth n-distribution over M; in other words, that in a neighborhood of every point the fibers of E are spanned by n smooth vector fields which are pointwise linearly independent in such neighborhood.

First note that since the assignment $a \mapsto X_a$ is linear, every set E_p is a subspace of T_pM. Furthermore, being $\mathcal{A}$ commutative for the Poisson brackets, it follows that:

$$\omega(X_a, X_b) = \{a, b\} = 0,$$

for every $a, b \in \mathcal{A}$. In particular, E_p is an isotropic subspace for ω and so has dimension at most n.

On the other hand, for every $p \in M$ we can choose smooth functions $a_1, \ldots, a_n \in \mathcal{A}$ whose differentials are linearly independent at p. Hence, it follows from Lemma 2.1 that the elements $(X_{a_1})_p, \ldots, (X_{a_n})_p$ are also linearly independent at p, thus showing that E_p has dimension exactly n. By continuity, the chosen vector fields $X_{a_1}, \ldots, X_{a_n}$ are linearly independent in a neighborhood of p and so their values span E_q for every q in such neighborhood. Hence, E is indeed an n-distribution and our proof shows that its fibers are Lagrangian.

By Lemma 2.2, the assignment $a \mapsto X_a$ is a homomorphism of Lie algebras, and so the commutativity of $\mathcal{A}$ with respect to the Poisson brackets implies that the vector fields X_a commute with each other for $a \in \mathcal{A}$. Since the latter span the distribution E we conclude that E is involutive and, by Frobenius' Theorem (see [9]), it is integrable to some foliation $\mathcal{F}$. Note that $\mathcal{F}$ is necessarily Lagrangian.

It is enough to prove that every $a \in \mathcal{A}$ is constant along the leaves of $\mathcal{F}$. But by hypothesis we have:

$$da(X_b) = \{a, b\} = 0,$$

for every $a, b \in \mathcal{A}$. Since the vector fields X_b ($b \in \mathcal{A}$) define the elements of E at the fiber level we conclude that for any given $a \in \mathcal{A}$ we have:

$$da(X) = 0$$

for every vector field X tangent to $\mathcal{F}$. This implies that every $a \in \mathcal{A}$ is constant along the leaves of $\mathcal{F}$. □

Finally, we establish the necessity of having a Lagrangian foliation underlying to every commutative algebra of Toeplitz operators with sufficiently rich (anti-Wick) symbols.

Proof of the Main Theorem. For every $h \in (0, 1)$, let us denote by $\widetilde{\mathcal{A}}_h(\mathcal{A})$ the algebra of Wick symbols corresponding to anti-Wick symbols $a \in \mathcal{A}$. By the first part of Theorem 3.1, the algebra $\widetilde{\mathcal{A}}_h(\mathcal{A})$ is commutative. Hence, by the correspondence principle stated in the second part of Theorem 3.1 it follows that $\mathcal{A}$ is commutative with respect to the Poisson brackets $\{\cdot, \cdot\}$. The result now follows from Theorem 1.1. □

References

[1] F.A. Berezin, *Quantization in complex bounded domains*, Soviet Math. Dokl. **14** (1973), 1209–1213

[2] A. Candel and L. Conlon, Foliations. I. Graduate Studies in Mathematics, 23. American Mathematical Society, Providence, RI, 2000.

[3] S. Grudsky, R. Quiroga-Barranco and N. Vasilevski, *Commutative C^*-algebras of Toeplitz operators and quantization on the unit disk*, J. Funct. Anal., **234** (2006), no. 1, 1–44.

[4] D. McDuff and D. Salamon, Introduction to symplectic topology. Second edition. Oxford Mathematical Monographs. The Clarendon Press, Oxford University Press, New York, 1998.

[5] R. Quiroga-Barranco and N. Vasilevski, *Commutative C^*-algebras of Toeplitz operators on the unit ball. I. Bargmann-type transforms and spectral representations of Toeplitz operators*, Integral Equations Operator Theory **59** (2007), no. 3, 379–419.

[6] R. Quiroga-Barranco and N. Vasilevski, *Commutative C^*-algebras of Toeplitz operators on the unit ball. II. Geometry of the level sets of symbols*, Integral Equations Operator Theory **60** (2008), no. 1, 89–132

[7] N.L. Vasilevski, *Toeplitz operators on the Bergman spaces: Inside-the-domain effects*, Contemp. Math., **289** (2001) 79–146.

[8] N.L. Vasilevski, *Bergmann space structure, commutative algebras of Toeplitz operators and hyperbolic geometry*, Integral Equations and Operator Theory **46** (2003), 235–251.

[9] F.W. Warner, Foundations of differentiable manifolds and Lie groups. Corrected reprint of the 1971 edition. Graduate Texts in Mathematics, 94. Springer-Verlag, New York-Berlin, 1983.

R. Quiroga-Barranco
Centro de Investigación en Matemáticas
Apartado Postal 402
36000, Guanajuato
Guanajuato, México
e-mail: quiroga@cimat.mx

Operator Theory:
Advances and Applications, Vol. 210, 203–216

Exponential Estimates of Eigenfunctions of Matrix Schrödinger and Dirac Operators

V. Rabinovich and S. Roch

Dedicated to Prof. N. Vasilevsky on the occasion of his 60th birthday.

Abstract. The paper is devoted to exponential estimates of eigenfunctions of the discrete spectrum of matrix Schrödinger operators with variable potentials, and Dirac operators for nonhomogeneous media with variable light speed and variable electric and magnetic potentials, For the study of exponential estimates we apply methods developed in our recent paper [25].

Mathematics Subject Classification (2000). Primary 58J10; Secondary 81Q10.
Keywords. Uniformly elliptic systems, exponential estimates, limit operators, Schrödinger and Dirac operators.

1. Introduction

In this paper, we consider exponential estimates of eigenfunctions of the discrete spectrum of matrix Schrödinger operators and Dirac operators for nonhomogeneous media with variable electric and magnetic potentials. Our approach is based on the results and methods developed in our recent paper [25].

The literature devoted to exponential estimates of solutions partial differential equations with applications to the Schrödinger operators is extensive; we would only like to mention [1, 2, 12, 13, 17]. Exponential estimates for solutions of pseudo-differential equations on $\mathbb{R}^n$ are also considered in [4, 5, 6, 16, 17, 19, 20, 21, 23, 24].

The contents of the paper is as follows. In Section 2 we introduce the necessary definitions and recall the main results from the paper [25]. The following sections are devoted to new applications of these results. We start in Section 3 with the essential spectrum and exponential estimates for eigenvectors of Schrödinger operators with matrix potentials. Note that the well-known Pauli operator is a Schrödinger operator of this kind (see, for instance, [7]). Moreover, such oper-

This work was partially supported by the German Research Foundation (DFG).

ators appear in the Born-Oppenheimer approximation for polyatomic molecules [15, 18, 22]. For potentials which are slowly oscillating at infinity we describe the location of the essential spectrum and give exact estimates of the behavior of eigenfunctions of the discrete spectrum at infinity. As example we consider the Pauli Hamiltonian with spin 1/2.

Note that in the paper [25] we considered the exponential estimates of eigenfunctions of scalar Schrödinger operators.

In the concluding Section 4 we consider the Dirac operator. The standard physical books are [8, 11, 27], see also mathematical books [9, 10] and the cited in this books literature. We consider the Dirac operator on $\mathbb{R}^3$ equipped with a Riemannian metric, under the assumption that the electric and magnetic potentials as well as the light speed are slowly oscillating at infinity. We give a description of the essential spectrum and exact exponential estimates of eigenfunctions of the discrete spectrum. Similar estimates for the Dirac operator in vacuum have been obtained in [25].

2. Exponential estimates of solutions of systems of partial differential equations

2.1. Essential spectrum

We will use the following standard notations.

- Given Banach spaces X, Y, $\mathcal{L}(X,Y)$ is the space of all bounded linear operators from X into Y. We abbreviate $\mathcal{L}(X,X)$ to $\mathcal{L}(X)$.
- $L^2(\mathbb{R}^n, \mathbb{C}^N)$ is the Hilbert space of all measurable functions on $\mathbb{R}^n$ with values in $\mathbb{C}^N$, provided with the norm
$$\|u\|_{L^2(\mathbb{R}^n,\mathbb{C}^N)} := \left(\int_{\mathbb{R}^n} \|u(x)\|^2_{\mathbb{C}^N} dx \right)^{1/2}.$$
- $H^s(\mathbb{R}^n, \mathbb{C}^N)$ is a Sobolev space of distributions with norm
$$\|u\|_{H^s(\mathbb{R}^n,\mathbb{C}^N)} := \left(\int_{\mathbb{R}^n} (1 + |\xi|^2)^s \, \|\hat{u}(\xi)\|^2_{\mathbb{C}^N} d\xi \right)^{1/2},$$
where $\hat{u}$ is the Fourier transform of u.
- The unitary operator V_h of shift by $h \in \mathbb{R}^n$ acts on $L^2(\mathbb{R}^n, \mathbb{C}^N)$ via
$$(V_h u)(x) := u(x - h), \quad x \in \mathbb{R}^n.$$
- $C_b(\mathbb{R}^n)$ is the C^*-algebra of all bounded continuous functions on $\mathbb{R}^n$.
- $C_b^u(\mathbb{R}^n)$ is the C^*-subalgebra of $C_b(\mathbb{R}^n)$ of all uniformly continuous functions.
- $SO(\mathbb{R}^n)$ is the C^*-subalgebra of $C_b^u(\mathbb{R}^n)$ which consists of all functions a which are slowly oscillating in the sense that
$$\lim_{x\to\infty} \sup_{y\in K} |a(x+y) - a(x)| = 0$$
for every compact subset K of $\mathbb{R}^n$.

- $SO^1(\mathbb{R}^n)$ is the set of all bounded differentiable functions a on $\mathbb{R}^n$ such that

$$\lim_{x\to\infty} \frac{\partial a(x)}{\partial x_j} = 0\ ,\ j = 1,\ \dots,\ n.$$

Evidently, $SO^1(\mathbb{R}^n) \subset SO(\mathbb{R}^n)$.

We also use the standard multi-index notation. Thus, $\alpha = (\alpha_1, \dots, \alpha_n)$ with $\alpha_j \in \mathbb{N} \cup \{0\}$ is a multi-index, $|\alpha| = \alpha_1 + \dots + \alpha_n$ is its length, and

$$\partial^\alpha := \partial_{x_1}^{\alpha_1} \dots \partial_{x_n}^{\alpha_n}, \qquad D^\alpha := (-i\partial_{x_1})^{\alpha_1} \dots (-i\partial_{x_n})^{\alpha_n}$$

are the operators of αth derivative. Finally, $\langle\xi\rangle := (1 + |\xi|^2)^{1/2}$ for $\xi \in \mathbb{R}^n$.

We consider matrix partial differential operators of order m of the form

$$(Au)(x) = \sum_{|\alpha|\le m} a_\alpha(x)(D^\alpha u)(x), \quad x \in \mathbb{R}^n \tag{2.1}$$

under the assumption that the coefficients a_α belong to the space

$$C_b^u(\mathbb{R}^n, \mathcal{L}(\mathbb{C}^N)) := C_b^u(\mathbb{R}^n) \otimes \mathcal{L}(\mathbb{C}^N).$$

The operator A in (2.1) is considered as a bounded linear operator from the Sobolev space $H^m(\mathbb{R}^n, \mathbb{C}^N)$ to $L^2(\mathbb{R}^n, \mathbb{C}^N)$. The operator A is said to be *uniformly elliptic* on $\mathbb{R}^n$ if

$$\inf_{x\in\mathbb{R}^n,\, \omega\in S^{n-1}} \left| \det \sum_{|\alpha|=m} a_\alpha(x)\omega^\alpha \right| > 0$$

where S^{n-1} refers to the unit sphere in $\mathbb{R}^n$.

The Fredholm properties of the operator A can be expressed in terms of its limit operators which are defined as follows. Let $h : \mathbb{N} \to \mathbb{R}^n$ be a sequence which tends to infinity. The Arzelà-Ascoli theorem combined with a Cantor diagonal argument implies that there exists a subsequence g of h such that the sequence of the functions $x \mapsto a_\alpha(x + g(k))$ converges as $k \to \infty$ to a limit function a_α^g uniformly on every compact set $K \subset \mathbb{R}^n$ for every multi-index α. The operator

$$A^g := \sum_{|\alpha|\le m} a_\alpha^g D^\alpha$$

is called the *limit operator of A defined by the sequence g*. Equivalently, A^g is the limit operator of A with respect to g if and only if, for every function $\chi \in C_0^\infty(\mathbb{R}^n)$,

$$\lim_{m\to\infty} V_{-g(m)} A V_{g(m)} \chi I = A^g \chi I$$

in the space $\mathcal{L}(H^m(\mathbb{R}^n, \mathbb{C}^N), L^2(\mathbb{R}^n, \mathbb{C}^N))$ and

$$\lim_{m\to\infty} V_{-g(m)} A^* V_{g(m)} \chi I = (A^g)^* \chi I$$

in $\mathcal{L}(L^2(\mathbb{R}^n, \mathbb{C}^N), H^{-m}(\mathbb{R}^n, \mathbb{C}^N))$. Here, I is the identity operator, and

$$A^* u = \sum_{|\alpha|\le m} D^\alpha(a_\alpha^* u), \qquad (A^g)^* u = \sum_{|\alpha|\le m} D^\alpha((a_\alpha^g)^* u)$$

refer to the adjoint operators of A, $A_g : H^m(\mathbb{R}^n, \mathbb{C}^N) \to L^2(\mathbb{R}^n, \mathbb{C}^N)$. Let finally $\mathrm{op}\,(A)$ denote the set of all limit operators of A obtained in this way.

Theorem 2.1. *Let A be a uniformly elliptic differential operator of the form* (2.1). *Then $A : H^m(\mathbb{R}^n, \mathbb{C}^N) \to L^2(\mathbb{R}^n, \mathbb{C}^N)$ is a Fredholm operator if and only if all limit operators of A are invertible as operators from $H^m(\mathbb{R}^n, \mathbb{C}^N)$ to $L^2(\mathbb{R}^n, \mathbb{C}^N)$.*

The uniform ellipticity of the operator A implies the a priori estimate

$$\|u\|_{H^m(\mathbb{R}^n,\mathbb{C}^N)} \le C\left(\|Au\|_{L^2(\mathbb{R}^n,\mathbb{C}^N)} + \|u\|_{L^2(\mathbb{R}^n,\mathbb{C}^N)}\right).$$

This estimate allows one to consider the uniformly elliptic differential operator A as a closed unbounded operator on $L^2(\mathbb{R}^n, \mathbb{C}^N)$ with dense domain $H^m(\mathbb{R}^n, \mathbb{C}^N)$. It turns out (see [3], pp. 27–32) that A, considered as an unbounded operator in this way, is a (unbounded) Fredholm operator if and only if A, considered as a bounded operator from $H^m(\mathbb{R}^n, \mathbb{C}^N)$ to $L^2(\mathbb{R}^n, \mathbb{C}^N)$, is a (common bounded) Fredholm operator.

We say that $\lambda \in \mathbb{C}$ belongs to the *essential spectrum* of A if the operator $A - \lambda I$ is not Fredholm as an unbounded differential operator. As above, we denote the essential spectrum of A by $\mathrm{sp}_{\mathrm{ess}}\, A$ and the common spectrum of A (considered as an unbounded operator) by $\mathrm{sp}\, A$. Then the assertion of Theorem 2.1 can be stated as follows.

Theorem 2.2. *Let A be a uniformly elliptic differential operator of the form* (2.1). *Then*

$$\mathrm{sp}_{\mathrm{ess}}\, A = \bigcup_{A_g \in \mathrm{op}\, A} \mathrm{sp}\, A_g. \tag{2.2}$$

2.2. Exponential estimates

Let w be a positive measurable function on $\mathbb{R}^n$, which we call a weight. By $L^2(\mathbb{R}^n, \mathbb{C}^N, w)$ we denote the space of all measurable functions on $\mathbb{R}^n$ such that

$$\|u\|_{L^2(\mathbb{R}^n,\mathbb{C}^N,w)} := \|wu\|_{L^2(\mathbb{R}^n,\mathbb{C}^N)} < \infty.$$

In what follows we consider weights of the form $w = \exp v$ where $\partial_{x_j} v \in C_b^\infty(\mathbb{R}^n)$ for $j = 1, \dots, n$ and

$$\lim_{x\to\infty} \partial^2_{x_i x_j} v(x) = 0\,, \quad 1 \le i, j \le n.$$

We call weights with these properties *slowly oscillating* and let $\mathcal{R}$ stand for the class of all slowly oscillating weights. Examples of slowly oscillating weights can be constructed as follows. For $l > 0$, let $v(x) := l\langle x\rangle$ and consider the weight $w := e^v$. It is clear that

$$\frac{\partial v(x)}{\partial x_i} = l\frac{x_i}{\langle x\rangle}$$

and that $\lim_{r\to\infty} \nabla v(r\omega) = l\omega$ for $\omega \in S^{n-1}$.

Theorem 2.3. *Let A be a uniformly elliptic differential operator of the form (2.1), and let $w = \exp v$ be a weight in $\mathcal{R}$ such that $\lim_{x\to\infty} w(x) = +\infty$. For $t \in [-1,1]$, set*

$$A_{w,t} := \sum_{|\alpha|\le m} a_\alpha (D + it\nabla v)^\alpha,$$

and assume that

$$0 \notin \bigcup_{t\in[-1,1]} \mathrm{sp}_{\mathrm{ess}}\, A_{w,t} = \bigcup_{t\in[-1,1]} \bigcup_{A^g_{w,t}\in \mathrm{op}\,(A_{w,t})} \mathrm{sp}\, A^g_{w,t}. \tag{2.3}$$

If u is a function in $H^m(\mathbb{R}^n, \mathbb{C}^N, w^{-1})$ for which Au is in $L^2(\mathbb{R}^n, \mathbb{C}^N, w)$, then u already belongs to $H^m(\mathbb{R}^n, \mathbb{C}^N, w)$.

Corollary 2.4. *Let A be a uniformly elliptic differential operator of the form (2.1), and let $w = \exp v$ be a weight in $\mathcal{R}$ with $\lim_{x\to\infty} w(x) = +\infty$. Let $\lambda \in \mathrm{sp}_{\mathrm{dis}}\, A$ and moreover $\lambda \notin \mathrm{sp}_{\mathrm{ess}}\, A_{tw}$ for all $t \in (0,1]$. Then every eigenfunction of A associated with λ belongs to the space $H^m(\mathbb{R}^n, \mathbb{C}^N, w)$.*

3. Schrödinger operators with matrix potentials

3.1. Essential spectrum

We consider the Schrödinger operator

$$\mathcal{H} := (i\partial_{x_j} - a_j)\rho^{jk}(i\partial_{x_k} - a_k)E + \Phi$$

where E is the $N \times N$ unit matrix, $a = (a_1, \dots, a_n)$ is referred to as the magnetic potential, and $\Phi = (\Phi_{pq})_{p,q=1}^N$ is a matrix potential on $\mathbb{R}^n$, the latter equipped with a Riemannian metric $\rho = (\rho_{jk})_{j,k=1}^n$ which is subject to the positivity condition

$$\inf_{x\in\mathbb{R}^n,\, \omega\in S^{n-1}} \rho_{jk}(x)\omega^j\omega^k > 0, \tag{3.1}$$

where $\rho_{jk}(x)$ refers to the matrix inverse to $\rho^{jk}(x)$. Here and in what follows, we make use of Einstein's summation convention.

In what follows we suppose that ρ^{jk} and a_j are real-valued functions in $SO^1(\mathbb{R}^n)$ and that $\Phi_{pq} \in SO(\mathbb{R}^n)$. Under these conditions, $\mathcal{H}$ can be considered as a closed unbounded operator on $L^2(\mathbb{R}^n, \mathbb{C}^N)$ with domain $H^2(\mathbb{R}^n, \mathbb{C}^N)$. If Φ is a Hermitian matrix-valued function, then $\mathcal{H}$ is a self-adjoint operator.

The limit operators of $\mathcal{H}$ are the operators with constant coefficients

$$\mathcal{H}^g = (i\partial_{x_j} - a^g_j)\rho^{jk}_g(i\partial_{x_k} - a^g_k)E + \Phi^g$$

where

$$a^g := \lim_{m\to\infty} a(g(m)), \quad \rho_g := \lim_{m\to\infty} \rho(g(m)), \quad \Phi^g := \lim_{m\to\infty} \Phi(g(m)). \tag{3.2}$$

The operator $\mathcal{H}^g$ is unitarily equivalent to the operator

$$\mathcal{H}^g_1 := -\rho^{jk}_g \partial_{x_j}\partial_{x_k}E + \Phi^g$$

(with the unitary equivalence given by the operator of multiplication by the function $x \mapsto \exp(-i\langle a^g, x\rangle)$), which in turn is unitarily equivalent to the operator $\widehat{\mathcal{H}_1^g}$ of multiplication by the matrix-function

$$\widehat{\mathcal{H}_1^g}(\xi) := (\rho_g^{jk}\xi_j\xi_k)E + \Phi^g$$

acting on $L^2(\mathbb{R}^n, \mathbb{C}^N)$. Evidently,

$$\mathrm{sp}\,\widehat{\mathcal{H}_1^g} = \bigcup_{j=1}^N \Gamma_j^g$$

where $\Gamma_j^g := \mu_j^g + \mathbb{R}^+$ and the μ_j^g, $1 \le j \le N$, run through the eigenvalues of the matrix Φ^g. Thus, specializing (2.2) to the present context we obtain the following.

Theorem 3.1. *The essential spectrum of the Schrödinger operator $\mathcal{H}$ is given by*

$$\mathrm{sp}_{\mathrm{ess}}\,\mathcal{H} = \bigcup_g \bigcup_{j=1}^N \Gamma_j^g \tag{3.3}$$

where the union is taken with respect to all sequences g for which the limits in (3.2) exist.

The description (3.3) of the essential spectrum becomes much simpler if Φ is a Hermitian matrix function, in which case $\mathcal{H}$ is a self-adjoint operator.

Theorem 3.2. *Let the potential Φ be a Hermitian and slowly oscillating matrix function. Then*

$$\mathrm{sp}_{\mathrm{ess}}\,\mathcal{H} = [d_\Phi, +\infty)$$

where

$$d_\Phi := \liminf_{x\to\infty} \inf_{\|\varphi\|=1} \langle \Phi(x)\varphi, \varphi\rangle.$$

Proof. Since Φ^g is Hermitian,

$$\gamma_g := \inf_{\|\varphi\|=1} \langle \Phi^g\varphi, \varphi\rangle$$

is the smallest eigenvalue of Φ^g. Hence, $\mathrm{sp}\,\mathcal{H}^g = [\gamma_g, +\infty)$ and, according to (2.2),

$$\mathrm{sp}_{\mathrm{ess}}\,\mathcal{H} = \bigcup_g [\gamma_g, +\infty) = [\inf_g \gamma_g, +\infty).$$

It remains to show that

$$\inf_g \gamma_g = d_\Phi. \tag{3.4}$$

Let g be a sequence tending to infinity for which the limit

$$\Phi^g := \lim_{m\to\infty} \Phi(g(m))$$

exists. Then, for each unit vector $\varphi \in \mathbb{C}^N$,

$$\langle \Phi^g\varphi, \varphi\rangle = \lim_{m\to\infty} \langle \Phi(g(m))\varphi, \varphi\rangle \ge \liminf_{x\to\infty} \langle \Phi(g(m))\varphi, \varphi\rangle \ge d_\Phi,$$

whence $\gamma_g \geq d_\Phi$. For the reverse inequality, note that there exist a sequence g_0 tending to infinity and a sequence φ in the unit sphere in $\mathbb{C}^N$ with limit φ_0 such that

$$d_\Phi = \lim_{m\to\infty} (\Phi(g_0(m))\varphi_m, \varphi_m) = (\Phi^{g_0}\varphi_0, \varphi_0) \geq \gamma_{g_0}.$$

Thus, $\gamma_{g_0} = d_\Phi$, whence (3.4). □

3.2. Exponential estimates of eigenfunctions of the discrete spectrum

Again we suppose that the components ρ^{jk} of the Riemannian metric and the coefficients a_α are real-valued functions in $SO^1(\mathbb{R}^n)$, that the weight w belongs to $\mathcal{R}$, and that Φ is a Hermitian slowly oscillating matrix function with components in $SO(\mathbb{R}^N)$. Every limit operator $(w^{-1}\mathcal{H}w)_g$ of $w^{-1}\mathcal{H}w$ is unitarily equivalent to the operator

$$\mathcal{H}_w^g := \rho_g^{jk}(D_{x_j} + i(\nabla v)_j^g)(D_{x_k} + i(\nabla v)_k^g) + \Phi^g E,$$

where ρ_g^{jk} and Φ^g are the limits defined by (3.2) and

$$(\nabla v)^g := \lim_{k\to\infty} (\nabla v)(g(k)) \in \mathbb{R}^n.$$

We set

$$|(\nabla v)|_\rho^2 := \rho^{jk}(\nabla v)_j(\nabla v)_k \quad \text{and} \quad |(\nabla v)^g|_{\rho_g}^2 := \rho_g^{jk}(\nabla v)_j^g(\nabla v)_k^g.$$

The operator $\mathcal{H}_w^g$ is unitarily equivalent to the operator of multiplication by the matrix-valued function

$$\widehat{\mathcal{H}_w^g}(\xi) := \rho_g^{jk}(\xi_j + i(\nabla v)_j^g)(\xi_k + i(\nabla v)_k^g) + \Phi^g E, \quad \xi \in \mathbb{R}^n,$$

the real part of which is

$$\Re(\widehat{\mathcal{H}_w^g}) = \rho_g^{jk}\xi_j\xi_k + (\Phi^g - |(\nabla v)^g|_{\rho_g}^2)E.$$

Corollary 2.4 implies the following.

Theorem 3.3. *Let $\lambda \in \mathrm{sp}_{\mathrm{dis}}\,\mathcal{H}$, and let $w = \exp v$ be a weight in $\mathcal{R}$ for which*

$$\limsup_{x\to\infty} |(\nabla v)(x)|_{\rho(x)} < \sqrt{d_\Phi - \lambda}.$$

Then every λ-eigenfunction of $\mathcal{H}$ belongs to $H^2(\mathbb{R}^n, w)$.

Proof. One has

$$\Re(\widehat{\mathcal{H}_{w^t}^g}(\xi) - \lambda I) = \left(\rho_g^{jk}\xi_j\xi_k - t|(\nabla v)^g|_{\rho_g}^2 - \lambda\right) I + \Phi_g.$$

The hypotheses of the theorem imply that there exists $\varepsilon > 0$ such that, for every $\xi \in \mathbb{R}^n$,

$$\Re(\widehat{\mathcal{H}_{w^t}^g}(\xi) - \lambda I) \geq \varepsilon E.$$

This estimate together with condition (3.1) yields $\lambda \notin \mathrm{sp}\,\mathcal{H}_{w^t}^g$ for every $t \in [0,1]$ and for every sequence g which defines a limit operator. □

Corollary 3.4. *Let $\lambda \in \mathrm{sp}_{\mathrm{dis}}\,\mathcal{H}$, and let $\gamma \in \mathbb{R}$ satisfy*

$$0 < \gamma < \frac{\sqrt{d_\Phi - \lambda}}{\rho^{\mathrm{sup}}}$$

where

$$\rho^{\mathrm{sup}} := \liminf_{x\to\infty} \sup_{\omega \in S^{n-1}} (\rho^{jk}(x)\omega_j\omega_k)^{1/2}.$$

Then the every λ-eigenfunction of $\mathcal{H}$ belongs to the space $H^2(\mathbb{R}^n, \mathbb{C}^N, w)$ with weight $w(x) = e^{\gamma\langle x\rangle}$.

As example we consider the Pauli Hamiltonian which describes particles with spin-$\frac{1}{2}$ (see for instance [7, p. 114], [14, p. 249]). Let

$$\mathcal{H} := (i\partial_{x_j} - a_j)\rho^{jk}(i\partial_{x_k} - a_k) + \Phi$$

be a Schrödinger operator on $\mathbb{R}^3$ with magnetic potential $\overrightarrow{a} = (a_1, a_2, a_3)$ and the scalar electric potential Φ. As above we suppose that $\rho^{jk}, a_j, \Phi \in SO(\mathbb{R}^3)$. We consider the Pauli Hamiltonian of the form

$$\mathcal{P} = \mathcal{H}E + \overrightarrow{\sigma} \cdot \overrightarrow{B},$$

where $\overrightarrow{B} = \nabla \times \overrightarrow{a}$ is the magnetic field, $\overrightarrow{\sigma} = (\sigma_1, \sigma_2, \sigma_3)$ is the matrix-vector with the standard Pauli matrices

$$\sigma_1 = \begin{pmatrix} 0 & 1 \\ 1 & 0 \end{pmatrix}, \sigma_2 = \begin{pmatrix} 0 & -i \\ i & 0 \end{pmatrix}, \sigma_3 = \begin{pmatrix} 1 & 0 \\ 0 & -1 \end{pmatrix},$$

E is 2×2 unit matrix.

Note that the operator $\mathcal{P}$ is a self-adjoint operator on $L^2(\mathbb{R}^3, \mathbb{C}^2)$ with domain $H^2(\mathbb{R}^3, \mathbb{C}^2)$. Because $a_j \in SO(\mathbb{R}^3)$,

$$\lim_{x\to\infty} \overrightarrow{\sigma} \cdot \overrightarrow{B} = 0,$$

hence Theorem 3.2 implies that

$$\mathrm{sp}_{\mathrm{ess}}\mathcal{P} = \mathrm{sp}_{\mathrm{ess}}\mathcal{H} = [\Phi_{\mathrm{inf}}, +\infty].$$

Let $\lambda \in (-\infty, \Phi_{\mathrm{inf}})$ be an eigenvalue of $\mathcal{P}$, and $\gamma \in \mathbb{R}$ satisfy

$$0 < \gamma < \frac{\sqrt{d_\Phi - \lambda}}{\rho^{\mathrm{sup}}}$$

where

$$\rho^{\mathrm{sup}} := \liminf_{x\to\infty} \sup_{\omega \in S^2} (\rho^{jk}(x)\omega_j\omega_k)^{1/2}.$$

Then Corollary 3.4 implies that every λ-eigenfunction of $\mathcal{P}$ belongs to the space $H^2(\mathbb{R}^3, \mathbb{C}^2, e^{\gamma\langle x\rangle})$.

4. Dirac operators

4.1. Essential spectrum of Dirac operators

In this section we consider the Dirac operator on $\mathbb{R}^3$ equipped with the Riemannian metric tensor (ρ_{jk}) depending on $x \in \mathbb{R}^3$ (for a general account on Dirac operators see, for example, [27]). We suppose that there is a constant $C > 0$ such that

$$\rho_{jk}(x)\xi^j\xi^k \geq C|\xi|^2 \quad \text{for every } x \in \mathbb{R}^3 \tag{4.1}$$

where we use the Einstein summation convention again. Let ρ^{jk} be the tensor inverse to ρ_{jk}, and let ϕ^{jk} be the positive square root of ρ^{jk}. The Dirac operator on $\mathbb{R}^3$ is the operator

$$\mathcal{D} := \frac{c}{2}\gamma_k(\phi^{jk}P_j + P_j\phi^{jk}) + \gamma_0 c^2 m - e\Phi \tag{4.2}$$

acting on functions on $\mathbb{R}^3$ with values in $\mathbb{C}^4$. In (4.2), the γ_k, $k = 0, 1, 2, 3$, are the 4×4 Dirac matrices, i.e., they satisfy

$$\gamma_j\gamma_k + \gamma_k\gamma_j = 2\delta_{jk}E \tag{4.3}$$

for all choices of $j, k = 0, 1, 2, 3$ where E is the 4×4 unit matrix,

$$P_j = D_j + \frac{e}{c}A_j, \qquad D_j = \frac{\hbar}{i}\frac{\partial}{\partial x_j}, \quad j = 1, 2, 3,$$

where $\hbar$ is the Planck constant, $\vec{A} = (A_1, A_2, A_3)$ is the vector potential of the magnetic field $\vec{H}$, that is $\vec{H} = \operatorname{rot}\vec{A}$, Φ is the scalar potential of the electric field $\vec{E}$, that is $\vec{E} = \operatorname{grad}\Phi$, and m and e are the mass and the charge of the electron and, finally, $c = c(x)$ is the light speed in the media.

We suppose that ρ^{jk}, A_j, Φ and c are real-valued functions which satisfy the conditions

$$\rho^{jk} \in SO^1(\mathbb{R}^3), \quad A_j \in SO^1(\mathbb{R}^3), \quad \Phi \in SO(\mathbb{R}^3), \quad c \in SO(\mathbb{R}^3), \tag{4.4}$$

and $c(x) \geq c_0 > 0$. We consider $\mathcal{D}$ as an unbounded operator on the Hilbert space $L^2(\mathbb{R}^3, \mathbb{C}^4)$ with domain $H^1(\mathbb{R}^3, \mathbb{C}^4)$. The conditions imposed on the magnetic and electric potentials $\vec{A}$ and Φ guarantee the self-adjointness of $\mathcal{D}$. The main symbol of $\mathcal{D}$ is $a_0(x,\xi) = c(x)\phi^{jk}(x)\xi_j\gamma_k$. Using (4.3) and the identity $\phi^{jk}\phi^{rt}\delta_{kt} = \rho^{jr}$ we obtain

$$\begin{aligned} a_0(x,\xi)^2 &= c^2(x)\hbar^2\phi^{jk}(x)\phi^{rt}(x)\xi_j\xi_r\gamma_k\gamma_t \\ &= c^2(x)\hbar^2\phi^{jk}(x)\phi^{rt}(x)\delta_{kt}\xi_j\xi_r \\ &= c^2(x)\hbar^2\rho^{jr}(x)\xi_j\xi_r E \geq c_0 C\hbar^2 \left|\xi\right|^2 E. \end{aligned}$$

This equality shows that $\mathcal{D}$ is a uniformly elliptic differential operator on $\mathbb{R}^3$.

Conditions (4.4) imply that limit operators $\mathcal{D}_g$ of $\mathcal{D}$ defined by sequences $g : \mathbb{N} \to \mathbb{Z}^3$ tending to infinity are operators with constant coefficients of the form

$$\mathcal{D}_g = c_g\gamma_k\phi_g^{jk}(D_j + \frac{e}{c_g}A_j^g) + \gamma_0 m c_g^2 - e\Phi^g$$

where

$$\phi_g^{jk} := \lim_{l\to\infty} \phi^{jk}(g_l), \qquad A_j^g := \lim_{l\to\infty} A_j(g_l)$$

and

$$\Phi^g := \lim_{l\to\infty} \Phi(g_l), \qquad c_g := \lim_{l\to\infty} c(g_l).$$

The operator $\mathcal{D}_g$ is unitarily equivalent to the operator

$$\mathcal{D}_g^1 = c_g\phi_g\gamma_{lg}^{jl} D_j + \gamma_0 mc_g^2 - e\Phi^g,$$

and the equivalence is realized by the unitary operator

$$T_{\vec{A}^g} : f \mapsto e^{i\frac{e}{c}\vec{A}^g\cdot x} f, \qquad \vec{A}^g := (A_1^g, A_2^g, A_3^g).$$

Let

$$\Phi^{\sup} := \limsup_{x\to\infty} \Phi(x), \qquad \Phi^{\inf} := \liminf_{x\to\infty} \Phi(x),$$

$$c_{\sup} := \limsup_{x\to\infty} c(x), \qquad c_{\inf} := \liminf_{x\to\infty} c(x).$$

The intervals $[\Phi^{\inf}, \Phi^{\sup}]$ and $[c_{\inf}, c_{\sup}]$ are just the sets of all partial limits Φ^g and c_g of the functions Φ and c as $x \to \infty$, respectively.

Theorem 4.1. *Let conditions* (4.4) *be fulfilled. Then the Dirac operator*

$$\mathcal{D} : H^1(\mathbb{R}^3, \mathbb{C}^4) \to L^2(\mathbb{R}^3, \mathbb{C}^4)$$

is a Fredholm operator if and only if

$$[\Phi^{\inf}, \Phi^{\sup}] \subset \left(-\frac{mc_{\sup}^2}{e}, \frac{mc_{\inf}^2}{e} \right). \tag{4.5}$$

Proof. Set $\mathcal{D}_0^g(\xi) := c_g\hbar\gamma_k\phi_g^{jk}\xi_j + \gamma_0 mc_g^2$ and $\rho_g^{jk} := \lim_{m\to\infty} \rho^{jk}(g_m)$. Then

$$\begin{aligned}(\mathcal{D}_0^g(\xi) - e\Phi^g E)\,(\mathcal{D}_0^g(\xi) + e\Phi^g E)\\ = \left(c_g^2\hbar^2\rho_g^{jk}\xi_j\xi_k + m^2c_g^4 - (e\Phi^g)^2\right) E.\end{aligned} \tag{4.6}$$

Let condition (4.5) be fulfilled. Then every partial limit $\Phi^g = \lim_{k\to\infty} \Phi(g_k)$ of Φ lies in the interval $(-\frac{mc_{\sup}^2}{e}, \frac{mc_{\inf}^2}{e})$. The identity (4.6) implies that

$$\det(\mathcal{D}_0^g(\xi) - e\Phi^g E) \neq 0$$

for every $\xi \in \mathbb{R}^3$. Hence, the operator $\mathcal{D}_g^1 : H^1(\mathbb{R}^3, \mathbb{C}^4) \to L^2(\mathbb{R}^3, \mathbb{C}^4)$ is invertible and, consequently, so is $\mathcal{D}_g$. By Theorem 2.1, $\mathcal{D}$ is a Fredholm operator.

For the reverse implication, assume that condition (4.5) is not fulfilled. Then there exist a sequence g tending to infinity and a vector $\xi^0 \in \mathbb{R}^3$ such that

$$c_g^2\rho_g^{jk}\xi_j^0\xi_k^0 + m^2c_g^4 - (e\Phi^g)^2 = 0.$$

Given ξ^0 we find a vector $u \in \mathbb{C}^4$ such that $v := \left(\mathcal{D}_0^g(\xi^0) + (e\Phi^g)E\right) u \neq 0$. Then (4.6) implies that $\left(\mathcal{D}_0^g(\xi^0) - e\Phi^g E\right) v = 0$, whence $\det(\mathcal{D}_0^g(\xi^0) - e\Phi^g E) = 0$. Thus, the operator $\mathcal{D}_g$ is not invertible. By Theorem 2.1, $\mathcal{D}$ cannot be a Fredholm operator. □

Let

$$\lambda_+^{\inf} := \liminf_{x\to\infty} \left(-e\Phi(x) + mc^2(x)\right),$$
$$\lambda_-^{\sup} := \limsup_{x\to\infty} \left(-e\Phi(x) - mc^2(x)\right).$$

Theorem 4.2. *If condition* (4.4) *is satisfied, then*

$$\mathrm{sp}_{\mathrm{ess}}\,\mathcal{D} = (-\infty, \lambda_-^{\sup}] \cup [\lambda_+^{\inf}, +\infty).$$

Proof. Let $\lambda \in \mathbb{C}$. The symbol of the operator $\mathcal{D}_g - \lambda I$ is the function $\xi \mapsto \mathcal{D}_0^g(\xi) - (e\Phi^g + \lambda)E$. Invoking (4.6) we obtain

$$\begin{aligned}(\mathcal{D}_0^g(\xi) - (e\Phi^g + \lambda)E)\,(\mathcal{D}_0^g(\xi) + (e\Phi^g + \lambda)E)\\ = \left(c_g^2\rho_g^{jk}\xi_j\xi_k + m^2c_g^4 - (e\Phi^g + \lambda)^2\right)E.\end{aligned}$$

Repeating the arguments from the proof of Theorem 4.1, we find that the eigenvalues $\lambda_\pm^g(\xi)$ of the matrix $\mathcal{D}_0^g(\xi) - e\Phi_1^g E$ are given by

$$\lambda_\pm^g(\xi) := -e\Phi^g \pm (c_g^2\rho_g^{jk}\xi_j\xi_k + m^2c_g^4)^{1/2}. \tag{4.7}$$

From (4.7) we further conclude

$$\{\lambda \in \mathbb{R} : \lambda = \lambda_-^g(\xi),\, \xi \in \mathbb{R}^3\} = (-\infty, -e\Phi^g - mc_g^2],$$
$$\{\lambda \in \mathbb{R} : \lambda = \lambda_+^g(\xi),\, \xi \in \mathbb{R}^3\} = [-e\Phi^g + mc_g^2, +\infty).$$

Hence,

$$\mathrm{sp}\,\mathcal{D}^g = (-\infty, -e\Phi^g - mc_g^2] \cup [-e\Phi^g + mc_g^2, +\infty),$$

whence the assertion via Theorem 2.2. □

Thus, if

$$\Phi^{\sup} - \Phi^{\inf} \geq \frac{mc_{\sup}^2}{e} + \frac{mc_{\inf}^2}{e}$$

then $\mathrm{sp}_{\mathrm{ess}}\,\mathcal{D}$ is all of $\mathbb{R}$, whereas $\mathrm{sp}_{\mathrm{ess}}\,\mathcal{D}$ has a proper gap in the opposite case.

4.2. Exponential estimates of eigenfunctions of the Dirac operator

Theorem 4.3. *Let the conditions* (4.4) *be fulfilled. Let λ be an eigenvalue of $\mathcal{D}$ which lies in the gap $(\lambda_-^{\sup}, \lambda_+^{\inf})$ of the essential spectrum. Further, let $w = \exp v$ be a weight in $\mathcal{R}$ with $\lim_{x\to\infty} w(x) = \infty$ which satisfies*

$$\limsup_{x\to\infty} |\nabla v(x)|_{\rho(x)} < \frac{1}{\hbar} \liminf_{x\to\infty} \frac{1}{c(x)}\sqrt{m^2c^4(x) - (e\Phi(x) + \lambda)^2}. \tag{4.8}$$

Then every eigenfunction of $\mathcal{D}$ associated with λ belongs to $H^1(\mathbb{R}^3, \mathbb{C}^4, w)$.

Proof. Let $\lambda \in (\lambda_-^{\sup}, \lambda_+^{\inf})$ be an eigenvalue of $\mathcal{D}$. As above, we examine the spectra of the limit operators $(\mathcal{D}_{w,t})_g$ of $\mathcal{D}_{w,t} := w^t\mathcal{D}w^{-t}$ for t running through $[0, 1]$. Let $(\mathcal{D}_{w,t})_g$ be a limit operator of $\mathcal{D}_{w,t}$ with respect to a sequence g tending to infinity. One easily checks that $(\mathcal{D}_{w,t})_g$ is unitarily equivalent to the operator

$$(\mathcal{D}'_{w,t})_g := A_{t,g} - e\Phi^g E$$

where

$$A_{t,g} := c\gamma_k \phi_g^{jk}(D_j + it\hbar(\frac{\partial v}{\partial x_j})^g) + \gamma_0 mc^2.$$

The operator $A_{t,g}$ has constant coefficients, and its symbol is the function

$$\widehat{A_{t,g}}(\xi) = c\gamma_k \phi_g^{jk}(\hbar(\xi_j + it\hbar(\frac{\partial v}{\partial x_j})^g)) + \gamma_0 mc^2.$$

Further,

$$\begin{aligned}
&\mathfrak{R}\left[\left(\widehat{A_{t,g}}(\xi) - (e\Phi^g - \lambda)E\right)\left(\widehat{A_{t,g}}(\xi) + (e\Phi^g - \lambda)E\right)\right] \\
&= \mathfrak{R}\left[c^2\hbar^2\rho_g^{jk}\left(\xi_j + it\hbar(\frac{\partial v}{\partial x_j})^g\right)\left(\xi_k + it\hbar(\frac{\partial v}{\partial x_k})^g\right)\right] \\
&\quad + \mathfrak{R}\left[\left(m^2c^4 - (e\Phi^g + \lambda)^2\right)E\right] \\
&= \left[c^2\hbar^2\rho_g^{jk}\xi_j\xi_k - c^2\hbar^2t^2\rho_g^{jk}(\frac{\partial v}{\partial x_j})^g(\frac{\partial v}{\partial x_k})^g + \left(m^2c^4 - (e\Phi^g + \lambda)^2\right)\right]E \\
&=: \gamma_{g,t}(\xi,\lambda)E.
\end{aligned}$$

Assume that condition (4.8) is fulfilled. Then, since $c^2\hbar^2\rho_g^{jk}\xi_j\xi_k \geq 0$,

$$\inf_{\xi\in\mathbb{R}^n} \gamma_{g,t}(\xi,\lambda) > 0$$

for all $t \in [0,1]$ and for all sequences $g \to \infty$ for which the limit operators exist. Hence, (4.8) implies that the matrix $\widehat{A_{t,g}}(\xi) - (e\Phi^g - \lambda)E$ is invertible for every $\xi \in \mathbb{R}^3$. On the other hand, due to the uniform ellipticity of $A_{t,g}$ one has $\lambda \in \mathrm{sp}\left(\widehat{A_{t,g}}(\xi) - (e\Phi^g - \lambda)E\right)$ if and only if there exists a $\xi_0 \in \mathbb{R}^3$ such that the matrix $\widehat{A_{t,g}}(\xi_0) - (e\Phi^g + \lambda)E$ is not invertible. Thus, $\lambda \notin \mathrm{sp}\,(\mathcal{D}_{w,t})_g$ for every $t \in [0,1]$ and every sequence $g \to \infty$. Via Corollary 2.4, the assertion follows. □

We conclude by a special case. Let the conditions (4.4) be fulfilled, and let λ be an eigenvalue of $\mathcal{D}$ in $(\lambda_-^{\sup}, \lambda_+^{\inf})$ and u_λ an associated eigenfunction. If a satisfies the estimate

$$0 < a < \frac{1}{\hbar\rho^{\sup}} \liminf_{x\to\infty} \frac{1}{c(x)} \sqrt{m^2c^4(x) - (e\Phi(x) + \lambda)^2}$$

where

$$\rho^{\sup} := \limsup_{x\to\infty} \sup_{\omega\in S^2} \left(\rho^{jk}(x)\omega_j\omega_k\right)^{1/2},$$

then $u_\lambda \in H^1(\mathbb{R}^3, \mathbb{C}^4, e^{a\langle x\rangle})$.

References

[1] S. Agmon, *Spectral properties of Schrödinger operators and scattering theory.* Ann. Scuola Norm. Sup. Pisa, Cl. Sci. **4** (1975), 151–218.

[2] S. Agmon, *Lectures on Exponential Decay of Solutions of Second Order Elliptic Equations.* Princeton University Press, Princeton, 1982.

[3] M.S. Agranovich, *Elliptic Operators on Smooth Manifolds.* In: Itogi Nauki i Tekhniki, Sovremennie Problemi Matematiki, Fundamentalnie Napravlenia, V. 63, Partial Differential Equations-6 (Russian), 5–130.

[4] E. Buzano, *Super-exponential decay of solutions to differential equations in* $\mathbb{R}^r$. in: Modern Trends in Pseudo-Differential Operators, Operator Theory: Adv. Appl. **172**, Birkhäuser, Basel 2007, 117–133.

[5] M. Cappiello, T. Gramchev, L. Rodino, *Super-exponential decay and holomorphic extensions for semilinear equations with polynomial coefficients.* J. Functional Anal. **237** (2006), 2, 634–654.

[6] M. Cappiello, T. Gramchev, L. Rodino, *Gelfand–Shilov spaces, pseudo-differential operators and localization operators.* in: Modern Trends in Pseudo-Differential Operators, Operator Theory: Adv. Appl. **172**, Birkhäuser, Basel 2007, 297–312.

[7] H.L. Cycon, R.G. Froese, W. Kirsch, B. Simon, *Schrödinger Operators with Applications to Quantum Mechanics and Global Geometry.* Springer-Verlag, Berlin, Heidelberg, New York 1987.

[8] P.A.M. Dirac, *The Principles of Quantum Mechanics.* Oxford: Clarendon Press, 1958.

[9] G. Esposito, *Dirac Operators and Spectral Geometry*, Cambridge Lecture Notes in Physics, Cambridge, University Press, 1998.

[10] V. Ivrii, *Microlocal Analysis and Precise Spectral Asymptotics*, Springer, Berlin Heidelberg New York, 1998.

[11] V.B. Berestecskiy, E.M. Lifshitz, L.P. Pitaevsky, *Quantum Electrodynamics*, 3 Edition, Nauka, Moskow, 1989 (in Russian).

[12] R. Froese, I. Herbst, *Exponential bound and absence of positive eigenvalue for N-body Schrödinger operators.* Comm. Math. Phys. **87** (1982), 429–447.

[13] R. Froese, I. Herbst, M. Hoffman-Ostenhof, T. Hoffman-Ostenhof, *L^2-exponential lower bound of the solutions of the Schrödinger equation.* Comm. Math. Phys. **87** (1982), 265–286.

[14] J.M. Jauch, *Foundation of Quantum Mechanics*, Addison-Wesley, London, 1973

[15] M. Klein, A. Martinez, R. Seiler, X.P. Wang, *On the Born-Oppenheimer expansion for polyatomic molecules.* Comm. Math. Phys. **143** (1992), 3, 607–639.

[16] Ya.A. Luckiy, V.S. Rabinovich, *Pseudodifferential operators on spaces of functions of exponential behavior at infinity.* Funct. Anal. Prilozh. **4** (1977), 79–80.

[17] M. Măntoiu, *Weighted estimations from a conjugate operator.* Letter Math. Phys. **51** (2000), 17–35.

[18] A. Martinez, *Eigenvalues and resonances of polyatomic molecules in the Born-Oppenheimer approximation*, in: Schrödinger Operators. The Quantum Mechanical Many-Body Problem, Lecture Notes in Physics **403**, Springer-Verlag, Berlin, Heidelberg 1992, 145–152.

[19] A. Martinez, *Microlocal exponential estimates and application to tunnelling.* In: Microlocal Analysis and Spectral Theory, L. Rodino (Editor), NATO ASI Series, Series C: Mathematical and Physical Sciences **490**, 1996, 349–376.

[20] A. Martinez, *An Introduction to Semiclassical and Microlocal Analysis.* Springer, New York 2002.

[21] S. Nakamura, *Agmon-type exponential decay estimates for pseudodifferential operators.* J. Math. Sci. Univ. Tokyo **5** (1998), 693–712.

[22] L. Nedelec, *Resonances for matrix Schrödinger operators.* Duke Math. J. **106** (2001), 2, 209–236.

[23] V.S. Rabinovich, *Pseudodifferential operators with analytic symbols and some of its applications.* In: Linear Topological Spaces and Complex Analysis **2**, Metu-Tübitak, Ankara 1995, 79–98.

[24] V. Rabinovich, *Pseudodifferential operators with analytic symbols and estimates for eigenfunctions of Schrödinger operators.* Z. Anal. Anwend. (J. Anal. Appl.) **21** (2002), 2, 351–370.

[25] V.S. Rabinovich, S. Roch, *Essential spectrum and exponential decay estimates of solutions of elliptic systems of partial differential equations. Applications to Schrödinger and Dirac operators.* Georgian Math. J. **15** (2008), 2, 333–351.

[26] M.A. Shubin, *Pseudodifferential Operators and Spectral Theory*, Springer-Verlag, Berlin, Heidelberg, New York 2001.

[27] B. Thaller, *The Dirac Equation*, Springer Verlag, Berlin, Heidelberg, New York 1992.

V. Rabinovich
Instituto Politécnico Nacional
ESIME-Zacatenco
Av. IPN, edif.1
México D.F. 07738, México
e-mail: `vladimir.rabinovich@gmail.com`

S. Roch
Fachbereich Mathematik
Technische Universität Darmstadt
Schlossgartenstrasse 7
D-64289 Darmstadt, Germany
e-mail: `roch@mathematik.tu-darmstadt.de`

Operator Theory:
Advances and Applications, Vol. 210, 217–231

The Laplace-Beltrami Operator on a Rotationally Symmetric Surface

Nikolai Tarkhanov

Dedicated to N. Vasilevski on the occasion of his 60th birthday

Abstract. The aim of this work is to highlight a number of analytic problems which make the analysis on manifolds with true cuspidal points much more difficult than that on manifolds with conic points while such singularities are topologically equivalent. To this end we discuss the Laplace-Beltrami operator on a compact rotationally symmetric surface with a complete metric. Even though the symmetry assumptions made here lead to a simplified situation in which standard separation of variables works, it is hoped that the study of this example can nevertheless bring to light some features which may subsist in the more general framework of the calculus on compact manifolds with cusps due to V. Rabinovich et al. (1997).

Mathematics Subject Classification (2000). Primary 58-XX; Secondary 47-XX.

Keywords. Manifolds with singularities, pseudodifferential operators, ellipticity.

Introduction

Calculus of pseudodifferential operators on manifolds with cuspidal points was first elaborated in [17] although many concrete problems had been solved in the early 1970s, see *ibid.* It is based on the concept of slowly varying symbols which allows one to give a description of Fredholm operators in the calculus by using Simonenko's local principle. This latter reads that under properly chosen scale of function spaces the Fredholm property is actually equivalent to local invertibility of the operator.

One question still unanswered in [17] is whether this calculus is efficient enough to construct explicit asymptotic expansions near cusps for solutions to corresponding equations. This question is at present far from being solved even for calculi on curves. Such asymptotic expansions can be derived if the "coefficients" of the equations behave well at the singular points, for cuspidal singularities can be topologically reduced to conical ones. One may conjecture that in the case of general equations with slowly varying symbols only a structure theorem is available for solutions, see [19]. It is therefore of interest to know if asymptotic expansions can be obtained for solutions of natural geometric equations, such as the Laplace-Beltrami equation, cf. [13].

With this as our starting point, we examine in this paper a local calculus of pseudodifferential operators on a manifold $\mathcal{Z}$ with cuspidal points. We restrict our attention to rotation-symmetric cusps, and so we may think of $\mathcal{Z}$ as a rotation surface in $\mathbb{R}^n$, which is due to locality. As but one example we show the so-called "citrus" $z^3(1-z)^3 = x^2 + y^2$ in $\mathbb{R}^3$, where $0 \le z \le 1$. It contains two singular points, both the points being cusps, and the calculus developed is of independent interest, cf. [18].

We define a complete Riemannian metric on the smooth part of $\mathcal{Z}$ and study the corresponding Laplace-Beltrami operator. It is by no means obvious that there is a change of variables $t = \delta(r)$ which reduces the Laplace-Beltrami operator on $\mathcal{Z}$ to the operator $-\partial_t^2 + \Delta_{\mathbb{S}^{n-2}}$ on the cylinder $\mathcal{C} = \mathbb{R} \times \mathbb{S}^{n-2}$ over the unit sphere in $\mathbb{R}^{n-1}$. The standard change of coordinates fails to do so. Note that we have not been able to find such a change of coordinates for the Laplace-Beltrami on $\mathcal{Z}$ corresponding to the Riemannian metric which is induced by the Euclidean metric on $\mathbb{R}^n$.

The operator $-\partial_t^2 + \Delta_{\mathbb{S}^{n-2}}$ can be thought of as ordinary differential operator with constant operator-valued coefficients. There are many excellent treatments of such operators, the monograph [11] being among the latest. Our paper [18] published independently gives even more, namely it treats also pseudodifferential operators on the real axis whose symbols take their values in pseudodifferential operators on $\mathbb{S}^{n-2}$.

Pseudodifferential operators on a rotation surface in $\mathbb{R}^n$ can be specified within global Fourier operators on an interval whose symbols take their values in the space of parameter-dependent pseudodifferential operators on $\mathbb{S}^{n-2}$. Using this description, we have shown an index formula for Fredholm operators in weighted Sobolev spaces, cf. [18]. This formula has the advantage of being quite explicit, when compared with a sophisticated index formula for elliptic pseudodifferential operators on an edged spindle proved in [6]. The latter requires well-elaborated techniques of corner calculus, cf. *ibid.* The formula for the index of the Laplace-Beltrami operator on $\mathcal{Z}$ given in the present paper is proved by a direct computation. It agrees to [18].

We also mention an explicit index formula of [2]. It applies to second-order elliptic differential operators in unbounded cylinders $\mathcal{C}$ in $\mathbb{R}^n$, the zero Dirichlet conditions being posed on $\partial\mathcal{C}$. This formula easily extends to unbounded cylinders over smooth compact closed manifolds like $\mathbb{S}^{n-1}$. It evaluates the index of operators in Hölder spaces but falls short of providing an index formula in weighted Sobolev spaces.

The same arguments still go when we replace $\mathbb{S}^{n-2}$ by an arbitrary smooth compact manifold $\mathcal{X}$, called the link. No efficient pseudodifferential calculus is yet available if the link itself has singularities. Indeed, on multiplying these singularities with the end singularities of the interval $[0,1]$ one obtains edges, corners, etc. The symbols corresponding to higher-order singularities are pseudodifferential operators on noncompact manifolds with singularities. The Fredholm property requires the symbols to be invertible, which is a much more complicated problem than the Fredholm property of original operators under study. Hence, mere concrete problems may be rich in content, as is pointed out by V. Kondratiev whose paper of 1967 initiated the theory. One needs surprising examples rather than a general theory.

1. Geometry

Let f be a positive differentiable function on the interval $(0,1)$ with the property that $f(0+) = f(1-) = 0$.

We consider the rotation surface $\mathcal{Z}$ in the Euclidean space $\mathbb{R}^n$ given by $f(x_n) = |x'|$, where $x' = (x_1,\dots,x_{n-1})$. It can parametrised under cylindrical coordinates

$$\left\{\begin{array}{lllllll} x_1 & = & f(r) & \cos\omega_1 & \dots & \cos\omega_{n-3} & \cos\omega_{n-2}, \\ x_2 & = & f(r) & \cos\omega_1 & \dots & \cos\omega_{n-3} & \sin\omega_{n-2}, \\ x_3 & = & f(r) & \cos\omega_1 & \dots & \sin\omega_{n-3}, & \\ & & & & \dots & & \\ x_{n-1} & = & f(r) & \sin\omega_1, & & & \\ x_n & = & r, & & & & \end{array}\right.$$

where $r \in [0,1]$, $\omega_k \in [-\pi/2, \pi/2]$ for $k = 1,\dots,n-3$, and $\omega_{n-2} \in [0, 2\pi]$. The parameters $\omega = (\omega_1,\dots,\omega_{n-2})$ form a global coordinate system on the unit sphere $\mathbb{S}^{n-2}$ in $\mathbb{R}^{n-1}$ which however fails to be differentiable everywhere. This surface is of spindle type, so it is smooth outside the points $r = 0$ and $r = 1$ which are singular in general.

The surface $\mathcal{Z}$ bears the Riemannian metric induced by the Euclidean metric $(dx_1)^2 + \dots + (dx_n)^2$ in $\mathbb{R}^n$. In order to emphasize the particular character of the points $r = 0$ and $r = 1$, we give $\mathcal{Z}$ its own Riemannian metric which may or not be complete according to the singularities at $r = 0$ and $r = 1$. It is the product of the Euclidean one by the factor $1/(f(r))^2$. In other words, this metric is defined as

$$\Big(g_{ij}\Big)_{\substack{i=1,\dots,n-1\\ j=1,\dots,n-1}} = \begin{pmatrix} \dfrac{1+(f')^2}{f^2} & 0 & 0 & \dots & 0 \\ 0 & 1 & 0 & \dots & 0 \\ 0 & 0 & \cos^2\omega_1 & \dots & 0 \\ 0 & 0 & 0 & \dots & 0 \\ 0 & 0 & 0 & \dots & \cos^2\omega_1 \dots \cos^2\omega_{n-3} \end{pmatrix}.$$

Lemma 1.1. *The surface measure on $\mathcal{Z}$ corresponding to the Riemannian metric is*

$$ds = \sqrt{g_{11}} dr\, d\sigma$$

where $d\sigma = \Big(\cos^{n-3} \omega_1 \dots \cos \omega_{n-3} \Big) d\omega_1 \dots d\omega_{n-2}$ *is the standard surface measure on* $\mathbb{S}^{n-2}$.

Proof. As is known, the surface measure is given by

$$ds = \sqrt{\det \Big(g_{ij} \Big)} dr\, d\omega_1 \dots d\omega_{n-2},$$

see for instance [16, p. 159]. On substituting the determinant of $\Big(g_{ij} \Big)$ we arrive at the desired formula. □

By the above, we get

$$\sqrt{g_{11}} = \frac{\sqrt{1 + (f')^2}}{f}.$$

2. Laplace-Beltrami operator

Any Riemannian metric on $\mathcal{Z}$ gives rise to a class of differentiable mappings of $\mathcal{Z}$ that multiply at each point ds^2 by some scalar. They are called conformal and can moreover be characterised by the property that their pullbacks on functions commute with an elliptic second-order partial differential operator on $\mathcal{Z}$ called the Laplace-Beltrami operator. On manifolds with singularities this is usually understood on the smooth part, i.e., outside the singularities.

Write g for the determinant and $\Big(g^{ij} \Big)$ for the inverse matrix of the metric tensor. The Laplace-Beltrami operator on $\mathcal{Z}$ is defined by

$$\Delta u = -\frac{1}{\sqrt{g}} \sum_{i,j=1}^{n-1} \partial_i \Big(\sqrt{g}\, g^{ij} \partial_j u \Big),$$

see for instance [16, p. 164]. Since the metric tensor is diagonal, this expression can be further simplified.

Lemma 2.1. *The Laplace-Beltrami operator on $\mathcal{Z}$ is explicitly given by*

$$\Delta u = -\Big(\frac{1}{\sqrt{g_{11}}} \partial_1 \Big)^2 u + \Delta_{\mathbb{S}^{n-2}} u$$

where $\Delta_{\mathbb{S}^{n-2}}$ is the Laplace-Beltrami operator on $\mathbb{S}^{n-2}$.

Proof. A trivial verification shows that

$$\begin{aligned}
\Delta u &= -\sum_{j=1}^{n-1} g^{jj}\partial_j^2 u + \Big(\frac{1}{2}\frac{\partial_j g}{g}\,g^{jj} + \partial_j g^{jj}\Big)\partial_j u \\
&= -\sum_{j=1}^{n-1} \frac{1}{g_{jj}}\partial_j^2 u + \frac{1}{2}\frac{1}{g_{jj}}\Big(\frac{\partial_j g_{11}}{g_{11}} + \cdots - \frac{\partial_j g_{jj}}{g_{jj}} + \cdots + \frac{\partial_j g_{n-1,n-1}}{g_{n-1,n-1}}\Big)\partial_j u,
\end{aligned}$$

only the jth term in parentheses being reduced.

The first summand is of key interest, for it is singular at $r = 0$ and $r = 1$ while the other terms are actually independent of the variable r. Hence it follows that

$$\Delta u = \Big(-\frac{1}{g_{11}}\partial_1^2 u + \frac{1}{2}\frac{1}{g_{11}}\frac{\partial_1 g_{11}}{g_{11}}\partial_1 u\Big) + \Delta_{\mathbb{S}^{n-2}} u, \tag{2.1}$$

$\Delta_{\mathbb{S}^{n-2}} u$ being the sum over $j = 2, \ldots, n-1$. This is precisely the Laplace-Beltrami operator on the sphere $\mathbb{S}^{n-2}$.

Since

$$\frac{1}{g_{11}}\partial_1^2 u - \frac{1}{2}\frac{1}{g_{11}}\frac{\partial_1 g_{11}}{g_{11}}\partial_1 u = \Big(\frac{1}{\sqrt{g_{11}}}\partial_1\Big)^2 u,$$

the lemma follows. □

On using Lemma 1.1 one easily sees that the Laplace-Beltrami operator on $L^2(\mathcal{Z})$ is self-adjoint, if given an appropriate domain.

3. Weighted spaces

Lemma 2.1 implies that the Laplace-Beltrami equation on a general rotation surface reduces to a canonical second-order ordinary differential equation with constant operator-valued coefficients on $\mathbb{R}$.

Namely, choose a mapping $t = \delta(r)$ of $(0,1)$ to the real axis with the property that

$$\frac{\partial}{\partial t} = \frac{1}{\sqrt{g_{11}}}\frac{\partial}{\partial r} \tag{3.1}$$

for all $r \in (0,1)$. This is equivalent to $\delta'(r) = \sqrt{g_{11}(r)}$ whence

$$\delta(r) = \int_{r_0}^{r} \frac{\sqrt{1 + (f'(s))^2}}{f(s)}\,ds,$$

for $r \in (0,1)$, where $r_0 \in (0,1)$ is an arbitrary fixed point. From the assumptions on $f(r)$ it follows that $t = \delta(r)$ is a diffeomorphism of the interval $(0,1)$ onto the interval $(\delta(0), \delta(1))$.

Example. Let $f(r) = r^p(1-r)^q$ with $p, q > 0$. This is a positive differentiable function on $(0,1)$ which is continuous up to $r = 0$ and $r = 1$. It vanishes at the end points and

$$\frac{\sqrt{1+(f'(s))^2}}{f(s)} = \frac{\sqrt{s^{2(1-p)}(1-s)^{2(1-q)} + p^2(1-s)^{2q} - 2pqs(1-s) + q^2 s^{2p}}}{s(1-s)}$$

for all $s \in (0,1)$. Hence we deduce that $\delta(0) = -\infty$ and $\delta(1) = +\infty$ whenever p and q are positive.

This example shows that $\delta(0) = -\infty$ happens not only in the case when the singular point at $r = 0$ is a sharp cusp ($p > 1$) but also in the case if it a conic point ($p = 1$) or even a smoothness point ($p < 1$) of $\mathcal{Z}$. An analogous fact is true for the singular point at $r = 1$. Note that for the commonly used change of coordinates $f(r)\, \partial/\partial r = \partial/\partial t$ the situation is quite different. Namely, the value of $t = \delta(r)$ at $r = 0$ or $r = 1$ is infinite only in the case of true cuspidal singularities at these points.

We shall make two standing assumptions on the functions $f(r)$ under consideration, namely $\delta(0) = -\infty$ and $\delta(1) = +\infty$. By the above, they are fulfilled in most interesting cases.

Lemma 3.1. *Under the change of variables $t = \delta(r)$, the Laplace-Beltrami operator on $\mathcal{Z}$ transforms into*

$$\delta_* \Delta = -\partial_t^2 + \Delta_{\mathbb{S}^{n-2}}.$$

Proof. Combine Lemma 2.1 and equality (3.1). □

We are thus led to a second-order ordinary differential operator with constant coefficients on the real axis, the coefficients taking their values in the space of differential operators on the sphere $\mathbb{S}^{n-2}$. This topic is nowadays well understood, cf. [4], [11], etc.

Let

$$\left(h_{j,k}\right)_{\substack{j=0,1,\dots \\ k=1,\dots,k_j}} \tag{3.2}$$

be an orthonormal basis in the space $L^2(\mathbb{S}^{n-2})$ consisting of the restrictions of homogeneous harmonic polynomials in $\mathbb{R}^{n-1}$. The index j means the degree of homogeneity, and the index k runs through the number of polynomials of degree j in the basis. Every $h_{j,k}$ is an eigenfunction of the Laplace-Beltrami operator on the unit sphere corresponding to the eigenvalue $\lambda_j = j(j+n-3)$, i.e., it satisfies

$$\Delta_{\mathbb{S}^{n-2}} h_{j,k} = \lambda_j h_{j,k},$$

and $k_j = \dfrac{(2j+n-3)(j+n-4)!}{(n-3)!j!}$.

We look for a solution to the inhomogeneous equation $\delta_* \Delta u = f$ whose Fourier expansion is

$$u(t) = \sum_{j=0}^{\infty} \sum_{k=1}^{k_j} u_{j,k}(t)\, h_{j,k}$$

for each $t \in \mathbb{R}$. Assuming $u_{j,k}(t)$ smooth enough, we apply $\delta_* \Delta$ termwise to the right-hand side and equate the Fourier coefficients of $\delta_* \Delta u$ and f, thus obtaining an infinite system of ordinary differential equations for the unknown functions $u_{j,k}(t)$ on the real axis, namely

$$-u''_{j,k}(t) + \lambda_j u_{j,k}(t) = f_{j,k}(t) \tag{3.3}$$

for all $j = 0, 1, \ldots$ and $k = 1, \ldots, k_j$.

The characteristic equation for (3.3) is $\tau^2 + \lambda_j = 0$ with roots $\tau_\pm = \pm\imath\sqrt{\lambda_j}$. The solution of the initial value problem for (3.3) with data $u_{j,k}(0) = u'_{j,k}(0) = 0$ is

$$(I_{j,k} f_{j,k})(t) = -\int_0^t \frac{\sinh\sqrt{\lambda_j}(t-s)}{\sqrt{\lambda_j}} f_{j,k}(s)ds,$$

and so the general solution to the inhomogeneous equation (3.3) on $\mathbb{R}$ can be represented by

$$u_{j,k}(t) = (I_{j,k} f_{j,k})(t) + c^+_{j,k}\, e^{\sqrt{\lambda_j}t} + c^-_{j,k}\, e^{-\sqrt{\lambda_j}t},$$

$c^\pm_{j,k}$ being arbitrary constants.

We have thus proved that the equation $\delta_* \Delta u = f$ on the real axis possesses a formal solution

$$u(t) = \sum_{j=0}^{\infty} \sum_{k=1}^{k_j} \Big((I_{j,k} f_{j,k})(t) + c^+_{j,k}\, e^{\sqrt{\lambda_j}t} + c^-_{j,k}\, e^{-\sqrt{\lambda_j}t} \Big) h_{j,k} \tag{3.4}$$

containing infinitely many arbitrary constants $c^\pm_{j,k}$. If we pose a proper restriction on the growth of $u(t)$ at $t = -\infty$ then only finitely many constants $c^-_{j,k}$ may be different from zero, for $\lambda_j \to +\infty$ as $j \to \infty$. On posing a restriction on the growth of $u(t)$ at $t = +\infty$ we obtain that all constants $c^+_{j,k}$ vanish but finitely many. Hence, the operator $\delta_* \Delta$, when acting in spaces of functions with growth restrictions at both points $\pm\infty$, is Fredholm. Formula (3.4) makes it also evident what function spaces are to be used.

Note that we use separation of variables here to merely write out the formal solution (3.4) to $\delta_* \Delta u = f$. To study the convergence of the series one requires further arguments like those in [5, Ch. 6]. We use (3.4) just to motivate our choice of functions spaces.

The functions $u(t)$ under study are defined on all of $\mathbb{R}$ and take their values in functions on the sphere $\mathbb{S}^{n-2}$. Hence, they can actually be thought of as functions on the cylinder $\mathcal{C} = \mathbb{R} \times \mathbb{S}^{n-2}$ over the sphere. For $s = 0, 1, \ldots$ and any real number γ, we denote by $H^{s,\gamma}(\mathcal{C})$ the completion of the space $C^\infty_{\mathrm{comp}}(\mathcal{C})$ with respect to the norm

$$\|u\|_{H^{s,\gamma}(\mathcal{C})} = \Big(\int_{\mathbb{R}} \sum_{j+A \le s} \|\partial^j (e^{-\gamma t} u)\|^2_{H^A(\mathbb{S}^{n-2})}\, dt \Big)^{1/2}. \tag{3.5}$$

Obviously, this space is Hilbert. If s is a negative integer, we define $H^{s,\gamma}(\mathcal{C})$ to be the dual of $H^{-s,-\gamma}(\mathcal{C})$. For fractional s, the space $H^{s,\gamma}(\mathcal{C})$ is defined by (complex) interpolation.

Lemma 3.2. *The operator* $\delta_* \Delta$ *extends to a continuous map* $H^{s,\gamma}(\mathcal{C}) \to H^{s-2,\gamma}(\mathcal{C})$ *for each* $s \in \mathbb{R}$.

Proof. For $s = 2, 3, \ldots$, this is proved by direct computation. It remains then to apply duality and interpolation results. □

The operator $\delta_* \Delta$ actually belongs to the calculus of pseudodifferential operators on the cylinder $\mathcal{C}$ elaborated in [18]. For functions $u \in H^{2,\gamma}(\mathcal{C})$, it can be written in the form

$$\delta_* \Delta u\,(t) = \frac{1}{2\pi} \int_{\Gamma_{-\gamma}} e^{\imath t z} \left(z^2 + \Delta_{\mathbb{S}^{n-2}} \right) \mathcal{F}u(z)\, dz \tag{3.6}$$

where $\Gamma_{-\gamma} = \{z \in \mathbb{C} : \Im z = -\gamma\}$ and $\mathcal{F}u$ stands for the Fourier transform of u. The holomorphic function $\sigma(\Delta)(z) = z^2 + \Delta_{\mathbb{S}^{n-2}}$ taking its values in pseudodifferential operators with parameter on the sphere $\mathbb{S}^{n-2}$ is sometimes called the conormal symbol of Δ. A more general designation is operator pencil.

4. Resolvent

Write $a(z) = z^2 + \Delta_{\mathbb{S}^{n-2}}$ for the conormal symbol of Δ. This is a polynomial of $z \in \mathbb{C}$ with values in pseudodifferential operators on $\mathbb{S}^{n-2}$ which are elliptic with parameter $\tau = \Re z$. Given any $s \in \mathbb{R}$, one can think of $a(z)$ as an operator $H^s(\mathbb{S}^{n-2}) \to H^{s-2}(\mathbb{S}^{n-2})$. The particular choice of s is not important for both the kernel and cokernel of $a(z)$ consist of C^∞ functions on the sphere. Thus, the results of [8] apply.

A point $z_0 \in \mathbb{C}$ is said to be a characteristic point of $a(z)$ if there exists a holomorphic function $u(z)$ in a neighbourhood of z_0 with values in $H^s(\mathbb{S}^{n-2})$, such that $u(z_0) \neq 0$ but $a(z)u(z)$ is holomorphic at z_0 and vanishes at this point. We call $u(z)$ a root function of $a(z)$ at z_0.

To find the characteristic points of $a(z)$, we invoke the orthonormal basis (3.2) in $L^2(\mathbb{S}^{n-2})$ and write

$$u(z) = \sum_{j=0}^{\infty} \sum_{k=1}^{k_j} u_{j,k}(z)\, h_{j,k}$$

on the sphere. Then,

$$a(z)u(z) = \sum_{j=0}^{\infty} \sum_{k=1}^{k_j} \left(z^2 + \lambda_j \right) u_{j,k}(z)\, h_{j,k}$$

which immediately shows that z_0 is a characteristic point of $a(z)$ if and only if $z_0 = \pm \imath \sqrt{\lambda_j}$ for some j. Any function $u_{j,k}(z)\, h_{j,k}$ holomorphic in a neighbourhood of $\pm \imath \sqrt{\lambda_j}$ and different from zero at this point is actually a root function of $a(z)$ at $\pm \imath \sqrt{\lambda_j}$.

Suppose z_0 is a characteristic point of $a(z)$ and $u(z)$ is a corresponding root function. The order of z_0 as a zero of $a(z)u(z)$ is called the multiplicity of $u(z)$, and $u(z_0) \in H^s(\mathbb{S}^{n-2})$ an eigenfunction of $a(z)$ at z_0. If supplemented by the zero function on $\mathbb{S}^{n-2}$, the eigenfunctions of $a(z)$ at z_0 form a vector space. It is called the kernel of $a(z)$ at z_0 and denoted by $\ker a(z_0)$. By the rank of an eigenfunction $u_0 \in H^s(\mathbb{S}^{n-2})$ we mean the supremum of the multiplicities of all root functions $u(z)$ with $u(z_0) = u_0$.

The kernel of $a(z)$ at $\pm\imath\sqrt{\lambda_j}$ is spanned by $h_{j,1}, \dots, h_{j,k_j}$. The rank of the eigenfunction $h_{0,1}$ at 0 is equal to 2. For $j \geq 1$, the rank of the eigenfunction $h_{j,k}$ at $\pm\imath\sqrt{\lambda_j}$ is 1.

Suppose that $\ker a(z_0)$ is of dimension K. By a canonical system of eigenfunctions of $a(z)$ at z_0 is meant any system of eigenfunctions $u_{0,1}, \dots, u_{0,K}$ with the property that the rank of $u_{0,1}$ is the maximum of the ranks of all eigenfunctions of $a(z)$ at z_0 and the rank of $u_{0,k}$ is the maximum of the ranks of all eigenfunctions in a direct complement in $\ker a(z_0)$ of the linear span of the vectors $u_{0,1}, \dots, u_{0,k-1}$, for $k = 2, \dots, K$. Let r_k be the rank of $u_{0,k}$, for $k = 1, \dots, K$. It is a simple matter to see that the rank of any eigenfunction of $a(z)$ at the characteristic point z_0 is always equal to one of the r_k. Hence it follows that the numbers r_k are determined uniquely by the function $a(z)$. Note that a canonical system of eigenfunctions is not, in general, uniquely determined. The numbers r_k are said to be partial multiplicities of the characteristic point z_0 of $a(z)$. Their sum $n(a(z_0)) = r_1 + \dots + r_K$ is called the multiplicity of z_0. If $a(z)$ has no root function at z_0, we naturally set $n(a(z_0)) = 0$.

Thus, $h_{j,1}, \dots, h_{j,k_j}$ is a canonical system of eigenfunctions of $a(z)$ at both characteristic points $\pm\imath\sqrt{\lambda_j}$ and the multiplicity of these characteristic points of $a(z)$ just amounts to k_j, if $j \geq 1$, and 2, if $j = 0$.

For $z \in \mathbb{C}$ away from the characteristic points of $a(z)$ this operator-valued function is invertible. The inverse $a^{-1}(z)$ called the resolvent of $a(z)$ takes its values in the space of pseudodifferential operators of order -2 with parameter on the unit sphere $\mathbb{S}^{n-2}$, cf. [18]. We now rehearse the expansion of the principal part of $a^{-1}(z)$ which is actually due to [8].

Theorem 4.1. *For each characteristic point* $z_0 = \pm\imath\sqrt{\lambda_j}$ *of* $a(z)$ *with* $j \geq 1$, *the principal part of* $a^{-1}(z)$ *at* z_0 *is*

$$\text{p.p.}\, a^{-1}(z) = \frac{1}{2z_0} \frac{\sum_{k=1}^{k_j} (\cdot, h_{j,k})_{L^2(\mathbb{S}^{n-2})}\, h_{j,k}}{z - z_0}. \tag{4.1}$$

Proof. Cf. Proposition 4.2 in [18]. □

In the case $j = 0$ the rank of the only eigenfunction $h_{0,1}$ of $a(z)$ at $z_0 = 0$ is equal to 2, and so (4.1) needs a modification. It becomes

$$\text{p.p.}\, a^{-1}(z) = \frac{(\cdot, h_{0,1})_{L^2(\mathbb{S}^{n-2})}\, h_{0,1}}{z^2},$$

as is easy to check.

We are now in a position to use the theory of [18] on the real axis and pull it back to the interval $(0,1)$ by the diffeomorphism $t = \delta(r)$.

5. Fredholm theory

Denote by $H^{s,\gamma}(\mathcal{Z}) = \delta^* H^{s,\gamma}(\mathcal{C})$ the Hilbert space consisting of the pullbacks $\delta^* u = u \circ \delta$ of functions $u \in H^{s,\gamma}(\mathcal{C})$. It is easy to see that the norm in this space is given by

$$\|u\|_{H^{s,\gamma}(\mathcal{Z})} = \Big(\int_0^1 \sum_{j+A \leq s} \| \Big(\frac{1}{\sqrt{g_{11}}} \partial_1 \Big)^j (e^{-\gamma\delta(r)} u) \|^2_{H^A(\mathbb{S}^{n-2})} \sqrt{g_{11}} dr \Big)^{1/2}, \qquad (5.1)$$

provided that $s = 0, 1, \ldots$.

The factor $e^{-\gamma\delta}$ entering into (5.1) can not control independently the behaviour of functions at both points $r = 0$ and $r = 1$. To this end, we introduce yet another scale of weighted Sobolev spaces on $\mathcal{Z}$, which includes two weight parameters. Namely, let $w = (w_0, w_1)$ be a pair of real numbers to inspect the growth of functions at $r = 0$ and $r = 1$. Fix an excision function $\chi(r)$ for the point $r = 1$ on $[0, 1]$, i.e., χ is a C^∞ function on $[0, 1]$ vanishing close to $r = 1$ and equal to 1 near $r = 0$. For $s \in \mathbb{R}$, set

$$H^{s,w}(\mathcal{Z}) = \chi\, H^{s,w_0}(\mathcal{Z}) + (1 - \chi)\, H^{s,-w_1}(\mathcal{Z}), \qquad (5.2)$$

the right-hand side being understood in the sense of nondirect sum of Fréchet spaces. In particular, on choosing $w = (\gamma, -\gamma)$ we recover $H^{s,w}(\mathcal{Z}) = H^{s,\gamma}(\mathcal{Z})$ for all $s, \gamma \in \mathbb{R}$.

The additive group structure of $\mathbb{R}$ induces under the diffeomorphism $t = \delta(r)$ a group structure on the interval $[0, 1]$. This is given by $r \circ s = \delta^{-1}(\delta(r) + \delta(s))$ for $r, s \in [0, 1]$. The corresponding invariant measure is a multiple of $\sqrt{g_{11}} dr$, and the abstract Fourier transform is

$$Fu\,(\varrho) = \int_0^1 e^{-\imath\varrho\delta(r)} u(r) \sqrt{g_{11}} dr$$

for $\varrho \in \mathbb{R}$.

Lemma 5.1. *Given any $u \in H^{s,w}(\mathcal{Z})$, the abstract Fourier transform $Fu(z)$ is holomorphic in the strip $-w_0 < \Im z < w_1$.*

Proof. See Lemma 6.3 in [18]. □

Combining (5.2) and Lemma 3.2 we deduce that the Laplace-Beltrami operator Δ extends to a continuous mapping $H^{s,w}(\mathcal{Z}) \to H^{s-2,w}(\mathcal{Z})$ for all $s \in \mathbb{R}$ and $w \in \mathbb{R}^2$.

If $-w_0 < w_1$ then the mapping $\Delta : H^{s,w}(\mathcal{Z}) \to H^{s-2,w}(\mathcal{Z})$ is specified in the pseudodifferential calculus on $\mathcal{Z}$ as

$$\Delta u\,(r) = \frac{1}{2\pi} \int_{\Gamma_{-\gamma}} e^{\imath\delta(r)\tau} \left(\varrho^2 + \Delta_{\mathbb{S}^{n-2}}\right) Fu(\varrho)\, d\varrho$$

for $r \in (0, 1)$, the integral being independent on the particular choice of the exponent $-\gamma$ in the interval $(-w_0, w_1)$, cf. (3.6).

Theorem 5.2. *Suppose w_0 and w_1 are different from $\pm\sqrt{\lambda_j}$ for all j. Then the mapping $\Delta : H^{s,w}(\mathcal{Z}) \to H^{s-2,w}(\mathcal{Z})$ is Fredholm for all $s \in \mathbb{R}$.*

Proof. The hypotheses of the theorem are equivalent to saying that the conormal symbol $\sigma(\Delta)(z)$ is invertible on the lines Γ_{-w_0} and Γ_{w_1}. Since the Laplace-Beltrami operator is a second-order elliptic operator on the smooth part of $\mathcal{Z}$, the Fredholm property follows for instance from [17]. □

If $-w_0 < w_1$, a sharper result is given by Corollary 6.8 in [18]. Namely, the operator $\Delta : H^{s,w}(\mathcal{Z}) \to H^{s-2,w}(\mathcal{Z})$ is injective and has a closed range of finite codimension,

$$\operatorname{codim} \operatorname{im} \Delta = \sum_{\substack{j=0,1,\ldots \\ k=1,\ldots,k_j \\ -w_0<\pm\sqrt{\lambda_j}<w_1}} (1+\delta_{0,j}) . \tag{5.3}$$

Note that the constant functions of $r \in [0,1]$ with values in $H^s(\mathbb{S}^{n-2})$ belong to the null-space of $\Delta : H^{s,w}(\mathcal{Z}) \to H^{s-2,w}(\mathcal{Z})$ provided that both w_0 and w_1 are negative.

6. Asymptotics

In this section we derive an explicit description of solutions of the homogeneous equation $\Delta u = 0$.

Theorem 6.1. *To every nonzero characteristic point $\pm\imath\sqrt{\lambda_j}$ of $\sigma(\Delta)(z)$ lying in the strip $w_1 < \Im z < -w_0$ there correspond k_j linearly independent solutions of $\Delta u = 0$, namely*

$$\left(e^{\mp\sqrt{\lambda_j}\delta(r)} h_{j,k}\right)_{k=1,\ldots,k_j} . \tag{6.1}$$

It is a simple matter to verify that each function of the form (6.1) belongs to $H^{\infty,w}(\mathcal{Z})$, provided that $w_1 < \pm\sqrt{\lambda_j} < -w_0$.

Proof. For the proof, fix any integer $k = 1, \ldots, k_j$. We have

$$e^{\mp\sqrt{\lambda_j}\delta(r)} h_{j,k} = \left(e^{\imath z\delta(r)} h_{j,k}\right)_{z=\pm\imath\sqrt{\lambda_j}},$$

as is easy to check. When applying the operator Δ to the right-hand side of this equality, we may interchange Δ and the evaluation in z. This implies

$$\begin{aligned} \Delta\left(e^{\mp\sqrt{\lambda_j}\delta(r)} h_{j,k}\right) &= \left(e^{\imath z\delta(r)} a(z) h_{j,k}\right)_{z=\pm\imath\sqrt{\lambda_j}} \\ &= 0, \end{aligned}$$

the right-hand side being zero because $a(z)h_{j,k}$ vanishes at the point $\pm\imath\sqrt{\lambda_j}$. To complete the proof it is sufficient to note that $\Delta : H^{-s+2,-w}(\mathcal{Z}) \to H^{-s,-w}(\mathcal{Z})$ is the transpose of $\Delta : H^{s,w}(\mathcal{Z}) \to H^{s-2,w}(\mathcal{Z})$. By Theorem 5.2, both the mappings are Fredholm, and so the range of the former is the dual of the null-space of the latter. We now use formula (5.3) to conclude that the dimension of the null-space of $\Delta : H^{s,w}(\mathcal{Z}) \to H^{s-2,w}(\mathcal{Z})$ is

$$\dim \ker \Delta = \sum_{\substack{j=0,1,\ldots \\ k=1,\ldots,k_j \\ w_0<\pm\sqrt{\lambda_j}<-w_1}} (1+\delta_{0,j}) \,. \tag{6.2}$$

Hence, the system (6.1) actually encompasses the whole contribution of $\pm\imath\sqrt{\lambda_j}$ to the null-space of Δ. □

Note that to the zero characteristic point of $\sigma(\Delta)(z)$ there correspond two linearly independent solutions of the Laplace-Beltrami equation on the surface $\mathcal{Z}$, namely $h_{0,1}$ and $\delta(r)h_{0,1}$.

Not only highlight (6.1) the asymptotic behaviour at singular points of solutions to the homogeneous equation $\Delta u = 0$ belonging to $H^{s,w}(\mathcal{Z})$ but they also describe the explicit structure of these solutions. The theorem gains in interest if we realise that the Laplace-Beltrami operator on $\mathcal{Z}$ has variable coefficients, in which case no asymptotics have been known for general cuspidal points. The explanation of this phenomenon lies obviously in the fact that there is a surprising change of variables which reduces the Laplace-Beltrami operator to an ordinary differential operator with constant operator-valued coefficients. No attempt has been made here to develop any general theory, we were rather going to motivate [18].

In [9] asymptotics of solutions to the Laplace-Beltrami equation on a rotation surface with an incomplete metric are studied by the method of Green-Liouville approximations, see [14].

The calculus of [17] encompasses much more general operators on manifolds with singular points. Close to a singular point they are of the form

$$Au\,(r) = \frac{1}{2\pi}\int_{\Gamma_{-\gamma}} e^{\imath\delta(r)z} a(r,z) Fu(z)\, dz \tag{6.3}$$

for $r > 0$, where r is the coordinate along an axis of the singular point corresponding to $r = 0$, and Fu is an abstract Fourier transform. In contrast to (3.6), the symbol $a(r,z)$ is a nonconstant smooth function of $r > 0$ with values in the space of parameter-dependent pseudodifferential operators on the link $\mathcal{X}$, the parameter being $\varrho = \Re z$. The behaviour of $a(r,z)$ at $r = 0$ is of crucial importance. In [17] we assume that $(\partial/\partial\delta)^k\, a(r,z)$ tends to zero as $r \to 0+$, for every $k = 1, 2, \ldots$. Operators (6.3) act in weighted Sobolev spaces with two weight functions, $e^{-\gamma\delta(r)}$ and $(\delta'(r))^{\mu}$, where $\gamma, \mu \in \mathbb{R}$. Under this assumptions, the local principle of Simonenko works to characterise Fredholm operators within the calculus. The lack of smoothness at $r = 0$ does not allow one to gain in the weight exponent γ. Hence, one needs a scale of weaker weight factors which control "smallness" of remainders. This is just $(\delta'(r))^{\mu}$ or $(-\delta(r))^{\nu}$, with $\nu \in \mathbb{R}$. For a recent contribution to the local principle we refer to [20] where a local principle is presented which describes C^*-algebras in terms of continuous sections of C^*-bundles. In the case of conical points we get $\delta(r) = \log r$, and so the scales $e^{-\gamma\delta(r)}$ and $(\delta'(r))^{\mu}$ coincide. In the case of true cusps the scale $(-\delta(r))^\nu$ provides us with stronger "glasses" to observe the gain in behaviour at $r = 0$. This demonstrates rather strikingly that asymptotics of solutions to $Au = 0$ near cuspidal points are very subtle. Obstructions arise when one looks for an asymptotic formula for the inverse symbol $a^{-1}(r, z)$ as $r \to \infty$. It is precisely the way at which small parameter problems enter into analysis on manifolds with singularities. It is therefore surprising that the solutions to the Laplace-Beltrami equation on a rotation hypersurface in $\mathbb{R}^n$ have very transparent structure (6.1) independently of the nature of singular points at $r = 0$ and $r = 1$.

7. Index formula

Given any weight data $w = (w_0, w_1)$, the Laplace-Beltrami operator determines a bounded mapping $\Delta : H^{s,w}(\mathcal{Z}) \to H^{s-2,w}(\mathcal{Z})$, as described above. Theorem 5.2 states that if $a(z)$ has no characteristic points on the lines Γ_{-w_0} and Γ_{w_1}, then A is a Fredholm operator. Hence the index of A is well defined and independent of the choice of s. The following corollary provides us with an explicit formula for the index.

Corollary 7.1. *Suppose that $\sigma(\Delta)(z)$ is invertible on the lines Γ_{-w_0} and Γ_{w_1}. Then,*

$$\operatorname{ind} \Delta = \sum_{\substack{j=0,1,\ldots\\ k=1,\ldots,k_j\\ w_0<\pm\sqrt{\lambda_j}<-w_1}} (1+\delta_{0,j}) - \sum_{\substack{j=0,1,\ldots\\ k=1,\ldots,k_j\\ -w_0<\pm\sqrt{\lambda_j}<w_1}} (1+\delta_{0,j}). \tag{7.1}$$

Proof. It is sufficient to combine formulas (6.2) and (5.3). Since the inequalities $w_0 < -w_1$ and $-w_0 < w_1$ are incompatible, at most one sum on the right-hand side of (7.1) is different from zero. □

Formula (7.1) is a rather particular case of the index theorem from [18], cf. also a more general logarithmic residue formula of [7]. If S_w is the horizontal strip bounded by Γ_{w_1} and $-\Gamma_{-w_0}$ then

$$\operatorname{ind} \Delta = \operatorname{tr} \frac{1}{2\pi\imath} \int_{\partial S_w} a^{-1}(z) da(z). \tag{7.2}$$

We emphasize that the integrals over single lines $\Im z = w_1$ and $\Im z = -w_0$ on the right-hand of (7.2) are divergent while their sum makes sense. Moreover, the operators $a^{-1}(z)a'(z)$ are not of trace class on $\mathbb{S}^{n-2}$ unless $n = 2$. However, the operator-valued function $a^{-1}(z)a'(z)$ is holomorphic everywhere in the strip between Γ_{-w_0} and Γ_{w_1}, except possibly at a finite number of points which are characteristic points of $a(z)$. Thus, only the principal parts of Laurent expansions of this function near characteristic points contribute to the sum of the integrals, as is clear from the residue formula. To see that the integral in (7.2) is of trace class it suffices to observe that these principal parts take their values in the space of smoothing operators on $\mathbb{S}^{n-2}$.

The above theory still applies to the differential operator $\Delta - \lambda$, with λ a complex parameter. We are thus in a position to describe the spectrum of Δ acting in $L^{2,w}(\mathcal{Z}) := H^{0,w}(\mathcal{Z})$ with domain $H^{2,w}(\mathcal{Z})$. For all $-w_0 = w_1$, the spectrum of the Laplace-Beltrami operator in $L^{2,w}(\mathcal{Z})$ is empty. On the one hand, if $-w_0 \leq w_1$ then the point spectrum of Δ in $L^{2,w}(\mathcal{Z})$ is empty. For $-w_0 > w_1$, we look for an eigenfunction of the Laplace-Beltrami operator in $L^{2,w}(\mathcal{Z})$ which has the form $u(r)h(\omega)$. On separating the variables we arrive at the ordinary differential equation

$$-\Big(\frac{1}{\sqrt{g_{11}}}\partial_1\Big)^2 u + \lambda_j u = \lambda u$$

for determining u, where λ_j is an eigenvalue of the Laplace-Beltrami operator on the sphere $\mathbb{S}^{n-2}$, and $h = h_{j,k}$ for some $k = 1, \dots, k_j$. The equation possesses two solutions

$$u_j(r) = e^{\imath\sqrt{\lambda-\lambda_j}\,\delta(r)}$$

which correspond to two different branches of the multivalent analytic function $\sqrt{\lambda - \lambda_j}$. The function $u_j(r)h_{j,k}(\omega)$ belongs to $L^{2,w}(\mathcal{Z})$ if and only if λ satisfies $w_1 < \Im\sqrt{\lambda - \lambda_j} < -w_0$. Each $\lambda \in \mathbb{C}$ satisfying this condition for some j belongs to the point spectrum of Δ in $L^{2,w}(\mathcal{Z})$. On the other hand, for $-w_0 \geq w_1$ the continuous spectrum of Δ in $L^{2,w}(\mathcal{Z})$ is empty. If $-w_0 < w_1$ then the continuous spectrum of Δ in $L^{2,w}(\mathcal{Z})$ contains all $\lambda \in \mathbb{C}$ satisfying $-w_1 < \Im\sqrt{\lambda - \lambda_j} < w_0$ for some j.

In either case, the spectrum of Δ in $L^{2,w}(\mathcal{Z})$ is purely point or purely continuous. The papers [1], [3], [10], [12] contain necessary and sufficient conditions for discreteness of the spectrum of the Laplace-Beltrami operator in L^2 for a rather big class of Riemannian manifolds.

References

[1] Baider, A., *Noncompact Riemannian manifolds with discrete spectra.* J. Differential Geom. **14** (1979), no. 1, 41–58.

[2] Collet, J.F., and Volpert, V.A., *Computation of the index of linear elliptic operators in unbounded cylinders.* J. Funct. Anal. **164** (1999), 34–59.

[3] Dodziuk, J., *L^2-harmonic forms on rotationally symmetric Riemannian manifolds.* Proc. Amer. Math. Soc. **77** (1979), no. 3, 395–401.

[4] Daletskii, Yu.L., and Krein, M.G., *Stability of Solutions to Differential Equations in a Banach Space.* Nauka, Moscow, 1970.

[5] Dezin, A.A., *General Questions of the Theory of Boundary Problems.* Nauka, Moscow, 1980.

[6] Fedosov, B., Schulze, B.-W., and Tarkhanov, N., *Analytic index formulas for elliptic corner operators.* Annales de l'Institut Fourier (Grenoble) **52** (2002), no. 3, 899–982.

[7] Gokhberg, I.Ts., and Sigal, E.I., *An operator generalisation of the logarithmic residue theorem and the theorem of Rouché.* Mat. Sb. **84 (126)** (1971), no. 4, 607–629.

[8] Keldysh, M.V., *On the characteristic values and characteristic functions of certain classes of non-selfadjoint equations.* Dokl. Akad. Nauk SSSR **77** (1951), 11–14.

[9] Kiselev, O., and Shestakov, I., *Asymptotics of solutions to the Laplace-Beltrami equation on manifolds with cusps.* SIAM J. of Math. Anal. (in preparation).

[10] Kleine, R., *Discreteness conditions for the Laplacian on complete noncompact Riemannian manifolds.* Math. Z. **198** (1988), no. 1, 127–141.

[11] Kozlov, V., and Maz'ya, V., *Differential Equations with Operator Coefficients.* Springer-Verlag, Berlin et al., 1999.

[12] Kuz'minov, V.I., and Shvedov, I.A., *An addition theorem for manifolds with discrete spectrum of the Laplace operator.* Siberian Math. J. **47** (2006), no. 3, 557–574.

[13] Maz'ya, V.G., and Plamenevskii, B.A., *On asymptotics of solutions of the Dirichlet problem near an isolated singularity of the boundary.* Vestnik Leningrad. Univ., Mat. **13** (1977), 60–65.

[14] Olwer, F.W.T., *Asymptotics and Special Functions.* Academic Press, N.Y., 1974.

[15] Plamenevskii, B.A., *Algebras of Pseudodifferential Operators.* Kluwer Academic Publishers, Dordrecht NL, 1989.

[16] Postnikov, M.M., *Riemannian Geometry.* Encyclopaedia of Mathematical Sciences, Vol. 91, Springer, Berlin et al., 2001.

[17] Rabinovich, V., Schulze, B.-W., and Tarkhanov, N., *A calculus of boundary value problems in domains with non-Lipschitz singular points.* Math. Nachr. **215** (2000), 115–160.

[18] Schulze, B.-W., and Tarkhanov, N., *Euler solutions of pseudodifferential equations.* Integral Equations and Operator Theory **33** (1999), 98–123.

[19] Schulze, B.-W., and Tarkhanov, N., *Singular functions and relative index for elliptic corner operators.* Ann. Univ. Ferrara, Sez. VII, Sc. Mat., Suppl. Vol. **XLV** (1999), 293–310.

[20] Vasilevski, N.L., *On a general local principle for C^*-algebras.* Izv. VUZ. North-Caucasian Region, Natural Sciences, Special Issue "Pseudodifferential operators and some problems of mathematical physics," 2005, 34–42.

Nikolai Tarkhanov
Institut für Mathematik
Universität Potsdam
Am Neuen Palais 10
D-14469 Potsdam, Germany
e-mail: `tarkhanov@math.uni-potsdam.de`

Operator Theory:
Advances and Applications, Vol. 210, 233–248

On the Structure of Operators with Automorphic Symbols

André Unterberger

Dedicated to Nikolai Vasilevski on the occasion of his 60th birthday

Abstract. We consider automorphic distributions on $\mathbb{R}^2$, a concept slightly more precise than that of automorphic functions on the hyperbolic half-plane: these objects introduce themselves in a natural way in symbolic calculi of operators such as Weyl's or the horocyclic calculus, linked to the projective discrete series of $SL(2,\mathbb{R})$. We interpret in operator- theoretic terms various constructions of number-theoretic interest, some of which raise quite deep questions.

Mathematics Subject Classification (2000). 11F11, 11F27, 47G30.

Keywords. Automorphic distributions, horocyclic calculus, Berezin calculus.

1. Introduction

Consider the domain $\mathcal{X}$, in the present text either the plane or the hyperbolic half-plane: the group $G = SL(2,\mathbb{R})$ acts on it by linear or fractional-linear transformations. There is a variety of so-called covariant symbolic calculi, each of which establishes a correspondence from a space of functions on $\mathcal{X}$ – hereafter called symbols – to a space of (generally unbounded) linear endomorphisms of some Hilbert space. Then, if H is a subgroup of G, it is natural to ask for a description of operators with H-invariant symbols.

When the symbolic calculus under consideration is the Toeplitz (or Berezin) calculus, and H is a one-parameter subgroup of G, one finds a commutative algebra of operators, and the question has been answered neatly by N.Vasilevski in [9] and related works. The case when $H = \Gamma{:}\ = SL(2,\mathbb{Z})$, or a more general arithmetic group, is quite different, and our understanding of it is limited to examples. However, these are connected to interesting questions of number theory; moreover, they provide some principles which should act as a guide towards a more general study.

Trying to bring to light some questions these principles raise, rather than giving genuine answers to them, is the limited object of the present paper. The examples described below have all been developed towards a different aim (*cf.* (4.3), (5.21)), namely, that of letting the zeta function with the Euler operator as an argument, or more general L-functions, participate in the spectral theory of interesting operators. Our emphasis on the concept of family of arithmetic coherent states, viewed as generalizing that which may serve as an introduction to the Berezin calculus, may, or so we hope, appeal to some readers of this volume in honour of Nikolai Vasilevski. We shall have to leave to possible distinct publications whatever answers we may find to some of the – always difficult – arithmetic questions raised here.

We first concentrate on the case of the Weyl calculus: then, $\mathcal{X} = \mathbb{R}^2$, and it is necessary to consider Γ-invariant distributions $\mathfrak{S}$ – to be called automorphic distributions – rather than functions. The Weyl operator $\mathrm{Op}(\mathfrak{S})$ acts from $\mathcal{S}(\mathbb{R})$ to $\mathcal{S}'(\mathbb{R})$, and understanding how the composition of two such operators can nevertheless be given some meaning (modulo the one-dimensional space of automorphic distributions homogeneous of degree -1), and sometimes computed, had been the main purpose of [7]. One interesting example of automorphic distribution we wish to report about here – the one denoted as $\mathfrak{T}_\infty$ – was obtained in [8], as a product of related, but different preoccupations: it is a good basis for doing away with some easy conjectures one might be willing to make about the structure of operators with automorphic symbols.

So as to link the present paper more closely to matters with which this introduction started, we shall also indicate some results in the same direction obtained in connection with calculi (it does not matter which, provided covariance is ensured) associated with the discrete series of representations of G.

2. Operators with automorphic symbols

We here consider the one-dimensional Weyl calculus Op: this is the rule that associates to a tempered distribution $\mathfrak{S} \in \mathcal{S}'(\mathbb{R}^2)$ in the plane the linear operator $\mathrm{Op}(\mathfrak{S})$ from $\mathcal{S}(\mathbb{R})$ to $\mathcal{S}'(\mathbb{R})$ defined by the equation

$$(\mathrm{Op}(\mathfrak{S})\,u)(x) = \int_{\mathbb{R}^2} \mathfrak{S}\left(\frac{x+y}{2},\,\eta\right)\, e^{2i\pi(x-y)\eta}\, u(y)\,dy\,d\eta\ : \tag{2.1}$$

this definition is to be taken in the weak sense, *i.e.*, after one has taken the scalar product with $v \in \mathcal{S}(\mathbb{R})$ on both sides. With $\Gamma = SL(2,\mathbb{Z})$, we call a distribution automorphic if it is invariant under the linear action of Γ in $\mathbb{R}^2$. A complete description of automorphic distributions is to be found in [7, Section 4]. It is essentially equivalent – actually, slightly more precise – to the theory of automorphic functions, with respect to the full modular group, on the hyperbolic half-plane Π.

The correspondence associates with $\mathfrak{S}$ the following pair of functions on the half-plane: the set of diagonal matrix elements of $\mathrm{Op}(\mathfrak{S})$ against the family $(u_z^1)_{z\in\Pi}$ defined in (3.7) below, and the one obtained in the same way, replacing this family by the appropriate family $(u_z)_{z\in\Pi}$ of even (Gaussian) functions.

One may call a linear operator A from $\mathcal{S}(\mathbb{R})$ to $\mathcal{S}'(\mathbb{R})$ automorphic if and only if its symbol $\mathfrak{S}$ (there is always a uniquely defined such object) is an automorphic distribution. Using the covariance of the Weyl calculus under the metaplectic representation, it is immediate (*cf.* [7, Section 2]) that this is the case if and only if the operator A commutes with the following two automorphisms of $\mathcal{S}'(\mathbb{R})$, at the same time automorphisms of $\mathcal{S}(\mathbb{R})$: (i) the operator $\mathcal{T}$ that multiplies functions of $x \in \mathbb{R}$ by $e^{i\pi x^2}$; (ii) the Fourier transformation $\mathcal{F}$, normalized in the way associated to the integral kernel $e^{-2i\pi xy}$.

Another important notion is that of homogeneous even distribution on $\mathbb{R}^2$, a substitute for that of (generalized or not) eigenfunction, on the hyperbolic half-plane, of the Laplacian, so that homogeneous automorphic distributions correspond to non-holomorphic modular forms, more precisely to pairs of such (this fits with the Lax-Phillips point of view [3], linked to scattering theory). A symbol $\mathfrak{S} \in \mathcal{S}'(\mathbb{R}^2)$ is homogeneous of degree $-1-\nu$ for some $\nu \in \mathbb{C}$ if and only if the associated operator A satisfies the equation

$$(x)\, A\, \frac{d}{dx} - \frac{d}{dx}\, A\, (x) = \nu A\,, \tag{2.2}$$

denoting as (x) the operator of multiplication by the variable x on the real line.

We have not yet transformed *all* concepts of importance in the classical theory of non-holomorphic modular forms (for the full modular group) into notions relative to operators A: the last operation which must be traced in this way is that of Hecke operator, which defines the fundamental concept of Hecke eigenform. Recall that, on automorphic functions f in the half-plane, the operator T_N $(N \geq 2)$ is defined by the equation

$$(T_N f)(z) = N^{-\frac{1}{2}} \sum_{\substack{ad=N,\, d>0 \\ b \bmod d}} f\left(\frac{az+b}{d}\right)\; ; \tag{2.3}$$

on automorphic distributions on $\mathbb{R}^2$, it transfers to the operator T_N^{dist}, which extends the operator defined on automorphic functions by the equation

$$(T_N^{\mathrm{dist}} h)(x,\, \xi) = N^{-\frac{1}{2}} \sum_{\substack{ad=N,\, d>0 \\ b \bmod d}} h\left(\frac{ax+b\xi}{\sqrt{N}},\, \frac{d\xi}{\sqrt{N}}\right). \tag{2.4}$$

Finally, using the covariance of the Weyl calculus under the metaplectic representation again, and introducing for $d>0,\, b \in \mathbb{R}$, the operator $H_{d,b}^{(N)}$ defined by

$$(H_{d,b}^{(N)}\, u)(x) = \left(\frac{d}{\sqrt{N}}\right)^{\frac{1}{2}} u\left(\frac{dx}{\sqrt{N}}\right) e^{\frac{i\pi b d x^2}{N}}\,, \tag{2.5}$$

one sees that, already assuming $\mathfrak{S}$ to be automorphic, the eigenvalue equation $T_N^{\text{dist}}\,\mathfrak{S} = \lambda\,\mathfrak{S}$ is equivalent, in terms of the operator A with symbol $\mathfrak{S}$, to

$$N^{-\frac{1}{2}} \sum_{\substack{ad=N,\, d>0 \\ b \bmod d}} H_{d,b}^{(N)}\; A\; \left(H_{d,b}^{(N)}\right)^{-1} = \lambda A\,. \tag{2.6}$$

Though the interpretation of this condition is certainly not as obvious as that of the previous ones regarding A, it shows the natural role played by the changes $x \mapsto \frac{x}{\sqrt{N}}$: these will occur again.

Automorphic operators do not quite make up an algebra, which is actually due to technicalities: they are too singular as distributions for this to be the case. But, even bypassing this difficulty as is possible [7], one finds that they do not commute, generally, which explains why a truly concrete description of these operators may probably be out of reach for some time to come. For instance, Eisenstein distributions (*cf. infra*), the complete linear superpositions of which make up a good part (cusp-distributions are still missing) of the space of automorphic distributions, are essentially familiar objects (not very different from the well-known non-holomorphic Eisenstein series): but we do not have a genuine understanding of the associated operators, even though *loc. cit.* contains an explicit formula for the composition, in the sense of the Weyl calculus, of any two Eisenstein distributions.

As this will be needed in the present paper, let us recall the definition of the Eisenstein distributions $\mathfrak{E}_\nu$. When $\nu \in \mathbb{C}$, $\operatorname{Re}\nu < -1$, we define this tempered distribution by the equation

$$\langle\, \mathfrak{E}_\nu\,,\, h\,\rangle = \frac{1}{2} \sum_{\substack{m,n\in\mathbb{Z} \\ |m|+|n|\neq 0}} \int_{-\infty}^{\infty} |t|^{-\nu}\, h(tn,\, tm)\, dt\,, \qquad h \in \mathcal{S}(\mathbb{R}^2)\,. \tag{2.7}$$

One then shows that the map $\nu \mapsto \mathfrak{E}_\nu$ extends as a meromorphic function (valued into $\mathcal{S}'(\mathbb{R}^2)$) in the whole plane, with poles only at ± 1: these are simple, and the residues there are given as $\operatorname{Res}_{\nu=-1}\mathfrak{E}_\nu = -1$ and $\operatorname{Res}_{\nu=1}\mathfrak{E}_\nu = \delta$, the unit mass at the origin of $\mathbb{R}^2$. One role of these distributions can be found in the fact that the standard Dirac comb $\mathfrak{D}_0$ on $\mathbb{R}^2$, the sum of unit masses at points with integral coordinates, can be written as

$$\mathfrak{D}_0 = 1 + \delta + \frac{1}{2\pi} \int_{-\infty}^{\infty} \mathfrak{E}_{i\lambda}\, d\lambda\,, \tag{2.8}$$

a decomposition into homogeneous components since $\mathfrak{E}_\nu$ is homogeneous of degree $-1-\nu$.

3. Some automorphic operators

Because of the covariance of the Weyl calculus under the metaplectic representation Met, one obvious way to build an automorphic operator is to consider the rank-one operator $u \mapsto (\mathfrak{d}\,|\,u)\,\mathfrak{d}$ associated to a distribution $\mathfrak{d}$ possessing the property that, given any point $\tilde{g}$ of the metaplectic group (recall that this is the twofold cover of G) lying above some element g of Γ, one should have $\mathrm{Met}(\tilde{g})\,\mathfrak{d} = \omega\,\mathfrak{d}$ for some $\tilde{g}$-dependent $\omega \in \mathbb{C}$ with $|\omega| = 1$. One of the two simplest examples (there is one such of each parity) is obtained if one takes

$$\mathfrak{d}(x) = \sum_{m\in\mathbb{Z}} \chi^{(12)}(m)\,\delta\left(x - \frac{m}{\sqrt{12}}\right), \tag{3.1}$$

$\chi^{(12)}$ being the Dirichlet character mod 12 such that $\chi^{(12)}(m) = 1$ if $m \equiv \pm 1$ mod 12, and $\chi^{(12)}(m) = -1$ if $m \equiv \pm 5$ mod 12. Then, introducing on $\mathbb{R}^2$ the Euler operator $2i\pi\mathcal{E} = x\,\frac{\partial}{\partial x} + \xi\,\frac{\partial}{\partial \xi} + 1$ so that, for $t > 0$, the operator $t^{2i\pi\mathcal{E}}$ can be defined on tempered distributions $\mathfrak{S}$ by the equation

$$\langle t^{2i\pi\mathcal{E}}\mathfrak{S}\,,\, h\rangle = \langle \mathfrak{S}\,,\,(x,\xi) \mapsto t^{-1}\,h(t^{-1}x,\, t^{-1}\xi)\rangle\,, \qquad h \in \mathcal{S}(\mathbb{R})\,, \tag{3.2}$$

one can show [8, Theorem 4.2] that the Weyl symbol $W(\mathfrak{d},\,\mathfrak{d})$ of the above-defined rank one operator is the image of $\mathfrak{D}_0$ under the operator $12^{i\pi\mathcal{E}}\,[\,1 - 2^{-2i\pi\mathcal{E}}\,]\,[\,1 - 3^{-2i\pi\mathcal{E}}\,]$.

This example immediately brings forward two questions. The first one starts from the observation that the support of the measure $W(\mathfrak{d},\,\mathfrak{d})$ is the set of points $(\frac{j}{\sqrt{12}},\,\frac{k}{\sqrt{12}})$ with $(j,\,k) \in \mathbb{Z}^2$: could one, say, get rid of the factors $\sqrt{12}$, at the same time finding an operator with such a clear-cut interpretation ? The answer to this question is negative: even such a simple operation as a rescaling $\mathfrak{S} \mapsto \mathfrak{T}$, with $\mathfrak{T}(x,\xi) = e^{\lambda}\,\mathfrak{S}(e^{\lambda}x,\, e^{\lambda}\xi)$, of symbols, in other words a transformation $k_{\mathfrak{S}} \mapsto k_{\mathfrak{T}}$, with

$$k_{\mathfrak{T}}(x,\,y) = k_{\mathfrak{S}}(x\,\cosh\lambda + y\,\sinh\lambda,\; x\,\sinh\lambda + y\,\cosh\lambda)\,, \tag{3.3}$$

of integral kernels, would destroy any interpretation of a given operator as being associated in any simple way to the use of distributions possessing, just like $\mathfrak{d}$, any amount of "metaplectic automorphy".

The next question concerns the possibility of generalizing the previous example with the help of some algebra, substituting for the measure $\mathfrak{d}$ a finite-dimensional space of distributions on the line, globally invariant under the part of the metaplectic representation lying above Γ. The (positive, this time) answer starts with the consideration of an arbitrary finite set S of prime numbers including 2 and, with $N = 2\prod_{p\in S} p$, of the (unique) character χ of the group $\Lambda = \{\mu \in (\mathbb{Z}/N\mathbb{Z})^{\times} : \mu^2 = 1\}$ taking the value -1 on each of the following elements μ: (i) the one such that $\mu \equiv -1$ mod 4 and $\mu \equiv 1$ mod p for every $p \in S$ with $p \neq 2$; (ii) for every $p \in S$ with $p \neq 2$, the one such that $\mu \equiv -1$ mod p, $\mu \equiv 1$ mod 4 and $\mu \equiv 1$ mod q for every $q \in S,\ q \neq 2,\ q \neq p$. Then, for every $\rho \in (\mathbb{Z}/N\mathbb{Z})^{\times}$, consider the measure

$$\varpi_{\rho}(x) = \sum_{\mu\in\Lambda} \chi(\mu) \sum_{\ell\in\mathbb{Z}} \delta\left(x - \frac{N\ell + \rho\mu}{\sqrt{N}}\right) \tag{3.4}$$

on the line. The linear space generated, for given N, by the distributions ϖ_{ρ}, satisfies the invariance condition aimed at. As a basis, one can take the set of ϖ_{ρ}'s with ρ describing any set R_N of representatives of $\mathbb{Z}/N\mathbb{Z})^\times$ mod Λ.We then consider the operator

$$u \mapsto \sum_{\rho \in R_N} (\varpi_\rho \,|\, u)\, \varpi_\rho \tag{3.5}$$

which, again, commutes with the part of the metaplectic representation above Γ. Its Weyl symbol W_N is given by the equation, obtained after some algebra:

$$W_N = N^{i\pi\mathcal{E}} \prod_{p \in S} (1 - p^{-2i\pi\mathcal{E}})\, \mathfrak{D}_0\,, \tag{3.6}$$

a Γ-invariant measure supported in the set $\frac{1}{\sqrt{N}}\,(\mathbb{Z} \times \mathbb{Z})$.

Many readers of this volume are probably familiar with the way it is possible, starting from the odd eigenfunction with lowest eigenvalue of the harmonic oscillator and using the metaplectic representation, to build in association with the odd part of $L^2(\mathbb{R})$ – a variant of course exists for the even part as well – a "family of coherent states", a.k.a. an overcomplete set, parametrized by the homogeneous space G/K, $K = SO(2)$, a model of which is the hyperbolic upper half-plane Π with invariant measure $d\mu$. Indeed, if one sets

$$u_z^1(t) = 2^{\frac{5}{4}}\, \pi^{\frac{1}{2}}\, (\mathrm{Im}\ z)^{\frac{3}{4}}\ t\ e^{i\pi z\, t^2}\,, \tag{3.7}$$

one has for every *odd* square-integrable function on the line the equation

$$\int_\Pi |(u_z^1 \,|\, u)|^2\, d\mu(z) = 8\pi\, \|\, u\, \|^2_{L^2(\mathbb{R})}\,. \tag{3.8}$$

Now, as shown in [8, Theorem 12.4], denoting as ϖ_ρ^g, for $g \in G$, the image of ϖ_ρ under the metaplectic transformation associated with any point of the metaplectic group lying above g^{-1} (phase factors have no importance here), one has whenever $u \in L^2(\mathbb{R})$ has the parity associated with the number of distinct prime divisors of N the equation

$$\sum_{\rho \in R_N} \int_{\Gamma\backslash G} |\langle \varpi_\rho^g\,,\, u\rangle|^2\, dg = \frac{2\pi}{3}\, N^{-\frac{1}{2}}\, \phi(N)\ \|\, u\, \|^2_{L^2(\mathbb{R})}\,, \tag{3.9}$$

in which ϕ is Euler's indicator function.

Comparing (3.8) with (3.9), one might look at the latter equation as the starting point of an arithmetic counterpart of Berezin's (or Toeplitz's) quantization theory, a space $\Gamma\backslash G$ of lattices taking the place of Π: however, this is rather removed from our current point of view. Indeed, we consider that what is important in a symbolic calculus is its *abstract* covariance group, not its explicit defining equation or even its phase space (always a homogeneous space of the covariance group): for all calculi with the same (sufficiently large) group of covariance are related in an explicit way. This will show up, again, in Section 5 of the present paper.

4. The distribution $\mathfrak{T}_\infty$

Here is our favourite example: as will be seen, it is quite interesting as a distribution, while the associated operator is mysterious.

Definition 4.1. The distribution $\mathfrak{T}_\infty$ in the plane is defined by the equation

$$\mathfrak{T}_\infty(x,\,\xi) = \sum_{\substack{j,k\in\mathbb{Z}\\ |j|+|k|\neq 0}} \Gamma_{jk}^{(\infty)}\,\delta(x-j)\,\delta(\xi-k)\,, \tag{4.1}$$

where

$$\Gamma_{jk}^{(\infty)} = \prod_{\substack{p \text{ prime}\\ p|j,\, p|k}} (1-p)\,. \tag{4.2}$$

It is not difficult to see, as proved in [8, Theorem 12.5], that the decomposition into homogeneous components of the distribution $\mathfrak{T}_\infty$ is given by the equation

$$\mathfrak{T}_\infty = \frac{1}{2\pi}\int_{-\infty}^{\infty} (\zeta(-i\lambda))^{-1}\,\mathfrak{E}_{i\lambda}\,d\lambda + \sum_j \mathrm{Res}_{\mu=\mu_j}\left(\frac{\mathfrak{E}_{-\mu}}{\zeta(\mu)}\right) : \tag{4.3}$$

here, μ_j is to vary over the set of non-trivial zeros of the Riemann zeta function ζ. Performing a change of contour, one sees that the Riemann hypothesis just means that it is possible to write a decomposition of $\mathfrak{T}_\infty$ as a generalized integral (principal values, in Cauchy's sense, are needed because of the critical zeros of zeta) of automorphic distributions homogeneous of degrees $-1-\nu$ supported by the line $\mathrm{Re}\,\nu = \frac{1}{2}$.

Of course, using the fact that the operator with symbol $\frac{1}{2}\,\delta(x)\,\delta(\xi)$ is the operator $u \mapsto \check{u}$, $\check{u}(x) = u(-x)$, together with the covariance of the Weyl calculus under the Heisenberg representation (which turns translations in the phase space into an operator-theoretic notion), one can write the operator $A_\infty = \mathrm{Op}(\mathfrak{T}_\infty)$ as a series of "symmetry" operators: this certainly does not lead to any deep understanding of A_∞, and we now formulate the question in a completely different way.

Let N be any positive integer. Consider the distribution

$$\mathfrak{T}_N(x,\,\xi) = \sum_{\substack{j,k\in\mathbb{Z}\\ |j|+|k|\neq 0}} \Gamma_{jk}^{(N)}\,\delta(x-j)\,\delta(\xi-k)\,, \tag{4.4}$$

with

$$\Gamma_{jk}^{(N)} = \prod_{\substack{p \text{ prime}\,|\,N\\ p|j,\, p|k}} (1-p)\,. \tag{4.5}$$

Clearly, $\mathfrak{T}_N \to \mathfrak{T}_\infty$ in $\mathcal{S}'(\mathbb{R}^2)$ if $N\to\infty$ in such a way that every finite set of prime numbers, from a certain point on, lies in the set of divisors of N. Let us indicate the way $\mathfrak{T}_N$ decomposes into homogeneous components. As soon as N is even, it is no loss of generality, since only the collection of prime factors of N is relevant in the definition of this distribution, to assume that N is the product of 4 by a squarefree odd integer. Recalling the Eulerian expansion of zeta, to wit $\zeta(\nu) = \prod_{p \text{ prime}}(1-p^{-\nu})^{-1}$, Re $\nu > 1$, set

$$\zeta_N(\nu) = \prod_{p \text{ prime} \,|\, N} (1-p^{-\nu})^{-1}. \tag{4.6}$$

Then [8, Section 12]

$$\mathfrak{T}_N = \frac{1}{2\pi} \int_{-\infty}^{\infty} (\zeta_N(-i\lambda))^{-1} \, \mathfrak{E}_{i\lambda} \, d\lambda + \mathrm{Res}_{\mu=1} \left(\frac{\mathfrak{E}_{-\mu}}{\zeta_N(\mu)} \right) \; : \tag{4.7}$$

comparing (4.3) and (4.7), note that the pole at $\mu = 1$ contributing to the right-hand side of (4.7) disappears in the limit because, contrary to ζ_N, ζ has a pole there. But infinitely many new poles, to wit the non-trivial zeros of zeta, make their appearance.

In view of the spectral decomposition of the measure $\mathfrak{T}_\infty$, a true understanding of the operator with symbol $\mathfrak{T}_N$ would be interesting. However, it is only the operator with symbol $N^{i\pi\mathcal{E}} \, \mathfrak{T}_N$ that has a close relation to the metaplectic representation: indeed, one has

$$N^{i\pi\mathcal{E}} \, \mathfrak{T}_N = W_N - N^{-\frac{1}{2}} \prod_{p \in S} (1-p) \; \times \; \delta \tag{4.8}$$

in terms of the measure W_N in Section 3 of the present paper, while, as has already been recalled, $\frac{1}{2}\,\delta$ is the symbol of the operator $u \mapsto \check{u}$. As pointed out in the paragraph around (3.3), a rescaling transformation of symbols such as $N^{i\pi\mathcal{E}}$ does not have any simple interpretation in terms of the associated operator.

5. Other symbolic calculi

Consider the measure $\mathfrak{d}$ in (3.1). Identifying even functions u on the real line with functions v on the half-line with the help of the quadratic change of variable $\mathrm{Sq}_{\mathrm{even}} \colon v \mapsto u$, where $u(x) = 2^{-\frac{3}{4}}\, |x| \, v(\frac{x^2}{2})$, one may identify $\mathfrak{d}$ with a measure $\mathfrak{s}$ on the half-line: the invariance of $\mathfrak{d}$, up to phase factors (*i.e.*, up to multiplication by complex numbers of absolute value 1), under the part of the metaplectic representation lying above Γ, becomes under this transfer the fact that the measure $\mathfrak{s}$ is invariant, up to phase factors, under the part of the representation $\mathcal{D}_{\frac{1}{2}}$ from the (prolongation of the) holomorphic discrete series of the twofold cover of G. We presently generalize the definition of such a measure $\mathfrak{s}$ to the case when the subscript $\frac{1}{2}$ is replaced by a more general positive number.

Before recalling the definition of the unitary representations to be considered, let us observe that we do not really care about phase factors, since we are only going, here, to consider scalar products in which these will appear on both sides. As a consequence, we do not have to substitute for the group G its universal cover, since we are satisfied with so-called projective representations, in which the basic homomorphism property is only assumed to be valid up to arbitrary phase factors, depending on the two elements of G under consideration. Then, for every

$\tau \in \mathbb{R}$, $\tau > -1$, there is an essentially unique unitary projective representation $\mathcal{D}_{\tau+1}$ of G in the space $L^2((0,\infty);\, s^{-\tau}\, ds)$ given, in the case when $b > 0$, by the equation (involving the Bessel function J_τ)

$$\left(\mathcal{D}_{\tau+1}\left(\left(\begin{smallmatrix} a & b \\ c & d \end{smallmatrix}\right)\right) v\right)(s) = \frac{2\pi}{b} \int_0^\infty v(t) \left(\frac{s}{t}\right)^{\frac{\tau}{2}} \exp\left(2i\pi \frac{ds+at}{b}\right) J_\tau\left(\frac{4\pi}{b}\sqrt{st}\right) dt\,. \tag{5.1}$$

Under a Laplace transformation, this representation transfers to the projective representation $\pi_{\tau+1}$ of G into an appropriate Hilbert space of holomorphic functions $\mathfrak{f}$ in the complex upper half-plane Π (that defined, only in the case when $\tau > 0$, by the condition $\mathfrak{f} \in L^2(\Pi,\, (\mathrm{Im}\, z)^{\tau+1} d\mu(z))$, with $d\mu(z) = (\mathrm{Im}\, z)^{-2}\, d\mathrm{Re}\, z\, d\mathrm{Im}\, z$), with which most readers are probably more familiar:

$$\left(\pi_{\tau+1}\left(\left(\begin{smallmatrix} a & b \\ c & d \end{smallmatrix}\right)\right) \mathfrak{f}\right)(z) = (-cz+a)^{-\tau-1}\, \mathfrak{f}\left(\frac{dz-b}{-cz+a}\right). \tag{5.2}$$

Note that the determination, in Π, of the power $(-cz+a)^{-\tau-1}$, has an arbitrary dependence on $\left(\begin{smallmatrix} a & b \\ c & d \end{smallmatrix}\right)$, but is of course assumed to be a continuous function of z.

It is customary, in the theory of modular forms, to set $q = e^{2i\pi z}$, so that $|q| < 1$ when $z \in \Pi$. A holomorphic function $\mathfrak{f}$ of the kind

$$\mathfrak{f}(z) = q^\kappa \sum_{m \geq 0} a_m\, q^m\,, \tag{5.3}$$

where $0 < \kappa \leq 1$ (of course, $q^\kappa = e^{2i\pi\kappa z}$ by definition) is always periodic of period 1 up to phase factors; it becomes a modular form of weight $\tau + 1$, for an unspecified multiplier system, if one assumes that it is invariant, up to some phase factor, under the transformation $\pi_{\tau+1}\left(\left(\begin{smallmatrix} 0 & 1 \\ -1 & 0 \end{smallmatrix}\right)\right)$, defined by extending (5.2), and if one also assumes that the sequence (a_m) of coefficients is controlled by some power of $m+1$. When $\kappa = 1$ and $\tau+1$ is an even integer (so that, in particular, the factor $(-cz+a)^{-\tau-1}$ ceases to be multi-valued), one gets back to the more classical concept of holomorphic (cusp-) form for the full modular group. Then, taking appropriate powers (when this is possible, *i.e.*, starting from a function without zeros in Π), one can find examples of such functions $\mathfrak{f}$ satisfying the required properties for any given $\tau > -1$. For instance, one may consider the function

$$\mathfrak{f}(z) = q^{\frac{\tau+1}{12}} \prod_{n \geq 1} (1-q^n)^{2\tau+2}\,, \tag{5.4}$$

the fractional power with exponent $\frac{\tau+1}{12}$ of the so-called Ramanujan Δ-function. In general, one associates with $\mathfrak{f}$ the measure $\mathfrak{s}_\tau$,

$$\mathfrak{s}_\tau(t) = \sum_{m \geq 0} a_m\, \delta(t-m-\kappa)\,, \tag{5.5}$$

of which $\mathfrak{f}$ is the Laplace transform. When $\tau = -\frac{1}{2}$ and $\mathfrak{f}$ is the function in (5.4), in this case called the Dedekind eta function, this would lead to a construction of the measure $\mathfrak{d}$ in (3.1).

Coming back to the general case, given $g \in G$, define the distribution $\mathfrak{s}_\tau^g$ (up to a phase factor) as being the image of $\mathfrak{s}_\tau$ under $\mathcal{D}_{\tau+1}(g^{-1})$. Setting

$$\| v \|_{\tau+1}^2 = \int_0^\infty |v(s)|^2 \, s^{-\tau} \, ds \,, \tag{5.6}$$

one has the following equation [8, Theorem 9.3], which shows again the role of the family $(\mathfrak{s}_\tau^g)_{g\in G}$ as an arithmetic family of coherent states for the representation $\mathcal{D}_{\tau+1}$:

$$\int_{\Gamma\backslash G} |(\mathfrak{s}_\tau^g \,|\, v)_{\tau+1}|^2 \, dg = \frac{(4\pi)^{\tau+1}}{\Gamma(\tau+1)} \, \| y^{\frac{\tau+1}{2}} \mathfrak{f} \|_{L^2(\Gamma\backslash\Pi)}^2 \, \| v \|_{\tau+1}^2 \,. \tag{5.7}$$

By the way, this equation gives the more serious of two reasons why, as the reader may have asked himself, one could not use, in place of the distribution $\mathfrak{d}$ in (3.1), the "more natural" one-dimensional Dirac comb $\sum_{m\in\mathbb{Z}} \delta(x-m)$: it is that the associated function $\mathfrak{f}$, a standard theta-function, would not be a cusp-form, *i.e.*, would not satisfy the case $\tau = -\frac{1}{2}$ of the condition

$$\| y^{\frac{\tau+1}{2}} \mathfrak{f} \|_{L^2(\Gamma\backslash\Pi)}^2 = \int_{\Gamma\backslash\Pi} y^{\tau+1} \, |\mathfrak{f}(z)|^2 \, d\mu(z) < \infty \,. \tag{5.8}$$

The other "reason", to wit that the one-dimensional Dirac comb is not invariant under multiplication by $e^{i\pi x^2}$, only under multiplication by $e^{2i\pi x^2}$, could be taken care of by substituting for Γ an appropriate congruence subgroup: details are to be found at the end of [8, Section 5].

Since this may be of some interest to practitioners of quantization theory, we shall spend some more time discussing symbolic calculi associated with the representation $\mathcal{D}_{\tau+1}$. The most popular way to obtain a covariant calculus in this context is Berezin's, which can be defined as follows. For every $z \in \Pi$, set

$$\phi_z^{\tau+1}(s) = s^\tau \, e^{2i\pi s \bar{z}^{-1}} \,, \qquad s > 0 \,, \tag{5.9}$$

so that, given $g \in G$,

$$\mathcal{D}_{\tau+1}\left(\left(\begin{smallmatrix} a & b \\ c & d \end{smallmatrix}\right)\right) \phi_z^{\tau+1} = \omega \left(a + \frac{b}{\bar{z}}\right)^{-\tau-1} \phi_{\frac{az+b}{cz+d}}^{\tau+1} \tag{5.10}$$

for some phase factor ω with $|\omega| = 1$ depending only on g. At least when $\tau > 0$, one can substitute for $\phi_z^{\tau+1}$ the normalized version

$$\psi_z^{\tau+1} = \frac{(4\pi)^{\frac{\tau+1}{2}}}{(\Gamma(\tau+1))^{\frac{1}{2}}} \left(\mathrm{Im}\,(-z^{-1})\right)^{\frac{\tau+1}{2}} \phi_z^{\tau+1} \,. \tag{5.11}$$

One then has the equation, to be compared to (3.8) (which is actually just the special case when $\tau = \frac{1}{2}$ of the identity that follows, rewritten after a quadratic change of variable)

$$\| v \|_{\tau+1}^2 = \frac{\tau}{4\pi} \int_\Pi |(\psi_z^{\tau+1} \,|\, v)|^2 \, d\mu(z) \tag{5.12}$$

for every $v \in L^2((0,\infty);\, s^{-\tau}\, ds)$. Given a linear operator A from the linear space generated by the functions $\psi_z^{\tau+1}$ to its algebraic dual – note that this is demanding very little from A – one defines the Berezin-covariant symbol $f^{\rm cov}$ of A as the function in Π such that

$$f^{\rm cov}(z) = (\psi_z^{\tau+1} \,|\, A\, \psi_z^{\tau+1}), \qquad z \in \Pi, \tag{5.13}$$

a definition with a natural extension to the case when $\tau > -1$. Sometimes, one uses the same definition with $\phi_z^{\tau+1}$ substituted for $\psi_z^{\tau+1}$: we here prefer the present one, for which the covariance property (a consequence of (5.10)) takes the following form: if $f^{\rm cov}$ is the Berezin-covariant symbol of some operator A, that of the operator $\mathcal{D}_{\tau+1}\left(\left(\begin{smallmatrix} a & b \\ c & d \end{smallmatrix}\right)\right) A\, \mathcal{D}_{\tau+1}\left(\left(\begin{smallmatrix} a & b \\ c & d \end{smallmatrix}\right)\right)^{-1}$ is the function $z \mapsto f\left(\frac{dz-b}{-cz+a}\right)$.

All this is extremely well known: also that, in conjunction with the covariant symbol, one must consider the contravariant symbol (which, in some sense, very seldom exists as a distribution), defined in such a way that the two maps operator $\mapsto$ covariant symbol and contravariant symbol $\mapsto$ operator are adjoint of each other, when the space $L^2(\Pi,\, d\mu)$ as a space of symbols, and the space of Hilbert-Schmidt operators on $L^2((0,\, \infty);\, s^{-\tau}\, ds)$ on the other hand, are considered. Some readers may also know that, in place of the pair of contravariant and covariant Berezin symbols, one can consider the pair of active and passive symbols: the advantage is that the gap between these two latter species is quite tractable, while that between the first two ones is too considerable for the Berezin calculus to be considered as a bona fide symbolic calculus of operators on the space $L^2((0,\, \infty);\, s^{-\tau}\, ds)$.

The point we wish to stress here, with a full realization that this point of view may not appeal to many practitioners of quantization theory, is that the hyperbolic upper half-plane Π does not seem to be the "right" phase space to use in order to define a symbolic calculus of operators on $L^2((0,\, \infty);\, s^{-\tau}\, ds)$. A much better choice, in two respects, is the plane $\mathbb{R}^2$, on which G acts under linear, rather than fractional-linear, transformations. Our first argument is based on the fact that it is impossible to build a manageable symbolic calculus, with Π as a phase space, which would establish an isometry between $L^2(\Pi,\, d\mu)$ and the space of Hilbert-Schmidt operators on $L^2((0,\, \infty);\, s^{-\tau}\, ds)$ (a desirable property as it would suppress the need to consider *a pair* of symbols). The situation, however, can be saved if one introduces the *isometric horocyclic* symbolic calculus, which we now proceed towards defining.

First, let us recall the definition of the Radon transformation V from functions on Π to functions on $\mathbb{R}^2$: provided convergence is ensured, one sets, given a function f on Π,

$$(Vf)(x,\, \xi) = \int_{-\infty}^{\infty} f\left(\left(\left(\begin{smallmatrix} x & b \\ \xi & d \end{smallmatrix}\right) \left(\begin{smallmatrix} 1 & t \\ 0 & 1 \end{smallmatrix}\right)\right) . i\right)\, dt, \qquad \left(\begin{smallmatrix} x & b \\ \xi & d \end{smallmatrix}\right) \in G, \tag{5.14}$$

where $\left(\begin{smallmatrix} p & q \\ r & s \end{smallmatrix}\right) . i = \frac{pi+q}{ri+s}$. It is immediate that, indeed, the right-hand side does not depend on the choice of the pair $(b,\, d)$ subject to the stated condition, and that the Radon transformation exchanges the two quasiregular actions of G on functions defined on Π or on $\mathbb{R}^2$. With the help of the Euler operator $2i\pi\mathcal{E}$ introduced in

Section 3, we also define, on functions on $\mathbb{R}^2$, the operator

$$T = \left(\frac{\pi}{2}\right)^{\frac{1}{2}} \frac{\Gamma(\frac{1}{2} - i\pi\mathcal{E})}{\Gamma(-i\pi\mathcal{E})} = \pi^{-\frac{1}{2}} (-i\pi\mathcal{E}) \int_0^\infty t^{-\frac{1}{2}} (1+t)^{-1+i\pi\mathcal{E}}\, dt\,, \tag{5.15}$$

and the first equation in [4, p. 91] gives a similar integral expression for the ratio of Gamma functions which occurs in (5.16) below. We then define the isometric horocyclic symbol h^{iso} of an operator A on $L^2((0,\infty);\, s^{-\tau}\, ds)$ in terms of its covariant symbol f^{cov} by the equation

$$h^{\text{iso}} = 2^{-\tau-\frac{1}{2}}\, \Gamma(\tau+1)\, \pi^{i\pi\mathcal{E}} \frac{\Gamma(\frac{\tau}{2} + \frac{1}{4} - \frac{i\pi}{2}\mathcal{E})}{\Gamma(\frac{\tau}{2} + \frac{3}{4} + \frac{i\pi}{2}\mathcal{E})}\,.$$

$$TV\left[\left(\Gamma\left(\frac{\tau}{2} + \frac{1}{4} + \frac{i}{2}\sqrt{\Delta - \frac{1}{4}}\right)\Gamma\left(\frac{\tau}{2} + \frac{1}{4} - \frac{i}{2}\sqrt{\Delta - \frac{1}{4}}\right)\right)^{-1} f^{\text{cov}}\right]. \tag{5.16}$$

Before coming to some technical points regarding this definition, let us emphasize the basic properties of the isometric horocyclic symbol. Besides the already mentioned fact that it establishes an isometry between a natural Hilbert space of symbols and a natural Hilbert space of operators – we shall have to state this with some more precision presently – it connects, in the case when $\tau = \mp\frac{1}{2}$, to the Weyl symbol of an operator on the line, as follows: given an operator A on $L^2((0,\infty);\, s^{\frac{1}{2}}\, ds)$, and identifying this latter space with the space $L^2_{\text{even}}(\mathbb{R})$ under the isometry $\mathrm{Sq}_{\text{even}}$ mentioned in the beginning of the present section (of course, a variant exists for the odd case), one can see that the isometric horocyclic symbol (considered with $\tau = -\frac{1}{2}$) of an operator $A\colon L^2_{\text{even}}(\mathbb{R}) \to L^2_{\text{even}}(\mathbb{R})$ coincides with the Weyl symbol of the extension of A as an operator on $L^2(\mathbb{R})$ which is zero on the space $L^2_{\text{odd}}(\mathbb{R})$.

Recall that the Gamma function is rapidly decreasing at infinity on vertical lines in the complex planes, so that the operator, the inverse of a product of Gamma functions in which the argument is close to $\pm\frac{i}{2}\sqrt{\Delta - \frac{1}{4}}$, which occurs on the right-hand side of (5.16), is extremely far from being a bounded operator in $L^2(\Pi,\, d\mu)$: this is not surprising since, as already mentioned, the covariant symbol of an operator can be defined under extremely large conditions: in other words, it is in some sense "too nice" a function. To give the right-hand side of (5.16) a meaning, one may start with the consideration of functions on Π the decomposition of which, into generalized eigenfunctions of Δ, is supported in a compact interval of $[\frac{1}{4}, \infty[$.

Now, the space of symbols to be considered in the isometric horocyclic calculus is not the whole of $L^2(\mathbb{R}^2)$: for instance, in the case when $\tau = -\frac{1}{2}$, it is only the space of even functions in $L^2(\mathbb{R}^2)$, invariant under the rescaled version $\mathcal{G}$ of the symplectic Fourier transformation defined by the equation

$$(\mathcal{G}\, h)(x,\, \xi) = 2\int_{\mathbb{R}^2} h(y,\, \eta)\, e^{4i\pi(x\eta - y\xi)}\, dy\, d\eta\,. \tag{5.17}$$

This can be explained by the fact that, in the Weyl calculus, these two properties characterize the symbols of operators which commute with the map $u \mapsto \check{u}$, $\check{u}(x) = u(-x)$, and which kill all odd functions. In the general case, the precise statement is that the isometric horocyclic symbols of Hilbert-Schmidt operators on $L^2((0, \infty); s^{-\tau}\, ds)$ exactly describe, in an isometric way, the subspace of $L^2_{\text{even}}(\mathbb{R}^2)$ consisting of functions invariant under the (unitary) symmetry

$$\frac{\Gamma(i\pi\mathcal{E})\,\Gamma(\tau + \frac{1}{2} - i\pi\mathcal{E})}{\Gamma(-i\pi\mathcal{E})\,\Gamma(\tau + \frac{1}{2} + i\pi\mathcal{E})}\, \mathcal{G}\,. \tag{5.18}$$

What is probably the main benefit of using symbols living on $\mathbb{R}^2$ rather than Π is that, under the Radon transformation, the operator $\Delta - \frac{1}{4}$ on functions defined in Π transfers to the operator $\pi^2\,\mathcal{E}^2$ on functions defined in the plane: this makes it possible to define a simple operator playing the role of a "square root" of $\Delta - \frac{1}{4}$. Then, one may substitute for the decomposition of functions on Π, as integral superpositions of generalized eigenfunctions of Δ, the simpler decomposition of functions (or distributions) in the plane, into their homogeneous components.

This makes it possible to perform computations which would hardly, if at all, have been feasible in the context of the half-plane, in particular when arithmetic is present. With

$$(\mathfrak{s}_\tau \,|\, w)_{\tau+1} = \langle\, \overline{\mathfrak{s}_\tau}\,,\; s \mapsto s^{-\tau}\, w(s)\,\rangle\,, \tag{5.19}$$

consider the operator $P_{\mathfrak{s}_\tau,\,\mathfrak{s}_\tau}$ defined by the equation

$$P_{\mathfrak{s}_\tau,\,\mathfrak{s}_\tau}\, w = (\mathfrak{s}_\tau \,|\, w)_{\tau+1}\, \mathfrak{s}_\tau\ : \tag{5.20}$$

the scalar product makes sense, for instance, if the function w on the real line lies in the space of C^∞ vectors of the representation $\mathcal{D}_{\tau+1}$. As a consequence of the assumptions made about the function $\mathfrak{f}$, the operator $P_{\mathfrak{s}_\tau,\,\mathfrak{s}_\tau}$ commutes with the operators $\mathcal{D}_{\tau+1}(g)$ with $g \in \Gamma$. In view of the covariance of the isometric horocyclic symbolic calculus, the symbol (in this calculus) $W(\mathfrak{s}_\tau,\, \mathfrak{s}_\tau)$ of this operator, a distribution in the plane, is invariant under linear changes of coordinates associated to elements of Γ.

The following theorem shows that, just like the symbol W_N from the Weyl calculus (*cf.* (3.6)), the symbol $W(\mathfrak{s}_\tau,\, \mathfrak{s}_\tau)$ is the image of the Dirac comb $\mathfrak{D}_0$ under some operator $\mathcal{B}$, a function (in the spectral-theoretic sense) of the Euler operator. In the case when $2N\kappa$ is an integer for some $N = 1, 2, \dots$, $\mathcal{B}$ is again the product of $N^{i\pi\mathcal{E}}$ by a Dirichlet series with the operator $i\pi\mathcal{E}$ as an argument.

Theorem 5.1. *For every* $\tau > -1$, *one has the equation*

$$W(\mathfrak{s}_\tau,\, \mathfrak{s}_\tau) = 2^{-i\pi\mathcal{E} - \frac{3}{2}}\, \frac{L(\bar{\mathfrak{f}} \otimes \mathfrak{f},\, \frac{1}{2} + i\pi\mathcal{E})}{\zeta(2i\pi\mathcal{E})}\, \mathfrak{D}_0\,, \tag{5.21}$$

where the function $s \mapsto L(\bar{\mathfrak{f}} \otimes \mathfrak{f},\, s)$ *is defined by analytic continuation, starting in the case when* $\operatorname{Re} s > 1$ *from the equation*

$$L(\bar{\mathfrak{f}} \otimes \mathfrak{f},\, s) = \sum_{m \geq 0} |a_m|^2\, (m+\kappa)^{-s-\tau}\,, \tag{5.22}$$

where κ *and the coefficients* a_m *have been defined in* (5.3).

Proof. The lengthy proof of a slightly different formulation of this theorem has been given in [8, Corollary 11.6], and we now provide the missing details. First, we must recall that the introduction of L-functions such as the one appearing here, properly called convolution L-functions, is an important subject in the theory of modular forms: their analytic extension is a classical application [1, 2] of the so-called Rankin-Selberg unfolding method; the method also shows that the function

$$L^*(\bar{\mathfrak{f}} \otimes \mathfrak{f},\, s) = \frac{\Gamma(s+\tau)}{(4\pi)^{s+\tau}}\, \pi^{-s}\, \Gamma(s)\, \zeta(2s)\, L(\bar{\mathfrak{f}} \otimes \mathfrak{f},\, s) \tag{5.23}$$

extends as a meromorphic function in the complex plane, with simple poles only at $s=0$ and $s=1$, invariant under the change $s \mapsto 1-s$. Denote as $\operatorname{Res}_{s=1}(L(\bar{\mathfrak{f}} \otimes \mathfrak{f},\, s))$ the residue of the L-function into consideration at $s=1$. Incidentally, this number can be connected to the coefficient on the right-hand side of (5.7) by the equation

$$\operatorname{Res}_{s=1}(L(\bar{\mathfrak{f}} \otimes \mathfrak{f},\, s)) = \frac{3}{\pi}\, \frac{(4\pi)^{\tau+1}}{\Gamma(\tau+1)}\, \| \, y^{\frac{\tau+1}{2}}\, \mathfrak{f} \, \|^2_{L^2(\Gamma\backslash\Pi)}\,. \tag{5.24}$$

The result of [8, Corollary 11.6] was the following spectral decomposition of the symbol $W(\mathfrak{s}_\tau,\, \mathfrak{s}_\tau)$:

$$W(\mathfrak{s}_\tau,\, \mathfrak{s}_\tau) = \frac{1}{2\pi} \int_{-\infty}^{\infty} 2^{\frac{i\lambda-3}{2}}\, \frac{L(\bar{\mathfrak{f}} \otimes \mathfrak{f},\, \frac{1-i\lambda}{2})}{\zeta(-i\lambda)}\, \mathfrak{E}_{i\lambda}\, d\lambda + \frac{1}{2} \operatorname{Res}_{s=1} (L(\bar{\mathfrak{f}} \otimes \mathfrak{f},\, s))\, (1 - \tau\, \delta)\,. \tag{5.25}$$

To obtain the present statement, we just have, starting from the decomposition (2.8), to consider the way the operator

$$\mathcal{B} = 2^{-i\pi\mathcal{E}-\frac{3}{2}}\, \frac{L(\bar{\mathfrak{f}} \otimes \mathfrak{f},\, \frac{1}{2}+i\pi\mathcal{E})}{\zeta(2i\pi\mathcal{E})} \tag{5.26}$$

acts on any of the distributions (all eigendistributions of the Euler operator) 1, δ and $\mathfrak{E}_{i\lambda}$. The eigenvalues corresponding to these three operators, in this order, are 1, -1 and $-i\lambda$ in view of their degrees of homogeneity. Hence, setting

$$B(s) = 2^{\frac{-s-3}{2}}\, \frac{L(\bar{\mathfrak{f}} \otimes \mathfrak{f},\, \frac{1+s}{2})}{\zeta(s)} \tag{5.27}$$

and observing that $B(-i\lambda)$ coincides with the coefficient of $\mathfrak{E}_{i\lambda}$ in the integrand on the right-hand side of (5.25), all we have to do is to check the pair of equations

$$B(1) = \frac{1}{2} \operatorname{Res}_{s=1}(L(\bar{\mathfrak{f}} \otimes \mathfrak{f},\, s))\,, \qquad B(-1) = -\frac{\tau}{2} \operatorname{Res}_{s=1}(L(\bar{\mathfrak{f}} \otimes \mathfrak{f},\, s))\,. \tag{5.28}$$

The first equation is immediate since the residue of the zeta function at $s=1$ is 1. On the other hand, the functional equation of the L-function under consideration shows that $L(\bar{\mathfrak{f}} \otimes \mathfrak{f},\, 0))$ is the value at $s = 1$ of the function

$$\frac{\Gamma(s+\tau)}{\Gamma(1-s+\tau)}\,\frac{(4\pi)^{1-s+\tau}}{(4\pi)^{s+\tau}}\,\pi^{1-2s}\,\Gamma(s)\,\frac{\zeta(2s)}{\zeta(2-2s)}\,\frac{L(\bar{\mathfrak{f}} \otimes \mathfrak{f},\, s)}{\Gamma(1-s)}\,, \tag{5.29}$$

hence (using $\zeta(0) = -\frac{1}{2}$)

$$L(\bar{\mathfrak{f}} \otimes \mathfrak{f},\, 0)) = \frac{\tau}{12}\,\mathrm{Res}_{s=1}(L(\bar{\mathfrak{f}} \otimes \mathfrak{f},\, s))\,. \tag{5.30}$$

Now, one has (as shown by the functional equation of the zeta function) $\zeta(-1) = -\frac{1}{12}$, so that the second equation (5.28) is proved as well. □

Remark 5.1. When $\tau = -\frac{1}{2}$ and $\mathfrak{s}$ coincides with the image, under $\mathrm{Sq}_{\mathrm{even}}^{-1}$, of the distribution $\mathfrak{d}$ on the line associated, in Section 3, to the non-trivial even Dirichlet character mod 12 (then, $\kappa = \frac{1}{24}$), one obtains

$$\mathcal{B} = 12^{i\pi\mathcal{E}}\,[1 - 2^{-2i\pi\mathcal{E}}\,]\,[1 - 3^{-2i\pi\mathcal{E}}\,]\,, \tag{5.31}$$

which fits with the result quoted right after (3.2). That, taking the first factor aside, the right-hand side appears as a Dirichlet "series" in the variable $2i\pi\mathcal{E}$, rather than $i\pi\mathcal{E}$, is due to the very special fact that, in this case, the only non-zero coefficients a_m (*cf.* (5.33) *infra*) are obtained in the case when $24m+1 = (6\ell-1)^2$ for some integer ℓ, as seen from the expansion of the Dedekind eta function.

Remark 5.2. For $r = 1, 2, \ldots$, set $\mu(r) = (-1)^{\nu(r)}$, where $\nu(r)$ is the number of distinct prime divisors of r, in the case when r is squarefree, while setting $\mu(r) = 0$ if r is divisible by the square of an integer ≥ 2: this is the Möbius function, which occurs in the formula

$$\frac{1}{\zeta(s)} = \sum_{r\geq 1} \frac{\mu(r)}{r^s}\,, \qquad \mathrm{Re}\ s > 1\,. \tag{5.32}$$

Hence, when $\mathrm{Re}\ s > 2$,

$$B(s) = 2^{-\frac{3}{2}-\frac{s}{2}} \sum_{r\geq 1} \mu(r)\, r^{-s} \sum_{m\geq 0} |a_m|^2\, (m+\kappa)^{-\tau-\frac{s+1}{2}}\,, \tag{5.33}$$

generally not a Dirichlet series in terms of the variable $\frac{s}{2}$, becomes one in the case when $\kappa = 1$. Indeed, one has in this case, when $\mathrm{Re}\ s > 2$,

$$B(s) = \sum_{\ell\geq 1} B_\ell\, (2\ell)^{-\frac{s}{2}} \tag{5.34}$$

with

$$B_\ell = 2^{-\frac{3}{2}} \sum_{\substack{m\geq 0,\, r\geq 1 \\ r^2(m+1)=\ell}} \mu(r)\ |a_m|^2\, (m+1)^{-\tau-\frac{1}{2}}\,. \tag{5.35}$$

It then follows from (3.2) that one has

$$W(\mathfrak{s}_\tau,\, \mathfrak{s}_\tau)(x,\, \xi) = \sum_{\ell\geq 1} (2\ell)^{\frac{1}{2}}\, B_\ell \sum_{(j,k)\in\mathbb{Z}^2} \delta(x - j\sqrt{2\ell})\, \delta(\xi - k\sqrt{2\ell})\,. \tag{5.36}$$

The expression of $W(\mathfrak{s}_\tau, \mathfrak{s}_\tau)$ is more complicated when $\kappa < 1$. It is to be noted that when $\kappa = 1$, $\tau + 1$ is an even integer and $\mathfrak{f}$ is a modular form for the full modular group in the usual sense, also an eigenfunction of every Hecke operator (from the *holomorphic* theory), the zeros of the function $L(\bar{\mathfrak{f}} \otimes \mathfrak{f}, s)$ on the spectral line Re $s = \frac{1}{2}$ include those on the same line of the zeta function (the so-called critical zeros): this, as explained in [2, p. 250], is as a consequence of results of Shimura [5].

As a conclusion, we wish to stress the fact that, though it would be possible to rephrase Theorem 5.1 (at the price of inserting some Gamma factors and substituting non-holomorphic Eisenstein series for Eisenstein distributions) in terms of any covariant symbolic calculus using the hyperbolic half-plane as a phase space, arithmetic results have both a neater formulation and an easier proof when expressed in terms of the horocyclic calculus, with the plane as a phase space.

References

[1] D. Bump, *Automorphic Forms and Representations*, Cambridge Series in Adv.Math. **55**, Cambridge, 1996.

[2] H. Iwaniec, *Topics in Classical Automorphic Forms*, Graduate Studies in Math. **17**, A.M.S., Providence, 1997.

[3] P.D. Lax, R.S. Phillips, *Scattering Theory for Automorphic Functions*, Ann. Math. Studies **87**, Princeton Univ. Press, 1976.

[4] W. Magnus, F.Oberhettinger, R.P. Soni, *Formulas and theorems for the special functions of mathematical physics*, 3rd edition, Springer-Verlag, Berlin, 1966.

[5] G. Shimura, *Modular Forms of half-integral weight*, Lecture Notes in Math. **320**, Springer-Verlag, Berlin–Heidelberg–New York, 1973.

[6] A. Unterberger, *Quantization and non-holomorphic modular forms*, Lecture Notes in Math. **1742**, Springer-Verlag, Berlin–Heidelberg, 2000.

[7] A. Unterberger, *Automorphic pseudodifferential analysis and higher-level Weyl calculi*, Progress in Math., Birkhäuser, Basel–Boston–Berlin, 2002.

[8] A. Unterberger, *Quantization and Arithmetic*, Progress in Math., Birkhäuser, Basel–Boston–Berlin, 2008.

[9] N. Vasilevski, *Bergman space structure, commutative algebras of Toeplitz operators, and hyperbolic geometry*, Integral Equations and Operator Theory **46**,2 (2003), 235–251.

André Unterberger
Mathématiques (FRE 3111)
Université de Reims
Moulin de la Housse, B.P.1039
F-51687 REIMS Cedex 2, France
e-mail: andre.unterberger@univ-reims.fr

Operator Theory:
Advances and Applications, Vol. 210, 249–272

Hilbert Bundles and Flat Connexions over Hermitian Symmetric Domains

Harald Upmeier

Dedicated to Nikolai Vasilevski on the occasion of his 60th birthday

Abstract. We study Hilbert spaces of holomorphic functions, generalizing the Fock spaces of entire functions, in the general setting of hermitian symmetric domains and Jordan algebras, using the concept of projectively flat connexion on a Hilbert subbundle.

Mathematics Subject Classification (2000). Primary 47B35; Secondary 32M15.

Keywords. Fock spaces, connexions, symmetric domains, Jordan algebras.

0. Introduction

In this paper we introduce the concept of Hilbert subbundle and associated *canonical connexion*, in the sense of differential geometry, and show that the unipotent "singular" holomorphic representations of semi-simple Lie groups, arising in the theory of hermitian symmetric domains and Jordan algebras, can be described by such a canonical Hilbert subbundle connexion which is *projectively flat*, i.e., its curvature is a scalar operator. This concept is well known in mathematical physics [2] and the theory of modular functions [11].

Of course, there is an extensive literature [3], [13], expressing these representations in geometric or Jordan algebraic terms. However, the approach presented here has some new features. Notably, the general construction (Section 4) is carried out in the bounded (Harish-Chandra) realization of a symmetric domain instead of the more familiar unbounded Siegel domains. In the bounded model, the representations are not given by measures on boundary orbits, but instead one needs distributions [1] or, as proposed here, projectively flat Hilbert bundles. Also, the bounded model treats all unbounded (Cayley transform) realizations on an equal

This work was completed with the support of the German-Israeli Foundation (GIF), I-415-023.06/95.

level. This is important for a study of geodesics and their behavior near the boundary.

In order to stay close to operator theory, we include a detailed discussion of the "classical" Fock space bundles over $\mathbb{K} = \mathbb{R}$, $\mathbb{C}$ and $\mathbb{H}$ (quaternions), the real case being the (even) metaplectic representation. These Hilbert spaces are given via representations of euclidean Jordan algebras of self-adjoint matrices over $\mathbb{K}$, as explained in Section 3. In Section 4 we outline the general construction of a Hilbert subbundle for any hermitian Jordan triple.

1. The canonical connexion on a Hilbert subbundle

For the basic setup, let D be a smooth manifold and H a complex Hilbert space. Let $\mathcal{L}(H)$ be the C^*-algebra of bounded operators on H, and $\mathcal{H}(H)$ be the subspace of all self-adjoint operators. Consider a smooth map

$$D \ni z \mapsto P_z \in \mathcal{H}(H) \tag{1.1}$$

such that

$$P_z = P_z\, P_z \tag{1.2}$$

is a (self-adjoint) projection for every $z \in D$. Since D is finite-dimensional, the notion of differentiability is the standard one, with values in a Banach space. For each $z \in D$, we consider the closed subspace

$$H_z := \text{Ran } P_z = P_z\, H \tag{1.3}$$

of H, and call the family of Hilbert spaces

$$(H_z)_{z \in D} \tag{1.4}$$

the *Hilbert subbundle* (of the trivial bundle $D \times H$) determined by the map (1.1). Of course, one has to impose additional conditions to ensure that (1.4) is locally trivial. For example, if the bundle is "covariant" under a transitive Lie group action, local triviality follows from Lie theoretic arguments, cf. the following Sections 2 and 3. We will only use the smooth field (1.1) of projections, not (local) trivializing maps.

By definition, a *section* of the Hilbert subbundle (1.4) is a smooth map

$$\Phi : D \to H, \quad z \mapsto \Phi_z \tag{1.5}$$

such that

$$\Phi_z = P_z\, \Phi_z \in H_z \tag{1.6}$$

for every $z \in D$. We write the z-variable as a subscript, since in applications H is itself a Hilbert function space and Φ_z depends on an extra variable. Note that, although Φ takes values in a Hilbert space, we do not endow D with a measure to obtain "square-integrable" sections. For every smooth function $f : D \to \mathbb{C}$, the pointwise product

$$z \mapsto (f \cdot \Phi)_z := f(z)\, \Phi_z \tag{1.7}$$

is again a (smooth) section of the Hilbert subbundle, since P_z is a linear operator on H. Let

$$T(D) = \{(z, \dot{z}) : z \in D,\ \ \dot{z} \in T_z D\} \tag{1.8}$$

be the tangent bundle of D. Consider the derivative

$$P'_z\, \dot{z} \in \mathcal{H}(H) \tag{1.9}$$

of the smooth map (1.1) at $(z, \dot{z}) \in T(D)$. Applying the product rule to (1.2), we obtain

$$P'_z\, \dot{z} = (P'_z\, \dot{z})\, P_z + P_z(P'_z\, \dot{z}) \tag{1.10}$$

as an operator relation in $\mathcal{L}(H)$. This implies

$$P_z(P'_z\, \dot{z})\, P_z = 0, \tag{1.11}$$

so that $P'_z\, \dot{z}$ maps H_z into $H_z^{\perp}$. Similarly, for a section Φ of the Hilbert subbundle, let

$$\Phi'_z\, \dot{z} \in H \tag{1.12}$$

denote the derivative at $(z, \dot{z}) \in T(D)$. Applying the product rule to (1.6), we obtain

$$\Phi'_z\, \dot{z} = P_z(\Phi'_z\, \dot{z}) + (P'_z\, \dot{z})\, \Phi_z. \tag{1.13}$$

This shows that

$$P_z(P'_z\, \dot{z})\, \Phi_z = 0 \tag{1.14}$$

or, equivalently,

$$(P'_z\, \dot{z})\, \Phi_z \in H_z^{\perp}. \tag{1.15}$$

Definition 1.1. The *canonical connexion* ∇ on the Hilbert subbundle (1.15) is defined by the *covariant derivative*

$$(\nabla_{\dot{z}}\Phi)_z := P_z(\Phi'_z\, \dot{z}) = \Phi'_z\, \dot{z} - (P'_z\, \dot{z})\, \Phi_z \tag{1.16}$$

at $(z, \dot{z}) \in T(D)$, applied to any (smooth) section Φ.

By construction, we have $(\nabla_{\dot{z}}\, \Phi)(z) \in H_z$ for all $(z, \dot{z}) \in T(D)$. In order to check the Leibniz rule, let $f : D \to \mathbb{C}$ be a smooth function and consider the product section $f \cdot \Phi$. Then

$$(f \cdot \Phi)'_z\, \dot{z} = (f'_z\, \dot{z})\, \Phi_z + f(z)(\Phi'_z\, \dot{z}) \tag{1.17}$$

as vectors in H, where

$$f'_z\, \dot{z} = (T_z\, f)\dot{z} \tag{1.18}$$

is the differential of f at z. Applying P_z to (1.17) we obtain

$$(\nabla_{\dot{z}}(f \cdot \Phi))_z := (f'_z\, \dot{z})\, \Phi_z + f(z)(\nabla_{\dot{z}}\, \Phi)_z \tag{1.19}$$

in view of (1.6). Finally, if X is a smooth tangent field on D, the associated covariant derivative is given by

$$(\nabla_X \Phi)_z := (\nabla_{X_z}\, \Phi)_z = P_z(\Phi'_z\, X_z) \tag{1.20}$$

for all $z \in D$, where $X_z \in T_z\, D$ is the value of X at z. Since (1.20) depends smoothly on z, it follows that $\nabla_X\, \Phi$ is a smooth H-valued mapping. In summary, we have shown that (1.16) (or (1.20)) define a (Hilbert) bundle connexion on D in the sense of differential geometry.

The *curvature* Ω of a connexion ∇ is a smooth operator-valued 2-form on D, denoted by

$$\Omega_z(\dot{z}, \ddot{z}) \in \mathcal{L}(H_z) \tag{1.21}$$

for all $z \in D$ and $\dot{z}, \ddot{z} \in T_z D$. It is defined via the formula

$$(\nabla_X \nabla_Y \Phi)_z - (\nabla_Y \nabla_X \Phi)_z - (\nabla_{[X,Y]}\Phi)_z = \Omega_z(X_z, Y_z)\, \Phi_z \tag{1.22}$$

for tangent fields X, Y on D, with commutator bracket

$$[X, Y]_z := Y'_z X_z - X'_z Y_z. \tag{1.23}$$

In our applications D is an open subset of a vector space Z, hence $T_z D \approx Z$ for all $z \in D$ and $X'_z \in \mathcal{L}(Z)$ is the usual derivative. Of course, one can give a more intrinsic definition which is, for example, important when studying quotients of D by discrete subgroups (moduli spaces [2]).

Our first result shows that the curvature of the canonical connexion of a Hilbert subbundle can be neatly expressed via the underlying field (1.1) of projections.

Proposition 1.2. *For any $z \in D$ and $\dot{z}, \ddot{z} \in T_z D$, the commutator $[P'_z \dot{z}, P'_z \ddot{z}] \in \mathcal{L}(H)$ commutes with P_z, and the canonical connexion* (1.16) *has the curvature 2-form*

$$\Omega_z(\dot{z}, \ddot{z}) = P_z[P'_z \dot{z}, P'_z \ddot{z}]\, P_z. \tag{1.24}$$

Proof. As a consequence of (1.10) it follows that the operator product $(P'_z \dot{z})(P'_z \ddot{z})$ (and a fortiori the commutator) commutes with P_z.

Now let X, Y be tangent fields on D, and let Φ be a section of the Hilbert subbundle. Applying ∇_X to the smooth section

$$(\nabla_Y \Phi)_z = \Phi'_z Y_z - (P'_z Y_z)\, \Phi_z \tag{1.25}$$

we obtain, denoting the second derivative by $''$,

$$\begin{aligned}(\nabla_X \nabla_Y \Phi)_z &= (\Phi' Y - (P'Y)\,\Phi)'_z X_z - (P'_z X_z)(\Phi'_z Y_z - (P'_z Y_z)\,\Phi_z) \\ &= \Phi''_z(Y_z, X_z) + \Phi'_z(Y'_z X_z) - P''_z(Y_z, X_z)\,\Phi_z - P'_z(Y'_z X_z)\,\Phi_z - (P'_z Y_z)(\Phi'_z X_z) \\ &\quad - (P'_z X_z)(\Phi'_z Y_z) + (P'_z X_z)(P'_z Y_z)\,\Phi_z.\end{aligned}$$

A similar formula holds for $(\nabla_Y \nabla_X \Phi)_z$. On the other hand, (1.23) implies that the tangent field $[X, Y]$ induces the covariant derivative

$$\begin{aligned}(\nabla_{[X,Y]}\Phi)_z &= \Phi'_z[X, Y]_z - (P'_z[X, Y]_z)\,\Phi_z \\ &= \Phi'_z(Y'_z X_z - X'_z Y_z) - P'_z(Y'_z X_z - X'_z Y_z)\,\Phi_z \\ &= \Phi'_z(Y'_z X_z) - \Phi'_z(X'_z Y_z) - P'_z(Y'_z X_z)\,\Phi_z + P'_z(X'_z Y_z)\,\Phi_z.\end{aligned}$$

The symmetry of the second derivatives

$$\begin{aligned}\Phi''_z(X_z, Y_z) &= \Phi''_z(Y_z, X_z) \in H, \\ P''_z(X_z, Y_z) &= P''_z(Y_z, X_z) \in \mathcal{H}(H)\end{aligned}$$

implies

$$\Omega_z(X_z, Y_z)\,\Phi_z = (P_z' X_z)(P_z' Y_z)\,\Phi_z - (P_z' Y_z)(P_z' X_z)\,\Phi_z = [P_z' X_z, P_z' Y_z]\,\Phi_z.$$

Putting $\dot{z} = X_z,\ \ddot{z} = Y_z$, the assertion follows. □

The above construction is easily adapted to include unitary group actions. More precisely, let T be a compact Lie group acting isometrically on H such that the projections (1.1) commute with the group action

$$t \ltimes (P_z\,\psi) = P_z(t \ltimes \psi) \tag{1.26}$$

for all $t \in T,\ \ z \in D$ and $\phi \in H$. Let

$$H^T := \{\psi \in H : t \ltimes \psi = \psi \quad \forall\, t \in T\}$$

denote the closed subspace of all T-invariant vectors in H. By (1.26), P_z commutes with the orthogonal projection

$$P^T\,\psi := \int_T dt\, t \ltimes \psi \tag{1.27}$$

from H onto H^T (here dt is the normalized Haar measure on T). Therefore the bundle

$$H_z^T := P^T\,P_z\,H = P^T\,H_z = P_z\,H^T \tag{1.28}$$

of T-invariant elements is a Hilbert subbundle of $D \times H^T$, with corresponding projection field

$$P_z^T := P_z\,P^T = P^T\,P_z = P^T\,P_z\,P^T \tag{1.29}$$

The canonical connexion ∇^T associated to (1.28), in the sense of Definition 1.1, is just the *restriction* of ∇ acting on T-invariant smooth sections

$$z \mapsto \Phi_z \in H_z^T. \tag{1.30}$$

This follows from the fact that $\Phi_z'\,\dot{z} \in H^T$ for such sections which implies

$$(\nabla_{\dot{z}}^T\,\Phi)_z := P_z^T\,(\Phi_z'\,\dot{z}) = P_z\,P^T\,(\Phi_z'\,\dot{z}) = P_z\,(\Phi_z'\,\dot{z}) = (\nabla_{\dot{z}}\,\Phi)_z.$$

Similarly, we have

$$[P^T,\ P_z'\,\dot{z}] = 0$$

for all $(z, \dot{z}) \in T(D)$ which, in view of (1.24), implies that the curvature 2-form (1.24) commutes with P^T. Thus the curvature 2-form Ω^T of ∇^T is just the restriction

$$\Omega_z^T\,(\dot{z}, \ddot{z}) = P^T\,\Omega_z\,(\dot{z}, \ddot{z})\,P^T \in \mathcal{H}(H_z^T)$$

of Ω. If $U : H \to \tilde{H}$ is an isometric isomorphism of Hilbert spaces, a Hilbert subbundle (H_z) of $D \times H$, with projection field (P_z), can of course be mapped to the Hilbert subbundle

$$\tilde{H}_z := U\,H_z \subset \tilde{H}$$

with corresponding projection field $\tilde{P}_z = U\, P_z\, U^*$. Since U is independent of $z \in D$, the canonical connexion and curvature of $(\tilde{H}_z)$ change only by a conjugation under U. As a special case, let

$$H = L^2(M,\lambda),\ \ \tilde{H} = L^2(M,\mu)$$

where λ and $\mu = \delta^{-1}\lambda$ are equivalent measures on M, with strictly positive Radon-Nikodym density function δ. Then $\psi \mapsto \tilde{\psi} = \sqrt{\delta}\,\psi$ defines an isometric isomorphism $U : H \to \tilde{H}$.

Lemma 1.3. *The integral kernels of* (P_z) *and* $(\tilde{P}_z)$ *are related by*

$$\tilde{P}_z(\xi,\eta) = \sqrt{\delta(\xi)}\ P_z(\xi,\eta)\,\sqrt{\delta(\eta)}$$

for all $\xi,\eta \in M$.

Proof. Since $\widetilde{P_z\,\psi} = \tilde{P}_z\,\tilde{\psi}$, we have

$$\int\limits_M d\mu(\eta)\,\tilde{P}_z(\xi,\eta)\,\tilde{\psi}(\eta) = (\tilde{P}_z\,\tilde{\psi})(\xi)$$

$$= \widetilde{P_z\psi}(\xi) = \sqrt{\delta(\xi)}\ (P_z\psi)(\xi) = \sqrt{\delta(\xi)}\int\limits_M d\lambda(\eta)\,P_z(\xi,\eta)\,\psi(\eta)$$

$$= \sqrt{\delta(\xi)}\int\limits_M d\mu(\eta)\,\delta(\eta)\,P_z(\xi,\eta)\,\psi(\eta) = \sqrt{\delta(\xi)}\int\limits_M d\mu(\eta)\,\sqrt{\delta(\eta)}\ P_z(\xi,\eta)\,\tilde{\psi}(\eta). \quad \square$$

2. The Fock bundle of entire functions

The classical example of a Hilbert subbundle endowed with a projectively flat connexion is the Fock bundle over a hermitian vector space E. Let $\zeta \mapsto \overline{\zeta}$ be a (conjugate-linear) conjugation in E, with real form

$$E_{\mathbb{R}} = \{\zeta \in E \,|\, \overline{\zeta} = \zeta\}.$$

Let $(\xi|\eta)$ denote the scalar product, anti-linear in η. Consider the convex unbounded domain

$$D := \{w = u + iv \in \mathcal{L}(E)\big|w^t = w,\ u \gg 0\} \tag{2.1}$$

in the space Z of complex symmetric endomorphisms w of E, where we put

$$u := \frac{w + w^*}{2}, \quad v := \frac{w - w^*}{2i} \tag{2.2}$$

and define

$$w^t\,\zeta := \overline{w^*\overline{\zeta}}$$

for all $\zeta \in E$. Consider the Hilbert space

$$H := L^2(E) \tag{2.3}$$

for the Lebesgue measure, with inner product conjugate-linear in the first variable. The *Fock subbundle* of $D \times L^2(E)$ is defined as follows: Let

$$H^2_F(E) = \left\{ \phi : E \to \mathbb{C} \text{ holomorphic} : \int_E d\xi\, e^{-(\xi|\xi)}\, |\phi(\xi)|^2 < +\infty \right\} \tag{2.4}$$

denote the standard multi-variable Fock space in the "complex wave" realization. For the unit matrix $e \in D$, the fibre $H_e \subset L^2(E)$ consists of all functions

$$\xi \mapsto e^{-(\xi|\xi)/2}\, \phi(\xi), \tag{2.5}$$

where $\phi \in H^2_F(E)$. The reproducing property

$$\phi(\xi) = \int_E d\sigma\, e^{(\xi|\sigma)}\, e^{-(\sigma|\sigma)}\, \phi(\sigma) \tag{2.6}$$

shows that the orthogonal projection $P_e : L^2(E) \to H_e$ has the integral kernel

$$P_e(\xi, \sigma) = \exp\left[(\xi|\sigma) - \frac{(\xi|\xi)}{2} - \frac{(\sigma|\sigma)}{2} \right] \tag{2.7}$$

for all $\xi, \sigma \in E$. In general, we have

Proposition 2.1. *For any $w \in D$, $P_w : L^2(E) \to L^2(E)$ is the operator with integral kernel*

$$P_w(\xi, \sigma) = \exp\left((\xi|\sigma)_w - \frac{(\xi|\xi)_w}{2} - \frac{(\sigma|\sigma)_w}{2} \right) \tag{2.8}$$

for $\xi, \sigma \in E$, where we put

$$4(\xi|\sigma)_w = (\xi + \overline{\xi} + \overline{w}(\xi - \overline{\xi}) \big| u^{-1}(\sigma + \overline{\sigma} + \overline{w}(\sigma - \overline{\sigma}))). \tag{2.9}$$

Proof. Consider the complex symplectic group

$$G^{\mathbb{C}} := Sp_{2r}(\mathbb{C}) = \left\{ \begin{pmatrix} a & b \\ c & d \end{pmatrix} \in GL_{2r}(\mathbb{C}) : \begin{pmatrix} a & b \\ c & d \end{pmatrix}^t \begin{pmatrix} 0 & 1 \\ -1 & 0 \end{pmatrix} \begin{pmatrix} a & b \\ c & d \end{pmatrix} = \begin{pmatrix} 0 & 1 \\ -1 & 0 \end{pmatrix} \right\}$$

written as 2×2 block matrices, and the real subgroup

$$G := \left\{ \begin{pmatrix} a & b \\ c & d \end{pmatrix} \in Sp_{2r}(\mathbb{C}) \,\middle|\, a, d \text{ real},\ b, c \text{ imaginary} \right\} \approx Sp_{2r}(\mathbb{R}).$$

This group acts on D by holomorphic (Möbius) transformations

$$\begin{pmatrix} a & b \\ c & d \end{pmatrix}(z) := (az + b)(cz + d)^{-1}. \tag{2.10}$$

(Note that D is a generalized "right" half-space instead of the usual "upper" half-space; therefore b and c are imaginary $r \times r$-matrices.) For any $w = u + iv \in D$, with $u > 0$, the matrix

$$\gamma_w := \begin{pmatrix} u^{1/2} & iv\, u^{-1/2} \\ 0 & u^{-1/2} \end{pmatrix}$$

belongs to G since

$$\gamma_w^t \begin{pmatrix} 0 & 1 \\ -1 & 0 \end{pmatrix} \gamma_w = \begin{pmatrix} u^{1/2} & 0 \\ iu^{-1/2}\,v & u^{-1/2} \end{pmatrix} \begin{pmatrix} 0 & 1 \\ -1 & 0 \end{pmatrix} \begin{pmatrix} u^{1/2} & iv\,u^{-1/2} \\ 0 & u^{-1/2} \end{pmatrix}$$
$$= \begin{pmatrix} 0 & u^{1/2} \\ -u^{-1/2} & iu^{-1/2}\,v \end{pmatrix} \begin{pmatrix} u^{1/2} & iv\,u^{-1/2} \\ 0 & u^{-1/2} \end{pmatrix} = \begin{pmatrix} 0 & 1 \\ -1 & 0 \end{pmatrix}.$$

Moreover, $\gamma_w(z) = (u^{1/2}\,z + iv\,u^{-1/2})\,u^{1/2} = u^{1/2}\,z\,u^{1/2} + iv$. Putting $z = e$, we obtain $\gamma_w(e) = u + iv = w$. Via the identification

$$\mathbb{C}^r \ni \zeta \equiv \begin{pmatrix} \frac{\zeta+\overline{\zeta}}{2} \\ \frac{\zeta-\overline{\zeta}}{2} \end{pmatrix} \in \begin{pmatrix} \mathbb{R}^r \\ i\,\mathbb{R}^r \end{pmatrix}$$

the $\mathbb{R}$-linear action

$$\begin{pmatrix} a & b \\ c & d \end{pmatrix} \begin{pmatrix} \xi \\ \eta \end{pmatrix} = \begin{pmatrix} a\xi + b\eta \\ c\xi + d\eta \end{pmatrix}$$

of $\left(\begin{smallmatrix} a & b \\ c & d \end{smallmatrix}\right) \in G$ can be written as

$$\begin{pmatrix} a & b \\ c & d \end{pmatrix} \cdot \zeta \equiv \begin{pmatrix} a & b \\ c & d \end{pmatrix} \begin{pmatrix} \frac{\zeta+\overline{\zeta}}{2} \\ \frac{\zeta-\overline{\zeta}}{2} \end{pmatrix} = \begin{pmatrix} a\,\frac{\zeta+\overline{\zeta}}{2} + b\,\frac{\zeta-\overline{\zeta}}{2} \\ c\,\frac{\zeta+\overline{\zeta}}{2} + d\,\frac{\zeta-\overline{\zeta}}{2} \end{pmatrix} \equiv (a+c)\,\frac{\zeta+\overline{\zeta}}{2} + (b+d)\,\frac{\zeta-\overline{\zeta}}{2}$$

Note that b and c have to be imaginary for the correct identification. Specializing to $g = \gamma_w^{-1} \in G$, for $w \in D$ fixed, we obtain $\gamma_w^{-1}(w) = e$ and

$$\gamma_w^{-1} \cdot \zeta = \begin{pmatrix} u^{-1/2} & -\,i\,u^{-1/2}\,v \\ 0 & u^{1/2} \end{pmatrix} \cdot \zeta$$
$$= u^{-1/2}\,\frac{\zeta+\overline{\zeta}}{2} + (u^{1/2} - i\,u^{-1/2}\,v)\,\frac{\zeta-\overline{\zeta}}{2} = u^{-1/2}\left(\frac{\zeta+\overline{\zeta}}{2} + \overline{w}\,\frac{\zeta-\overline{\zeta}}{2}\right),$$

since $\overline{w} = u - iv$. Thus we define for $\xi, \sigma \in E$

$$\begin{aligned}
(\xi|\sigma)_w &:= (\gamma_w^{-1} \cdot \xi \,|\, \gamma_w^{-1} \cdot \sigma)_e \\
&= \left(u^{-1/2}\left(\frac{\xi+\overline{\xi}}{2} + \overline{w}\,\frac{\xi-\overline{\xi}}{2}\right) \,\Big|\, u^{-1/2}\left(\frac{\sigma+\overline{\sigma}}{2} + \overline{w}\,\frac{\sigma-\overline{\sigma}}{2}\right)\right) \qquad (2.11) \\
&= \left(\frac{\xi+\overline{\xi}}{2} + \overline{w}\,\frac{\xi-\overline{\xi}}{2} \,\Big|\, u^{-1}\left(\frac{\sigma+\overline{\sigma}}{2} + \overline{w}\,\frac{\sigma-\overline{\sigma}}{2}\right)\right)
\end{aligned}$$

to obtain the G-invariant kernel

$$P_w(\xi,\sigma) = \exp\left((\xi|\sigma)_w - \frac{(\xi|\xi)_w}{2} - \frac{(\sigma|\sigma)_w}{2}\right) = P_e\,(\gamma_w^{-1} \cdot \xi,\ \gamma_w^{-1} \cdot \sigma). \qquad \square$$

Lemma 2.2. *For $g \in G$ we have the invariance property*

$$(g \cdot \xi | g \cdot \sigma)_{g(w)} = (\xi|\sigma)_w.$$

Proof. The stabilizer subgroup $K := \{k \in G | k(e) = e\}$ can be identified with the unitary group $U_r(\mathbb{C})$ under the map

$$U_r(\mathbb{C}) \ni a + ib \mapsto \begin{pmatrix} a & ib \\ -ib & a \end{pmatrix} \in G.$$

It follows that

$$(k \cdot \xi \,|\, k \cdot \sigma) = (\xi|\sigma) \tag{2.12}$$

for all $k \in K$ and $\xi, \sigma \in E$. Now let $g \in G$ and $w \in D$. Then $k := \gamma_{g(w)}^{-1}\, g\, \gamma_w \in G$ satisfies

$$k\,(e) = \gamma_{g(w)}^{-1}\, g\, \gamma_w(e) = \gamma_{g(w)}^{-1}\, g(w) = e.$$

Hence $k \in K$ and (2.12) yields

$$\begin{aligned}(g \cdot \xi \,|\, g \cdot \sigma)_{g(w)} &:= (\gamma_{g(w)}^{-1} \cdot g \cdot \xi \,|\, \gamma_{g(w)}^{-1} \cdot g \cdot \sigma) \\ &= (k \cdot \gamma_w^{-1} \cdot \xi \,|\, k \cdot \gamma_w^{-1} \cdot \sigma) = (\gamma_w^{-1} \cdot \xi \,|\, \gamma_w^{-1} \cdot \sigma) = (\xi|\sigma)_w. \qquad \square\end{aligned}$$

Applying Lemma 1.3 to the (G-invariant) Lebesgue measure λ on E, and the Gauss measure $d\mu(\eta) = d\lambda(\eta)\, e^{-(\eta|\eta)}$ with $\delta(\eta) = \exp(\eta|\eta)$, we obtain

$$\tilde{P}_e(\xi, \eta) = \sqrt{\delta(\xi)}\; P_e(\xi, \eta) \sqrt{\delta(\eta)} = \exp \frac{(\xi|\xi)}{2}\; P_e(\xi, \eta)\; \exp \frac{(\eta|\eta)}{2} = \exp(\xi|\eta)$$

and, more generally

$$\begin{aligned}\tilde{P}_w(\xi, \eta) &= \sqrt{\delta(\xi)}\; P_w(\xi, \eta) \sqrt{\delta(\eta)} = \sqrt{\delta(\xi)}\; P_e(\gamma_w^{-1} \cdot \xi,\; \gamma_w^{-1} \cdot \eta) \sqrt{\delta(\eta)} \\ &= \mu_w(\xi)\; \exp(\gamma_w^{-1} \cdot \xi \,|\, \gamma_w^{-1} \cdot \eta)\; \mu_w(\eta)\end{aligned}$$

where

$$\mu_w(\xi) = \frac{\delta(\xi)}{\delta(\gamma_w^{-1} \cdot \xi)}$$

is the Radon-Nikodym density of the image measure of μ under γ_w. This follows from the fact that $\lambda = \delta \cdot \mu$ is G-invariant and therefore

$$\begin{aligned}&\int_E d\mu(\xi)\, \frac{\delta(\xi)}{\delta(\gamma_w^{-1} \cdot \xi)}\, f(\xi) = \int_E d\lambda(\xi)\, \frac{f(\xi)}{\delta(\gamma_w^{-1} \cdot \xi)} = \int_E d\lambda(\eta)\, \frac{f(\gamma_w \cdot \eta)}{\delta(\gamma_w^{-1} \cdot \gamma_w \cdot \eta)} \\ &= \int_E d\lambda(\eta)\, \frac{f(\gamma_w \cdot \eta)}{\delta(\eta)} = \int_E d\mu(\eta)\, f(\gamma_w \cdot \eta) = \int_E d(\gamma_w \cdot \mu)(\xi)\, f(\xi).\end{aligned}$$

For any fixed $\xi, \sigma \in E$

$$\begin{aligned}(\xi|\sigma)_w &= \left(\frac{\xi + \overline{\xi}}{2} + \overline{w}\, \frac{\xi - \overline{\xi}}{2} \,\middle|\, u^{-1} \left(\frac{\sigma + \overline{\sigma}}{2} + \overline{w}\, \frac{\sigma - \overline{\sigma}}{2} \right) \right) \\ &= \left(\frac{\sigma + \overline{\sigma}}{2} + w\, \frac{\overline{\sigma} - \sigma}{2} \,\middle|\, u^{-1} \left(\frac{\xi + \overline{\xi}}{2} + w\, \frac{\overline{\xi} - \xi}{2} \right) \right)\end{aligned}$$

defines a "sesqui-polynomial" function on D, with derivative given as follows:

$$
\begin{aligned}
(\xi|\sigma)'_w\,\dot w &= \left(\dot w\frac{\overline{\sigma}-\sigma}{2}\,\Big|\,u^{-1}\left(\frac{\xi+\overline{\xi}}{2}+w\,\frac{\overline{\xi}-\xi}{2}\right)\right)\\
&-\left(\frac{\sigma+\overline{\sigma}}{2}+w\,\frac{\overline{\sigma}-\sigma}{2}\,\Big|\,u^{-1}\,\frac{\dot w+\dot w^*}{2}\,u^{-1}\left(\frac{\xi+\overline{\xi}}{2}+w\,\frac{\overline{\xi}-\xi}{2}\right)\right)\\
&+\left(\frac{\sigma+\overline{\sigma}}{2}+w\,\frac{\overline{\sigma}-\sigma}{2}\,\Big|\,u^{-1}\,\dot w\,\frac{\overline{\xi}-\xi}{2}\right).
\end{aligned}
\tag{2.13}
$$

Here we use the fact that the map $w \mapsto u^{-1} = (\frac{w+w^*}{2})^{-1}$ has the derivative

$$\dot w \mapsto -u^{-1}\left(\frac{\dot w+\dot w^*}{2}\right)u^{-1} \tag{2.14}$$

at $w \in D$. Specializing to $w = u = e$ we obtain

$$
\begin{aligned}
2(\xi|\sigma)'_e\,\dot w &= (\dot w(\overline{\sigma}-\sigma)|\overline{\xi}) - (\overline{\sigma}|(\dot w+\dot w^*)\,\overline{\xi}) + (\overline{\sigma}|\dot w(\overline{\xi}-\xi))\\
&= -(\dot w\sigma|\overline{\xi}) - (\overline{\sigma}|\dot w\xi).
\end{aligned}
\tag{2.15}
$$

In view of (2.8) it follows that

$$
\begin{aligned}
4\,\frac{(P'_e\dot w)(\xi,\sigma)}{P_e(\xi|\sigma)} &= 4\,(\xi|\sigma)'_e\,\dot w - 2\,(\xi|\xi)'_e\,\dot w - 2\,(\sigma|\sigma)'_e\,\dot w\\
&= -2(\dot w\sigma|\overline{\xi}) - 2(\overline{\sigma}|\dot w\xi) + (\dot w\xi|\overline{\xi}) + (\overline{\xi}|\dot w\xi) + (\dot w\sigma|\overline{\sigma}) + (\overline{\sigma}|\dot w\sigma)\\
&= (\overline{\xi}-\overline{\sigma}|\dot w(\xi-\sigma)) + (\dot w(\xi-\sigma)|\overline{\xi}-\overline{\sigma})
\end{aligned}
\tag{2.16}
$$

since $(\dot w\sigma|\overline{\xi}) = (\dot w\xi|\overline{\sigma})$ and $(\overline{\xi}|\dot w\sigma) = (\overline{\sigma}|\dot w\xi)$ because $\dot w$ is symmetric.

For any fixed $a \in E$, let

$$(\partial_a f)(\zeta) := \frac{1}{2}(f'(\zeta)\,a - i\,f'(\zeta)\,ia) \tag{2.17}$$

denote the holomorphic Wirtinger derivative of a smooth function f on E, and let

$$(\iota_a f)(\zeta) = (a|\zeta)\,f(\zeta) \tag{2.18}$$

be the multiplication operator with (anti-linear) symbol $\zeta \mapsto (a|\zeta)$. Note that both operators ∂_a and ι_a depend $\mathbb{C}$-linearly on $a \in E$. It follows that there exists a $\mathbb{C}$-linear map

$$E \otimes E \to \mathcal{D}_{\mathbb{C}}\,(E \otimes \mathbb{R}),\ \ A \mapsto D_A \tag{2.19}$$

into the complex Weyl algebra $\mathcal{D}_{\mathbb{C}}(E \otimes \mathbb{R})$ of all polynomial differential operators on the real vector space $E \otimes \mathbb{R}$, which is uniquely determined by

$$D_{a\otimes b} = (\iota_a - \partial_a)(\iota_b - \partial_b)$$

for all $a, b \in E$. Since $\iota_a(\zeta)$ is anti-linear in ζ, the operators $\iota_a - \partial_a$ and $\iota_b - \partial_b$ commute and we may restrict to symmetric tensors $E \vee E \subset E \otimes E$. Put

$$(a\,a^t)(\zeta) := a(\zeta|\overline{a}).$$

Then $E \vee E$ can be identified with the space

$$\mathcal{L}_s(E) = \{w \in \mathcal{L}(E)\,|\,\overline{w\zeta} = w^*\overline{\zeta}\}$$

of all symmetric endomorphisms of E, via the map $a\otimes a \mapsto a\,a^t$. Thus (2.19) yields a map

$$\mathcal{L}_s(E) \to \mathcal{D}_{\mathbb{C}}\,(E\otimes\mathbb{R}),\quad w\mapsto D_w \tag{2.20}$$

which is uniquely determined by

$$D_{a\,a^t} = (\iota_a - \partial_a)^2$$

for all $a\in E$.

Lemma 2.3. *For any $\xi,\eta\in E$ we have*

$$(D_w\, e^{(\cdot|\eta)})(\xi) = (w(\overline{\xi}-\overline{\eta})|\xi-\eta)\, e^{(\xi|\eta)}. \tag{2.21}$$

Proof. We may assume that $w = a\,a^t$ for some $a\in E$. Then

$$((\iota_a-\partial_a)\, e^{(\cdot|\eta)})(\sigma) = (a|\sigma)\, e^{(\sigma|\eta)} - (\partial_a\, e^{(\cdot|\eta)})(\sigma) = (a|\sigma-\eta)\, e^{(\sigma|\eta)} \tag{2.22}$$

and hence, since $(a|\sigma-\eta)$ is anti-holomorphic in σ,

$$\begin{aligned}((\iota_a-\partial_a)^2\, e^{(\cdot|\eta)})(\xi) &= (a|\xi)(a|\xi-\eta)\, e^{(\xi|\eta)} - (a|\xi-\eta)(\partial_a\, e^{(\cdot|\eta)})(\xi) \quad (2.23)\\ &= (a|\xi-\eta)^2\, e^{(\xi|\eta)} = (aa^t(\overline{\xi}-\overline{\eta})\,|\,\xi-\eta)\, e^{(\xi|\eta)}.\end{aligned}$$

□

Proposition 2.4. *The operator $P_e'\dot{w}$, restricted to H_e, has the form*

$$((P_e'\dot{w})(e^{-(\cdot|\cdot)/2}\phi))(\xi) = e^{-(\xi|\xi)/2}(D_{\overline{\dot{w}}}\,\phi)(\xi) \tag{2.24}$$

for all $\phi\in H^2_F(E)$ and $\dot{w}\in\mathcal{L}_s(E)$.

Proof. In view of (2.16) we have for any fixed $\xi\in E$

$$\begin{aligned}&e^{(\xi|\xi)/2}((P_e'\,\dot{w})(e^{-(\cdot|\cdot)/2}\phi))(\xi) \quad (2.25)\\ &= e^{(\xi|\xi)/2}\int_E d\sigma\, P_e(\xi,\sigma)\, e^{-(\sigma|\sigma)/2}\phi(\sigma)[(\dot{w}(\xi-\sigma)|\overline{\xi}-\overline{\sigma}) + (\overline{\xi}-\overline{\sigma}|\dot{w}(\xi-\sigma))]\\ &= \int_E d\sigma\, e^{-(\sigma|\sigma)}\, e^{(\xi|\sigma)}\phi(\sigma)[(\dot{w}(\xi-\sigma)|\overline{\xi}-\overline{\sigma}) + (\overline{\xi}-\overline{\sigma}|\dot{w}(\xi-\sigma))] = P((f+\overline{f})\,\phi)(\xi).\end{aligned}$$

Here P is the orthogonal projection onto $H^2_F(E)$, with respect to the measure $e^{-(\sigma|\sigma)}d\sigma$ on E, and $f(\sigma) = (\dot{w}(\xi-\sigma)|\overline{\xi}-\overline{\sigma})$ is a holomorphic polynomial vanishing at ξ. Since $\phi\in H^2_F(E)$, it follows that

$$P(f\phi)(\xi) = (f\phi)(\xi) = f(\xi)\,\phi(\xi) = 0. \tag{2.26}$$

The adjoint $T_f^*\,\phi = T_{\overline{f}}\,\phi = P(\overline{f}\,\phi)$ of the Toeplitz operator $T_f\,\phi = P(f\phi) = f\phi$ is the holomorphic differential operator induced by f via the scalar product [14]. Therefore, by Lemma 2.3

$$P(\overline{f}\, e^{(\cdot|\eta)})(\xi) = (T_f^*\, e^{(\cdot|\eta)})(\xi) = \overline{f(\eta)}\, e^{(\xi|\eta)} = (\overline{\xi}-\overline{\eta}|\dot{w}(\xi-\eta))\, e^{(\xi|\eta)} = (D_{\overline{\dot{w}}}\, e^{(\cdot|\eta)})(\xi)$$

whenever $\eta\in E$ is fixed. Using $e^{(\cdot|\eta)}$ as the reproducing kernel, we obtain

$$P(\overline{f}\phi)(\xi) = (D_{\overline{\dot{w}}}\,\phi)(\xi). \tag{2.27}$$

In view of (2.25) and (2.26), the assertion follows. □

In order to compute the curvature 2-form Ω of the canonical Fock subbundle connexion, consider first the base point $e \in D$.

Proposition 2.5. *Let $\dot{w}, \ddot{w} \in Z$. Then we have*

$$(P'_e\,\dot{w})(P'_e\,\ddot{w})\big|_{H_e} = \mathrm{tr}_E\,(\dot{w}\,\ddot{w}^*)\ \mathrm{id}\,. \tag{2.28}$$

Proof. Let $\eta \in E$ be fixed. Applying (2.24), we obtain for the $L^2(E)$-inner product

$$\begin{aligned}
&(e^{-(\cdot|\cdot)/2}\,e^{(\cdot|\eta)}|(P'_e\,\dot{w})(P'_e\,\ddot{w})(e^{-(\cdot|\cdot)/2}\,e^{(\cdot|\eta)}))\\
&= ((P'_e\,\dot{w})(e^{-(\cdot|\cdot)/2}\,e^{(\cdot|\eta)})|(P'_e\,\ddot{w})\,(e^{-(\cdot|\cdot)/2}\,e^{(\cdot|\eta)}))\\
&= (e^{-(\cdot|\cdot)/2}\,(D_{\overline{\dot{w}}}\,e^{(\cdot|\eta)})|e^{-(\cdot|\cdot)/2}\,(D_{\overline{\ddot{w}}}\,e^{(\cdot|\eta)}))\\
&= \int_E d\sigma\,e^{-(\sigma|\sigma)}\ \overline{(D_{\overline{\dot{w}}}\,e^{(\cdot|\eta)})(\sigma)}\ (D_{\overline{\ddot{w}}}\,e^{(\cdot|\eta)})(\sigma)\\
&= \int_E d\sigma\,e^{-(\sigma|\sigma)}\ (\dot{w}(\sigma-\eta)|\overline{\sigma}-\overline{\eta})\ e^{(\eta|\sigma)}\ (\overline{\sigma}-\overline{\eta}|\ddot{w}(\sigma-\eta))\ e^{(\sigma|\eta)}\\
&= e^{(\eta|\eta)}\int_E d\sigma\,e^{-(\sigma-\eta|\sigma-\eta)}\ (\dot{w}(\sigma-\eta)|\overline{\sigma}-\overline{\eta})\ (\overline{\sigma}-\overline{\eta}|\ddot{w}(\sigma-\eta))\\
&= e^{(\eta|\eta)}\int_E d\tau\,e^{-(\tau|\tau)}\ (\dot{w}\tau|\overline{\tau})\ (\overline{\tau}|\ddot{w}\tau) = e^{(\eta|\eta)}\ \mathrm{tr}_E(\dot{w}\ddot{w}^*)
\end{aligned}$$

as the Fock inner product of the quadratic polynomials $(\dot{w}\tau|\overline{\tau})$ and $(\ddot{w}\tau|\overline{\tau})$ in $\tau \in E$. Here we put $\tau = \sigma - \eta$, where η is fixed. By polarization, we obtain

$$\begin{aligned}
&(e^{-(\cdot|\cdot)/2}\,e^{(\cdot|\eta_1)}|(P'_e\dot{w})(P'_e\ddot{w})(e^{-(\cdot|\cdot)/2}\,e^{(\cdot|\eta_2)})) = e^{(\eta_1|\eta_2)}\ \mathrm{tr}_E(\dot{w}\ddot{w}^*)\\
&= \mathrm{tr}_E(\dot{w}\ddot{w}^*)\int_E e^{-(\sigma|\sigma)}\ e^{(\eta_1|\sigma)}\ e^{(\sigma|\eta_2)} = \mathrm{tr}_E(\dot{w}\ddot{w}^*)(e^{-(\cdot|\cdot)/2}\,e^{(\cdot|\eta_1)}|e^{-(\cdot|\cdot)/2}\,e^{(\cdot|\eta_2)}).
\end{aligned}$$

The reproducing kernel property implies

$$(e^{-(\cdot|\cdot)/2}\,\phi|(P'_e\dot{w})(P'_e\ddot{w})(e^{-(\cdot|\cdot)/2}\,\psi)) = \mathrm{tr}_E(\dot{w}\ddot{w}^*)(e^{-(\cdot|\cdot)/2}\,\phi|e^{-(\cdot|\cdot)/2}\,\psi)$$

for all $\phi, \psi \in H^2_F(E)$. Since the operator $(P'_e\dot{w})(P'_e\ddot{w})$ leaves H_e invariant, as a consequence of Proposition 1.2, the assertion

$$(P'_e\dot{w})(P'_e\ddot{w})(e^{-(\cdot|\cdot)/2}\,\psi) = \mathrm{tr}_E(\dot{w}\ddot{w}^*)\,e^{-(\cdot|\cdot)/2}\,\psi \tag{2.29}$$

follows. □

Theorem 2.6. *The Fock subbundle over D, with projection field (2.8), has a canonical connexion which is projectively flat. More precisely, the curvature form Ω_w is a scalar operator which coincides (up to a constant multiple) with the $Sp_{2r}(\mathbb{R})$-invariant symplectic form ω on D.*

Proof. By Proposition 2.5 the curvature 2-form Ω at the base point $e \in D$ is the scalar operator 2-form

$$\Omega_e(\dot{w}, \ddot{w}) = \mathrm{tr}_E(\dot{w}\ddot{w}^* - \ddot{w}\dot{w}^*)\ \mathrm{id}_{H_e} \tag{2.30}$$

on H_e. Since $\mathrm{tr}_E(\dot{w}\ddot{w}^* - \ddot{w}\dot{w}^*)$ coincides (up to a constant factor) with the $Sp_{2r}(\mathbb{R})$-invariant symplectic form ω on D, evaluated at e, and the projection field (2.8) is G-invariant, the assertion follows. □

3. Fock bundles for Jordan algebra representations

In this section we generalize the above construction in a Jordan algebraic setting, starting with the three "classical" cases $\mathbb{K} = \mathbb{R},\ \mathbb{C}$ or $\mathbb{H}$ (quaternions). Let $\mathbb{K}^{r\times r}$ denote the $(r \times r)$-matrices over $\mathbb{K}$ and consider the euclidean Jordan algebra of all hermitian matrices

$$X := \mathcal{H}_r(\mathbb{K}) = \{x \in \mathbb{K}^{r\times r} : x^* = x\}. \tag{3.1}$$

Then

$$D_{\mathbb{K}} := \{w = u + iv : u, v \in \mathcal{H}_r(\mathbb{K}),\ \ u > 0 \ \text{ positive definite}\} = G/K$$

is a well-known symmetric tube domain generalizing the Siegel half-space ($\mathbb{K} = \mathbb{R}$). Let S_1 denote the compact manifold of all rank 1 tripotents (partial isometries) u in $X^{\mathbb{C}}$. It carries a transitive K-action, with invariant probability measure du.

Consider a representation

$$X \times E_{\mathbb{R}} \to E_{\mathbb{R}},\quad (x, \zeta) \mapsto x\zeta \tag{3.2}$$

of X (in the sense of [5]) such that

$$(x\,\zeta_1|\zeta_2) = (\zeta_1|x\,\zeta_2) \tag{3.3}$$

for all $\zeta_1, \zeta_2 \in E_{\mathbb{R}}$. By complexification, (3.2) extends to a representation

$$X^{\mathbb{C}} \times E \to E,\quad (w, \zeta) \mapsto w\zeta$$

on $E = E_{\mathbb{R}}^{\mathbb{C}}$ such that $(w\,\zeta_1|\zeta_2) = (\zeta_1|w^*\zeta_2)$.

Lemma 3.1. *For all* $\dot{w} \in X^{\mathbb{C}}$ *and* $\xi, \sigma \in E$

$$(\dot{w}\,\overline{\xi}|\sigma) = (\dot{w}\,\overline{\sigma}|\xi) \tag{3.4}$$

and, hence,

$$(\xi|\dot{w}\,\overline{\sigma}) = (\sigma|\dot{w}\,\overline{\xi}). \tag{3.5}$$

Proof. Since (3.4) is conjugate-linear in $\dot{w} \in X^{\mathbb{C}}$ we may assume $\dot{w} \in X$. In this case $\dot{w}$ is hermitian and commutes with the conjugation. It follows that

$$(\dot{w}\,\overline{\xi}|\sigma) = (\overline{\xi}|\dot{w}\sigma) = (\overline{\dot{w}\sigma}|\xi) = (\dot{w}\overline{\sigma}|\xi).$$

The relation (3.5) follows by complex conjugation. □

For the Fock spaces over $\mathbb{K} = \mathbb{R}, \mathbb{C}, \mathbb{H}$ studied in Section 2 we will now construct a quadratic map

$$\begin{array}{ccc} E & \xrightarrow{Q} & X^{\mathbb{C}} \\ \cup & & \cup \\ E_{\mathbb{R}} & \longrightarrow & X \end{array} \tag{3.6}$$

where E is a complex representation space of dimension ar, $a = \dim_{\mathbb{R}} \mathbb{K}$, having a real form

$$E_{\mathbb{R}} = \{\zeta \in E : \zeta^\sharp = \zeta\}$$

with respect to a suitable conjugation $\zeta \mapsto \zeta^\sharp$.

For $\mathbb{K} = \mathbb{R}$, the euclidean Jordan algebra

$$X = \mathcal{H}_r(\mathbb{R}) = \{x \in \mathbb{R}^{r\times r} \,|\, x^t = x\} \subset X^{\mathbb{C}} = \{z \in \mathbb{C}^{r\times r} \,|\, z^t = z\}$$

has the representation space $E = \mathbb{C}^r$ and

$$E_{\mathbb{R}} = \mathbb{R}^r = \{\zeta \in E \,|\, \overline{\zeta} = \zeta\},$$

where $\overline{\zeta}$ denotes the usual conjugation. The quadratic map $Q(\zeta) := \zeta\zeta^t = \zeta\overline{\zeta}^*$ satisfies (3.6).

Proposition 3.2. *Let $f : X^{\mathbb{C}} \to \mathbb{C}$ be continuous with at most polynomial growth. Then*

$$\int_{\mathbb{C}^r} d\zeta\; e^{-(\zeta|\zeta)} f(\zeta\zeta^t) = \frac{\pi^r}{\Gamma(r)} \int_0^\infty \frac{dt}{t}\, t^r\, e^{-t} \int_{S_1} du\, f(tu).$$

Proof. Using polar coordinates $\zeta = s\sigma$, with $s > 0$ and $\sigma \in \mathbb{S}^{2r-1}$, we have according to [12, 1.4.3 and Remark on p. 17]

$$\int_{\mathbb{C}^r} d\zeta\, F(\zeta) = \frac{2\pi^r}{\Gamma(r)} \int_0^\infty \frac{ds}{s}\, s^{2r} \int_{\mathbb{S}^{2r-1}} d\sigma\, F(s\sigma) = \frac{\pi^r}{\Gamma(r)} \int_0^\infty \frac{dt}{t}\, t^r \int_{\mathbb{S}^{2r-1}} d\sigma\, F(t^{1/2}\,\sigma) \tag{3.7}$$

by making the change of variables $t = s^2$, with $\frac{dt}{t} = 2\,\frac{ds}{s}$. Since the map $Q : \mathbb{S}^{2r-1} \to S_1$, $\sigma \mapsto \sigma\sigma^t$, is equivariant under $K = U_r(\mathbb{C})$ and hence preserves the normalized Haar measures, it follows that

$$\int_{\mathbb{C}^r} d\zeta\, e^{-(\zeta|\zeta)}\, f(\zeta\zeta^t)$$

$$= \frac{\pi^r}{\Gamma(r)} \int_0^\infty \frac{dt}{t}\, t^r\, e^{-t} \int_{\mathbb{S}^{2r-1}} d\sigma\, f(t\,\sigma\sigma^t) = \frac{\pi^r}{\Gamma(r)} \int_0^\infty \frac{dt}{t}\, t^r\, e^{-t} \int_{S_1} du\, f(tu).$$

□

For $\mathbb{K} = \mathbb{C}$, the euclidean Jordan algebra

$$X = \mathcal{H}_r(\mathbb{C}) = \{x \in \mathbb{C}^{r\times r} \,|\, x^* = x\} \subset X^{\mathbb{C}} = \mathbb{C}^{r\times r}$$

has the representation space $E = \mathbb{C}^r \times \overline{\mathbb{C}}^r = \{(\xi, \overline{\eta}) : \xi, \eta \in \mathbb{C}^r\}$ and

$$E_{\mathbb{R}} = \{(\xi, \overline{\xi}) : \xi \in \mathbb{C}^r\} = \{\zeta \in E | \zeta^\# = \zeta\},$$

where

$$(\xi, \overline{\eta})^{\#} = (\eta, \overline{\xi}), \quad \xi, \eta \in \mathbb{C}^r$$

denotes the flip involution. The quadratic map $Q(\xi, \overline{\eta}) = \xi\eta^*$ satisfies (3.6).

Proposition 3.3. *Let $f : X^{\mathbb{C}} \to \mathbb{C}$ be continuous with at most polynomial growth. Then we have for the Bessel function K_0 [4]*

$$\int_{\mathbb{C}^r} d\xi \int_{\mathbb{C}^r} d\eta \; e^{-(\xi|\xi)-(\eta|\eta)} f(\xi\eta^*) = \frac{4\pi^{2r}}{\Gamma(r)^2} \int_0^\infty dx\, x^{2r-1} \; K_0(2x) \int_{S_1} du\, f(xu).$$

Proof. Using polar coordinates $\xi = s\sigma, \;\; \eta = t\tau$ with $s, t > 0$ and $\sigma, \tau \in \mathbb{S}^{2r-1}$ we obtain in view of (3.7)

$$\begin{aligned}
&\int_{\mathbb{C}^r} d\xi \int_{\mathbb{C}^r} d\eta \; e^{-(\xi|\xi)-(\eta|\eta)} f(\xi\eta^*) \\
&= \frac{4\pi^{2r}}{\Gamma(r)^2} \int_0^\infty ds\, s^{2r-1}\, e^{-s^2} \int_0^\infty dt\; t^{2r-1}\, e^{-t^2} \int_{\mathbb{S}^{2r-1}} d\sigma \int_{\mathbb{S}^{2r-1}} d\tau \; f(st\sigma\tau^*) \\
&= \frac{4\pi^{2r}}{\Gamma(r)^2} \int_0^\infty ds \; s^{2r-1} \; e^{-s^2} \int_0^\infty dt\, t^{2r-1} \; e^{-t^2} \int_{S_1} du\, f(stu)
\end{aligned}$$

since du is the unique probability measure on S_1 which is invariant under $K = U_r(\mathbb{C}) \times U_r(\mathbb{C})$. The change of variables $x = st > 0, \;\; y = s/t > 0$ with $xy = s^2, \;\; x/y = t^2$ and

$$dx\, dy = \Big| \det \begin{pmatrix} x_s & y_s \\ x_t & y_t \end{pmatrix} \Big| \; ds\; dt = 2y\, ds\, dt$$

yields

$$\begin{aligned}
&\int_0^\infty ds \; s^{2r-1} \; e^{-s^2} \int_0^\infty dt \; t^{2r-1} \; e^{-t^2} \; g(st) \\
&= \frac{1}{2} \int_0^\infty dx \; x^{2r-1} \int_0^\infty \frac{dy}{y} \; e^{-xy-\frac{x}{y}} \; g(x) = \int_0^\infty dx \; x^{2r-1} \; K_0(2x) \; g(x)
\end{aligned}$$

using formula (23) of [4, p. 82]. Putting

$$g(x) := \int_{S_1} du \; f(xu),$$

the assertion follows. □

For $\mathbb{K} = \mathbb{H}$, the euclidean Jordan algebra

$$X = \mathcal{H}_r(\mathbb{H}) = \{u + vj : \; u, v \in \mathbb{C}^{r\times r}, \;\; u^* = u, \;\; v^t = -v\}$$

has the complexification $X^{\mathbb{C}} = \{z \in \mathbb{C}^{2r\times 2r} : \; Jz^t = zJ\}$ via the map

$$u + vj \mapsto \begin{pmatrix} u & v \\ -\overline{v} & \overline{u} \end{pmatrix}.$$

Here $J := \begin{pmatrix} 0 & 1 \\ -1 & 0 \end{pmatrix}$. Put

$$E = \mathbb{C}^{2r\times 2} = \left\{ \begin{pmatrix} \alpha & \beta \\ \gamma & \delta \end{pmatrix} : \alpha, \beta, \gamma, \delta \in \mathbb{C}^r \right\}$$

and

$$E_{\mathbb{R}} = \left\{ \begin{pmatrix} \alpha & \beta \\ -\overline{\beta} & \overline{\alpha} \end{pmatrix} : \alpha, \beta \in \mathbb{C}^r \right\} = \{\zeta \in E | \zeta^{\sim} = \zeta\}.$$

Here we define the conjugation $\zeta^{\sim} = J\overline{\zeta}J^{-1}$, with components

$$\begin{pmatrix} \alpha & \beta \\ \gamma & \delta \end{pmatrix}^{\sim} = J \begin{pmatrix} \overline{\alpha} & \overline{\beta} \\ \overline{\gamma} & \overline{\delta} \end{pmatrix} J^{-1} = \begin{pmatrix} \overline{\delta} & -\overline{\gamma} \\ -\overline{\beta} & \overline{\alpha} \end{pmatrix}$$

for all $\alpha, \beta, \gamma, \delta \in \mathbb{C}^r$. Then the quadratic map $Q(\zeta) = \zeta\widetilde{\zeta}^* = \zeta J \zeta^t J^{-1}$, with components

$$Q \begin{pmatrix} \alpha & \beta \\ \gamma & \delta \end{pmatrix} = \begin{pmatrix} \alpha\,\delta^t - \beta\,\gamma^t & \beta\alpha^t - \alpha\beta^t \\ \gamma\,\delta^t - \delta\,\gamma^t & \delta\alpha^t - \gamma\beta^t \end{pmatrix}$$

satisfies (3.6). Note the analogy with the real case.

Proposition 3.4. *Let $f : X^{\mathbb{C}} \to \mathbb{C}$ be continuous with at most polynomial growth. Then we have for the Bessel function K_1 [4]*

$$\int\limits_{\mathbb{C}^{2r\times 2}} d\zeta \; e^{-tr\,\zeta\,\zeta^*} f(\zeta\,\tilde{\zeta}^*) = \frac{4\pi^{4r}}{\Gamma(2r)\,\Gamma(2r-1)} \int\limits_0^\infty dt \; t^{4r-2}\; K_1(2t) \int\limits_{S_1} du \; f(tu).$$

Proof. The space $E = \mathbb{C}^{2r\times 2}$ carries the Jordan algebra representation

$$E \times \mathcal{H}_2(\mathbb{C}) \to E, \quad \zeta, x \mapsto \zeta x$$

in the sense of [5, Section IV.4], with associated quadratic map $q_\zeta := \zeta^*\zeta \in \mathcal{H}_2(\mathbb{C})$ in view of the identity

$$(\zeta x|\zeta)_E := tr(\zeta x \zeta^*) = tr(x\,\zeta^*\zeta) = (x\,q_\zeta) = (x|q_\zeta)_{\mathcal{H}_2(\mathbb{C})}.$$

The action of $K = U_{2r}(\mathbb{C})$ on $X^{\mathbb{C}}$ is given by $[k\cdot(w\,J^{-1})]\,J = k\,w\,k^t$ for all $k \in U_{2r}(\mathbb{C}),\ w \in \mathbb{C}^{2r\times 2r}_{\text{asym}}$. Equivalently, $k\cdot z = k\,z\,J\,k^t J^{-1}$ for $z \in X^{\mathbb{C}}$. Since

$$\begin{aligned} Q_{k\zeta} &= (k\zeta)(\widetilde{k\zeta})^* = (k\zeta)(J\,\overline{k\zeta}\,J^{-1})^* = k\,\zeta\,(J\,\overline{k}\,\overline{\zeta}\,J^{-1})^* = k\,\zeta\,J\,\zeta^t\,k^t\,J^{-1} \\ &= k\,\zeta\,J\,\zeta^t\,J^{-1}\,J\,k^t\,J^{-1} = k\,Q_\zeta\,J\,k^t\,J^{-1} = k\cdot Q_\zeta \end{aligned}$$

we see that the map $Q : E \to X^{\mathbb{C}}$ is K-equivariant. The compact submanifold

$$S(E) := \{\sigma \in E : \; q_\sigma = \overset{*}{\sigma}\,\sigma = 1_2\} \subset E$$

is called the *Stiefel manifold* [5, p. 351]. For $k \in U_{2r}(\mathbb{C})$ we have

$$q_{k\zeta} = (k\zeta)^*k\,\zeta = \zeta^*k^*k\,\zeta = \zeta^*\zeta = q_\zeta$$

which shows that $U_{2r}(\mathbb{C})$ acts (transitively) on $S(E)$. Let $d\sigma$ denote the normalized $U_{2r}(\mathbb{C})$-invariant measure on $S(E)$. By [5, Proposition XVI.2.1] we have for any integrable function F on E

$$\int_E d\zeta\, F(\zeta) = \frac{\pi^{4r}}{\Gamma_\Omega(2r)} \int_{\mathcal{H}_2^+(\mathbb{C})} dx\ \det(x)^{2r-2} \int_{S(E)} d\sigma\, F(\sigma\, x^{1/2})$$

since $\dim_{\mathbb{R}} E = 8r$ and $\mathcal{H}_2(\mathbb{C})$ has dimension 4 and rank 2. The normalizing Γ-factor is computed as $\Gamma_\Omega(2r) = (2\pi)^{\frac{4-2}{2}}\,\Gamma(2r)\,\Gamma(2r-1) = 2\pi\,\Gamma(2r)\,\Gamma(2r-1)$. We have

$$tr(\sigma\, x^{1/2})(\sigma\, x^{1/2})^* = tr\,(\sigma\, x\, \sigma^*) = tr(x\, \sigma^*\, \sigma) = tr(x\, 1_2) = tr\, x$$

and

$$(\sigma\, x^{1/2})(\widetilde{\sigma\, x^{1/2}})^* = (\sigma\, x^{1/2})(\tilde\sigma\, \tilde x^{1/2})^* = \sigma\, x^{1/2}\, \tilde x^{1/2}\, \tilde\sigma^* = (\det\, x)^{1/2}\, \sigma\, \tilde\sigma^*.$$

Therefore

$$\begin{aligned}
&\int_E d\zeta\; e^{-tr\,\zeta\zeta^*} f(\zeta\,\tilde\zeta^*) \\
&= \frac{\pi^{4r-1}}{2\Gamma(2r)\,\Gamma(2r-1)} \int_{\mathcal{H}_2^+(\mathbb{C})} dx\ \det(x)^{2r-2}\, e^{-tr\,x} \int_{S(E)} d\sigma\, f((\det\, x)^{1/2}\, \sigma\, \tilde\sigma^*) \qquad (3.8)\\
&= \frac{\pi^{4r-1}}{2\Gamma(2r)\,\Gamma(2r-1)} \int_{\mathcal{H}_2^+(\mathbb{C})} dx\, (\det\, x)^{2r-2}\; e^{-tr\,x} \int_{S_1} du\, f((\det\, x)^{1/2} u)
\end{aligned}$$

since, by K-invariance, du is the image measure of $d\sigma$ under Q. By [5, Corollary VII.1.3] we have for $s > 1$

$$\int_{\mathcal{H}_2^+(\mathbb{C})} dx\; e^{-tr\,x}\, (\det\, x)^{s-2} = \Gamma_\Omega(s) = 2\pi\,\Gamma(s)\,\Gamma(s-1) = 8\pi \int_0^\infty dt\, t^{2s-2}\, K_1(2t)$$

using formula (27), p. 51, of [11]. A Stone-Weierstrass argument implies

$$\int_{\mathcal{H}_2^+(\mathbb{C})} dx\; e^{-tr\,x}\, (\det\, x)^{-2}\, g(\det\, x) = 8\pi \int_0^\infty dt\ t^{-2}\, K_1(2t)\, g(t^2) \qquad (3.9)$$

for suitably bounded continuous functions g on $\mathbb{R}_+$. Applying (3.9) to the function

$$g(t) = \int_{S_1} du\, f(t^{1/2}\, u)$$

we obtain from (3.8)

$$\int_E d\zeta\, e^{-tr\,\zeta\,\zeta^*}\, f(\zeta\,\tilde{\zeta}^*) = \frac{\pi^{4r-1}}{2\Gamma(2r)\,\Gamma(2r-1)}\, 8\pi \int_0^\infty dt\, t^{-2}\, K_1(2t)\, t^{4r} \int_{S_1} du\, f(tu)$$

$$= \frac{4\pi^{4r}}{\Gamma(2r)\,\Gamma(2r-1)} \int_0^\infty dt\, t^{4r-2}\, K_1(2t) \int_{S_1} du\, f(tu). \qquad \square$$

For $\mathbb{K} = \mathbb{R}, \mathbb{C}, \mathbb{H}$ consider the (right) action

$$E_{\mathbb{R}} \times U(\mathbb{K}) \to E_{\mathbb{R}}, \quad (\zeta, \vartheta) \mapsto \zeta\vartheta$$

of the compact Lie group $U(\mathbb{K}) \approx \mathbb{S}^{a-1}$ on $E_{\mathbb{R}}$, extended to a $\mathbb{C}$-linear action on E. By Theorem 2.6, the "invariant" Fock bundle

$$H_z^{U(\mathbb{K})} \subset \{\psi \in L^2(E) \,|\, \psi(\zeta\vartheta) = \psi(\zeta) \ \ \forall\, \vartheta \in U(\mathbb{K})\} \tag{3.10}$$

carries a projectively flat canonical connexion. Since

$$Q_{\zeta\vartheta} = Q_\zeta \qquad \forall\, \vartheta \in U(\mathbb{K})$$

and a *holomorphic* $U(\mathbb{K})$-invariant function on E factorizes through Q, it follows that this projectively flat Hilbert subbundle can also be realized, via the quadratic map Q, on the Jordan algebra $X^{\mathbb{C}}$, more precisely on the subvariety of rank 1 elements. This motivates the general construction of the next Section 4.

For $\mathbb{K} = \mathbb{R}$, we have $U(\mathbb{R}) = \{\pm 1\}$ and (3.10) consists precisely of the *even* functions. In the other cases, we have invariance under $U(\mathbb{C}) \approx \mathbb{T}$ and $U(\mathbb{H}) \approx SU_2(\mathbb{C})$, respectively.

4. The Jordan theoretic framework

In this section we present a general construction of projectively flat Hilbert subbundles in the framework of Jordan algebras, which generalize the Fock type bundles associated with the Jordan matrix algebras (3.1). The main result (Theorem 4.2) determines the underlying measure μ on the Jordan algebra, more precisely its rank 1 elements, and the projection P_0 at the "origin" 0 of the unit ball B. At the other points $z \in B$, the projections P_z are constructed by invariance under the group G of biholomorphic automorphism of B, similar as in Proposition 2.1. In order to identify the fibre H_0, we need the following result (compare also [16, Theorem 2.18].

Theorem 4.1. *Let Z be an irreducible hermitian Jordan triple of rank r and dimension d. Put*

$$\frac{d}{r} = 1 + \frac{a}{2}\,(r-1) + b$$

where a and b are the characteristic multiplicities of Z. Then, for any $1 \le \ell \le r-1$, the Hilbert space

$$H^2_\ell(Z) = H^2(S^{\mathbb{C}}_\ell) = \sum_{\boldsymbol{m}\in\mathbb{N}^\ell_+} \mathcal{P}_{\boldsymbol{m}}(Z) \tag{4.1}$$

of "ℓ-harmonic" holomorphic functions has the reproducing kernel

$$E_\ell(z,w) = \sum_{\boldsymbol{m}\in\mathbb{N}^\ell_+} \frac{(d/r)_{\boldsymbol{m}}\,(ra/2)_{\boldsymbol{m}}}{(\ell a/2)_{\boldsymbol{m}}}\, E^{\boldsymbol{m}}(z,w)$$

for all $z,w \in Z$, where $E^{\boldsymbol{m}}$ is the Fock reproducing kernel of

$$\mathcal{P}^{\boldsymbol{m}}(Z) \subset H^2(Z)$$

in the complex Segal-Bargmann space of Z.

Proof. Put

$$d_{\boldsymbol{m}} = \dim \mathcal{P}^{\boldsymbol{m}}(Z).$$

By [6, Theorem 2.1] there exists a unique polynomial $\phi_{\boldsymbol{m}} \in \mathcal{P}^{\boldsymbol{m}}(Z)$ which is invariant under the group

$$L := \{k \in K : \; ke = e\}$$

and is normalized by $\phi_{\boldsymbol{m}}(e) = 1$. Here $e = e_1 + \cdots + e_r$. By [6, Lemma 3.3], we have

$$\int_K dk\, |\phi_{\boldsymbol{m}}(k\cdot x)|^2 = \frac{1}{d_{\boldsymbol{m}}}\, \phi_{\boldsymbol{m}}(x^2) \tag{4.2}$$

for all $x = \sum_{j=1}^{r} x_j\, e_j$, and

$$E^{\boldsymbol{m}}(z,e) = \frac{d_{\boldsymbol{m}}}{(d/r)_{\boldsymbol{m}}}\, \phi_{\boldsymbol{m}}(z) \tag{4.3}$$

by [6, Lemma 3.1 and Theorem 3.4]. Therefore

$$E^{\boldsymbol{m}}(e,e) = \frac{d_{\boldsymbol{m}}}{(d/r)_{\boldsymbol{m}}}\, \phi_{\boldsymbol{m}}(e) = \frac{d_{\boldsymbol{m}}}{(d/r)_{\boldsymbol{m}}}.$$

Putting $x = u_\ell = e_1 + \cdots + e_\ell$ in (4.2), we obtain for partitions $\boldsymbol{m} \in \mathbb{N}^\ell_+$ of length $\le \ell$

$$\int_{S_\ell} du\, |\phi_{\boldsymbol{m}}(u)|^2 = \int_K dk\, |\phi_{\boldsymbol{m}}(k\cdot u_\ell)|^2 = \frac{1}{d_{\boldsymbol{m}}}\, \phi_{\boldsymbol{m}}(u_\ell^2) = \frac{1}{d_{\boldsymbol{m}}}\, \phi_{\boldsymbol{m}}(u_\ell) \tag{4.4}$$

since $S_\ell = K \cdot u_\ell$. By [1, Proposition 3.7], we have

$$\phi_{\boldsymbol{m}}(u_\ell) = \frac{(\ell a/2)_{\boldsymbol{m}}}{(ra/2)_{\boldsymbol{m}}}. \tag{4.5}$$

On the other hand, by (4.3), the Fischer-Fock norm $\|\cdot\|_F$ satisfies

$$\|\phi_{\boldsymbol{m}}\|^2_F = \frac{(d/r)^2_{\boldsymbol{m}}}{d^2_{\boldsymbol{m}}}\, \|E^{\boldsymbol{m}}(-,e)\|^2_F = \frac{(d/r)^2_{\boldsymbol{m}}}{d^2_{\boldsymbol{m}}}\, E^{\boldsymbol{m}}(e,e) = \frac{(d/r)_{\boldsymbol{m}}}{d_{\boldsymbol{m}}}.$$

Combining (4.4) and (4.5), it follows that

$$\int_{S_\ell} du\, |\phi_{\boldsymbol{m}}(u)|^2 = \frac{(\ell a/2)_{\boldsymbol{m}}}{d_{\boldsymbol{m}}(ra/2)_{\boldsymbol{m}}} = \frac{(\ell a/2)_{\boldsymbol{m}}}{(d/r)_{\boldsymbol{m}}(ra/2)_{\boldsymbol{m}}}\, \|\phi_{\boldsymbol{m}}\|_F^2. \tag{4.6}$$

Since $\mathcal{P}_{\boldsymbol{m}}(Z)$ is irreducible under K this implies

$$\int_{S_\ell} du\ \overline{q(u)}\ p(u) = \frac{(\ell a/2)_{\boldsymbol{m}}}{(d/r)_{\boldsymbol{m}}\,(ra/2)_{\boldsymbol{m}}}\ (q|p)_F$$

for all $p, q \in \mathcal{P}_{\boldsymbol{m}}(Z)$. In particular, putting $q = E^{\boldsymbol{m}}(-, z)$ for some fixed $z \in Z$, we have

$$\int_{S_\ell} du\, \overline{E^{\boldsymbol{m}}(u,z)}\ p(u) = \frac{(\ell a/2)_{\boldsymbol{m}}}{(d/r)_{\boldsymbol{m}}\,(ra/2)_{\boldsymbol{m}}}\ (E^{\boldsymbol{m}}(-,z)|p)_F = \frac{(\ell a/2)_{\boldsymbol{m}}}{(d/r)_{\boldsymbol{m}}\,(ra/2)_{\boldsymbol{m}}}\ p(z)$$

whenever $p \in \mathcal{P}^{\boldsymbol{m}}(Z)$ and $\boldsymbol{m} \in \mathbb{N}_+^\ell$. Since (4.6) is an orthogonal sum, the assertion follows. □

Theorem 4.2. *Let Z be an irreducible hermitian Jordan algebra with unit ball $B = G/K$, having rank r, dimension d and characteristic multiplicity a. Let S_1 be the compact manifold of all rank 1 tripotents in Z, with K-invariant probability measure du, and let*

$$Z_1 = \{t \cdot u \,|\, t \in \mathbb{R}_+,\ u \in S_1\}$$

denote the space of all rank 1 elements in Z. Then the Hilbert space $H = L^2(Z_1, d\mu)$ on Z_1, with respect to the measure

$$d\mu(t\cdot u) = \frac{4\,dt\,du}{\Gamma(\frac{d}{r})\,\Gamma(\frac{ra}{2})}\ t^{a(r-\frac{1}{2})}\, K_{\frac{a}{2}-1}(2t) \tag{4.7}$$

on Z_1, has a G-equivariant Hilbert subbundle $(H_z)_{z\in B}$ defined over the unit ball $B \subset Z$. The fibre at the base point $0 \in B$ consists of holomorphic functions

$$H_0 = \sum_{m=0}^{\infty} \mathcal{P}_{m0\ldots0}(Z) \quad \text{[Hilbert sum]} \tag{4.8}$$

and the associated projection, realized as an integral operator on Z_1, has the kernel

$$P_0(u,v) = \sum_{m=0}^{\infty} \frac{(u|v)^m}{m!(\frac{a}{2})_m} = {}_0F_1(\frac{a}{2})((u|v)). \tag{4.9}$$

The canonical connexion on this field is projectively flat, with curvature at 0 given by

$$\Omega_0(\dot z, \ddot z) = (\dot z|\ddot z) - (\ddot z|\dot z)$$

for all $\dot z,\ \ddot z \in Z = T_0(B)$. Here $(z|w)$ is the normalized K-invariant measure on Z.

Proof. Specializing Theorem 4.1 to $\ell = 1$, and choosing polynomials $p(u) = (u|e_1)^m$, $q(u) = (u|v)^m$ for fixed $v \in Z_1$ we have

$$\int_{S_1} du\,(v|u)^m\,(u|e_1)^m = \frac{(a/2)_m}{(d/r)_m\,(ra/2)_m}\,(q|p)_F = \frac{(a/2)_m\,m!}{(d/r)_m\,(ra/2)_m}\,(v|e_1)^m \quad (4.10)$$

for all $m \geq 0$. This follows from the fact that $p, q \in \mathcal{P}_{m0\ldots0}(Z)$, and the well-known value

$$((-|v)^m \,|\, (-|e_1)^m)_F = m!(v|e_1)^m$$

for the Fock inner product [14]. By (4.8), the fibre $H_0 \subset L^2(Z_1, d\mu)$ over $0 \in B$ is generated by the functions $(-|v)^m$, for $v \in Z_1$, or equivalently by the K-translates of $(-|e_1)^m$, since $S_1 = K \cdot e_1$. Expressing the underlying measure on Z_1 in the form

$$d\mu(t \cdot u) = \mathcal{K}(t)\,dt\,du$$

for some density function $\mathcal{K}$ on $\mathbb{R}_+$, it follows that the orthogonal projection P_0 onto H_0 satisfies

$$\begin{aligned}(v|e_1)^m &= \int_0^\infty dt\,\mathcal{K}(t) \int_{S_1} du\,P_0(v,\,t\cdot u)(t\cdot u|e_1)^m \\ &= \int_0^\infty dt\,\mathcal{K}(t)\,t^m \int_{S_1} du\,P_0(v,\,t\cdot u)(u|e_1)^m.\end{aligned}$$

Writing

$$P_0(v,u) = \sum_{k\geq 0} \frac{(v|u)^k}{c_k}$$

as a power series in $(v|u)$, we obtain

$$\begin{aligned}(v|e_1)^m &= \sum_{k\geq 0} \frac{1}{c_k} \int_0^\infty dt\,\mathcal{K}(t)\,t^{m+k} \int_{S_1} du\,(v|u)^k\,(u|e_1)^m \\ &= \tfrac{1}{c_m} \int_0^\infty dt\,\mathcal{K}(t)\,t^{2m} \int_{S_1} du\,(v|u)^m\,(u|e_1)^m\end{aligned} \quad (4.11)$$

since different powers are orthogonal over S_1. Using (4.10) for the S_1-integral we obtain

$$(v|e_1)^m = \frac{(v|e_1)^m}{c_m}\,\frac{(a/2)_m\,m!}{(d/r)_m\,(ra/2)_m} \int_0^\infty dt\,\mathcal{K}(t)\,t^{2m}$$

and therefore

$$c_m = \frac{(a/2)_m\,m!}{(d/r)_m\,(ra/2)_m} \int_0^\infty dt\,\mathcal{K}(t)\,t^{2m}.$$

By (4.7) the density function $\mathcal{K}(t)$ has the form

$$\mathcal{K}(t) = \frac{4}{\Gamma(\frac{d}{r})\,\Gamma(\frac{ra}{2})}\; t^{a(r-\frac{1}{2})}\, K_{\frac{a}{2}-1}(2t).$$

Since $\frac{d}{r} = 1 + \frac{a}{2}(r-1)$, formula (16) in [8, p. 668] implies

$$\int_0^\infty dt\, \mathcal{K}(t)\, t^{2m} = \frac{4}{\Gamma(\frac{d}{r})\,\Gamma(\frac{ra}{r})} \int_0^\infty dt\, t^{2m+a(r-\frac{1}{2})}\, K_{\frac{a}{2}-1}(2t)$$
$$= \frac{\Gamma(m+\frac{ar}{2})\,\Gamma(m+\frac{d}{r})}{\Gamma(\frac{d}{r})\,\Gamma(\frac{ra}{2})} = \left(\frac{ar}{2}\right)_m \left(\frac{d}{r}\right)_m.$$

Therefore $c_m = m!\,(a/2)_m$, as asserted □

For the classical Fock type spaces over $\mathbb{K} = \mathbb{R}, \mathbb{C}, \mathbb{H}$, our general result yields the correct measure as well as the projection formula (4.9), as will now be shown. If $\mathbb{K} = \mathbb{R}$, we have $a = 1$ and hence $d = \frac{r(r+1)}{2}$ (symmetric $r \times r$-matrices). Therefore

$$P_0(u,v) = \sum_{m=0}^\infty \frac{(u|v)^m}{(\frac{1}{2})_m\, m!} = \sum_{m=0}^\infty \frac{4^m (u|v)^m}{(2m)!}$$

and the measure μ on Z_1 is given by $d\mu(t\cdot u) = t^{r-1}\, e^{-t}$ as shown in Lemma 3.2. This case (the classical one!) is slightly exceptional since $\frac{a}{2} < 1$ and $K_{-1/2}(2t)$ is replaced by $t^{-1/2}\, e^{-t}$. If $\mathbb{K} = \mathbb{C}$, we have $a = 2$ and hence $d = r^2$ (all $r \times r$-matrices). Applying Theorem 4.2 we obtain

$$P_0(u,v) = \sum_{m=0}^\infty \frac{(u|v)^m}{(1)_m\, m!} = \sum_{m=0}^\infty \frac{(u|v)^m}{m!^2}$$

and $d\mu(t\cdot u) = dt\, du\, t^{2r-1}\, K_0(2t)$ in accordance with Lemma 3.3. If $\mathbb{K} = \mathbb{H}$, we have $a = 4$ and hence $d = r(2r-1)$ (skew-symmetric $2r \times 2r$-matrices). In this case Theorem 4.2 yields

$$P_0(u,v) = \sum_{m=0}^\infty \frac{(u|v)^m}{(2)_m\, m!} = \sum_{m=0}^\infty \frac{(u|v)^m}{m!^2 (m+1)}$$

and $d\mu(t\cdot u) = dt\, du\, t^{4r-2}\, K_1(2t)$ in accordance with Proposition 3.4.

In order to construct the field (H_z) at all points $z \in B$, and compute its connection and curvature, we use invariance under a suitable G-action on $Z_1 = \mathbb{R}_+ S_1$, similar as in Proposition 2.1. In order to explain the G-action, we need some Jordan algebraic concepts explained in [10, Chaper 6]. For every tripotent $u \in S_1$, the Jordan triple Z admits a *Peirce decomposition* [8]

$$Z = Z_u^1 \oplus Z_u^{1/2} \oplus Z_u^0 \tag{4.12}$$

and, putting $B_u^0 := B \cap Z_u^0 = \{z \in B : \{uu^*z\} = 0\}$, the set

$$u + B_u^0 \subset \partial B \tag{4.13}$$

is called the *boundary component* with center u. The (disjoint) union

$$\partial_1 B = \dot{\bigcup_{u \in S_1}} u + B_u^0 \tag{4.14}$$

is a G-invariant stratum of ∂B, which is open in ∂B, cf. [10]. It follows that for every $g \in G$ we have

$$g(u + B_u^0) = \tilde{g}(u) + B_{\tilde{g}(u)}^0 \tag{4.15}$$

for a unique tripotent $\tilde{g}(u) \in S_1$. In this way we obtain a real analytic action

$$\begin{array}{lcl} G \times S_1 & \to & S_1 \\ g, u & \mapsto & \tilde{g}(u). \end{array} \tag{4.16}$$

For a careful study of this action, more precisely its infinitesimal version, we refer to [15].

References

[1] J. Arazy, H. Upmeier, *Boundary measures for symmetric domains and integral formulas for the discrete Wallach points.* Int. Equ. Op. Th. **47** (2003), 375–434.

[2] S. Axelrod, S. Della Pietra, E. Witten, *Geometric quantization of Chern-Simons gauge theory.* J. Diff. Geom. **33** (1991), 787–902.

[3] H. Ding, K. Gross, R. Kunze, D. Richards, *Bessel functions on boundary orbits and singular holomorphic representations.* Proc. Symp. Pure Math. **68** (2000), 225–254.

[4] A. Erdelyi et al., *Higher Transcendental Functions II.* McGraw-Hill, 1953.

[5] J. Faraut, A. Korányi, *Analysis on Symmetric Cones.* Clarendon Press, Oxford 1994.

[6] J. Faraut, A. Korányi, *Function spaces and reproducing kernels on bounded symmetric domains.* J. Funct. Anal. **88** (1990), 64–89.

[7] W. Kirwin, S. Wu, *Geometric quantization, parallel transport and the Fourier transform.* Commun. Math. Phys. **266** (2006), 577–594.

[8] I.S. Gradshteyn, I.M. Ryzhik, *Table of Integrals, Series and Products.* Acad. Press 2000.

[9] W. Kirwin, S. Wu, *Geometric quantization, parallel transport and the Fourier transform.* Commun. Math. Phys. **266** (2006), 577–594.

[10] O. Loos, *Bounded Symmetric Domains and Jordan Pairs.* Univ. of California, Irvine, 1977.

[11] J. Peetre, *The Fock bundle.* Preprint, Lund University.

[12] W. Rudin, *Function Theory in the Unit Ball of* $\mathbb{C}^n$. Springer Grundlehren Vol. 241, 1980.

[13] S. Sahi, *Explicit Hilbert spaces for certain unipotent representations.* Invent. Math. **110** (1992), 409–418.

[14] H. Upmeier, *Toeplitz Operators and Index Theory in Several Complex Variables.* OT **81**, Birkhäuser 1996.

[15] H. Upmeier, *Projectively flat Hilbert bundles on Jordan varieties.* Preprint.

[16] Z. Yan, *Differential operators and function spaces.* Contemp. Math **142** (1993), 121–142.

Harald Upmeier
Fachbereich Mathematik
Philipps-Universität
Hans-Meerwein-Straße
D-35032 Marburg, Germany
e-mail: upmeier@mathematik.uni-marburg.de

MIX
Papier aus verantwortungsvollen Quellen
Paper from responsible sources
FSC® C105338

If you have any concerns about our products,
you can contact us on
ProductSafety@springernature.com

In case Publisher is established outside the EU,
the EU authorized representative is:
Springer Nature Customer Service Center GmbH
Europaplatz 3, 69115 Heidelberg, Germany

Printed by Libri Plureos GmbH
in Hamburg, Germany